Encounters With Life

SIXTH EDITION

Larry J. Scott, Ed.D.
Central Virginia Community College, Lynchburg, Virginia

Hans F. E. Wachtmeister, Ed.D.
Belmont Hill School, Belmont, Massachusetts

Morton Publishing Company

925 W. Kenyon, Unit 12
Englewood, Colorado 80110
www.morton-pub.com

ISBN: 0-89582-586-4

Printed in the United States of America

10 9 8 7 6 5 4 3 2 1

RECYCLED
PAPER

Preface

LABORATORY MANUAL

This laboratory guide was designed for use in a one-semester or a one-year introductory biology course at the college level and can be coordinated with general biology textbooks. Because each exercise is a self-contained unit containing clearly-stated objectives, a variety of learning experiences, and thought-provoking review questions, the order of use of the exercises may be varied by the instructor to suit the organization of his or her course.

TO THE STUDENT

You are encouraged to think for yourself as much as possible. The instructors are more than willing to help you in the lab. If you have any questions, feel free to ask. Sometimes learning is best accomplished by trying, then asking, because until you have attempted an exercise, what you do not understand is not always clear to you.

The exercises are written with the idea that you will read them in advance of the lab period. Most of the exercises are detailed and will require the entire lab period to complete, therefore it is imperative that you be prepared when you begin so that your time may be spent more fruitfully in doing. Before you come to the lab, you should definitely read the list of objectives that explain what you should get out of each lab exercise. The objectives will also be useful as a review for tests. At the end of each exercise is a set of review questions. You are encouraged to complete the answers to these questions before leaving the lab. Some of the questions may require additional outside reading on your part before you can answer them. Terms in bold face type are important and should become part of your daily vocabulary.

Regular attendance is extremely important for two reasons:

1. If you miss one lab, you are missing the equivalent of one week of work.
2. The materials used in each lab will not be available later, so it is impossible to make up lab work.

Each student is responsible for cleaning up the materials he or she has used during the lab and for returning equipment and materials to their original places. In this way, all students taking the lab will be able to begin with materials and equipment that are clean and in good order, just as you did.

TO THE INSTRUCTOR

The Instructor's Manual is available online at www.morton-pub.com. It contains information on equipment and materials, helpful hints to the instructor, and answers to all of the questions that appear in this manual.

Acknowledgments

We would like to thank the many people who have contributed to *Encounters with Life* over the years. From Tidewater Community College: Agnes Flemming, Larry Blevins, Tom Weikel, Jim Hatstat, Donna Wright, Deborah Miller, Andi Helfant, Rita Parisi, Wendell Taylor, Nell Baynard, Paul Aiken, Marion Cullins, Greg Frank, and Sharon Sheffield. Also, a special thanks to Molly Perry and John Joyce who helped with the text of the first editions, and to Sharon Lawhorne, Arlien Steiner, Anna Ferreri, and Carol Kasper for the development of many of the original exercises upon which this manual is based.

We would also like to thank the many colleges and biology faculty members who have given advice on ways to improve the manual as well as pointing out errors in our manuscript. Norman Tweed of Pierce College and Kristi Sather-Smith of Hinds Community College have been especially helpful in this regard.

For this sixth edition, we are particularly indebted to Gregory Garman of Centralia College who critically reviewed the text of the fifth edition as well as the additions to this edition. Also, thanks go out to Carolina Biological Supply Company for allowing us to use their experimental procedure on electrophoresis and transformation in our last two editions. Carolina also allowed us to use the photograph of their electrophoresis apparatus, and Kimball Glass Inc. and Ohaus Corporation graciously allowed us to photograph their measuring devices for Exercise One.

We wish to thank Joanne R. Saliger and the personnel at Ash Street Typecrafters, Inc., for their excellent work in the proofreading and typesetting of the manuscript.

A special thanks goes out to Doug Morton and everyone at Morton Publishing Company for their encouragement for this edition, and for their continuous support over the many years.

Contents

The Scientific Method and Metric Measurements

OBJECTIVES

After completion of this exercise, the student should be able to do each of the following:

- Define the terms *hypothesis, null hypothesis, independent variable, dependent variable, control variables, control group, experimental group.*

- List and describe the steps in the scientific method.

- Explain the difference between the experimental and non-experimental approaches to testing a hypothesis.

- Design a scientific experiment if given a simple question and hypothesis. Use all of the terms in the first objective.

- Explain why objectivity is important in analyzing new discoveries in science.

- List the standard metric units of measure for length, volume, and mass along with the prefixes for units that are less than the standard unit.

- Name each of the instruments used in a lab to measure length, volume, and mass, and be able to make accurate measurements with each.

- Distinguish between weight and mass.

THE SCIENTIFIC METHOD

WAYS OF KNOWING

We tend to compartmentalize human knowledge into fields of study such as art, music, literature, philosophy, or science. Each field has its own methods by which knowledge is acquired and classified. These fields of knowledge often begin by making observations and asking questions about these observations.

Science takes its observations from the natural world. Also, what sets science apart from other fields of knowledge is that in science, the possible answers to questions are tested for their validity. Those answers that are supported by the testing are tentatively added to the volume of knowledge about the natural world.

The procedure that science follows is called the **scientific method** or **scientific reasoning**. Although somewhat oversimplified, the method is usually described as a series of steps (there are several ways the method can be modified in use and still test possible answers).

STEPS IN THE SCIENTIFIC METHOD

1. *Making observations.* As in other fields of study, science begins with observation. Scientists, due to their training, make very careful, detailed, precise observations, often involving intricate measurements. Part of this precision comes because scientists are making observations of the natural world around them.

2. *Asking questions.* Scientists are not only precise; they are also very curious. They ask questions about what they have observed, and the questions are often ones that others have never thought of.

3. *Proposing answers.* Scientists propose possible answers, called **hypotheses** (singular: **hypothesis**), to the question. A hypothesis is also called an "educated guess" because the scientist has done a lot of background research before proposing an answer to the question. This research involves an extensive study of everything that has been published in the area related to the question, as well as talking to experts in that particular area.

4. Testing the hypothesis. At this point, the scientific method becomes distinctly different from other fields of knowledge. All fields of knowledge can make observations, ask questions, and propose

answers, but only in science are the hypotheses put to the test. Most people think that science tests hypotheses only by performing experiments but there are other forms of testing. Sometimes the testing of hypotheses can involve making a new different set of observations that would test the validity of the hypothesis. This testing method can be used when conditions cannot be manipulated by the researcher. Often this is because the conditions could be too huge to manipulate (for example, in astronomy or meteorology) or the event under study has already occurred in the past (as in paleontology or archaeology).

In experimental testing, the hypothesis is first used to make a prediction and this prediction can be tested. In testing the hypothesis, there are 3 variables (conditions that can change) involved: **independent variable, dependent variable,** and **control variables**. The independent variable is the one that can be manipulated or changed by the researcher. The dependent variable is the observed (measured or counted) outcome of the test. In other words, what happens to the dependent variable "depends" on what is manipulated in the independent variable. Control variables are all other variables that must remain constant during the test. If other variables are involved other than the independent variable then the results could have been due to any of these variables or a combination of them and not just the independent variable.

To conduct an experiment, two groups are set up. These two groups are called the **control group** (not to be confused with control variables), and the **experimental group**. These two groups are as identical as possible (all control variables are accounted for) and the one differing variable, the independent variable, is given only to the experimental group.

As an illustration, suppose a scientist wanted to know if the addition of calcium to the water would stimulate Coleus plants to grow taller (research question). The hypothesis to be tested would be: "Water containing calcium will stimulate Coleus plants to grow taller." A prediction from this hypothesis would be: "If Coleus plants are watered with water + calcium, then the plants will grow taller than they would with just water alone" (notice that the prediction is written in "if-then" form). If an experiment were set up, the independent variable would be the presence or absence of calcium, and the dependent variable would be the measured height of the plants. Some control variables that would have to remain constant would be the light intensity, temperature, amount of distilled water given (no other chemical factors in the water), and frequency of watering. Control and experimental groups of plants would be set up, and the independent variable (calcium) would be given in the water to the experimental group only.

5. *Accepting or rejecting the hypothesis.* Based on the results of the test, the hypothesis can be accepted or rejected. A hypothesis can be proven to be false (rejected) but a hypothesis can never be proven to be correct. The hypothesis is merely the best explanation that is available at the time. If the hypothesis is rejected, then the scientists go back to step 3 again.

6. *Repeat the test (make replications).* No one should ever base their results on only one series of tests since there could be unknown variables that have affected the experiment. If the test is repeated many times and the results are basically the same, then the test is more reliable. Also, the experiment should be able to be repeated by other scientists who would get the same results.

THE METHOD IN ACTION — A MEDICAL EXAMPLE

In early 2000, a disease appeared in the U.S. that had never been seen before. The disease produced a high fever, nasal congestion, and a very obvious feature — a bright red nose. The disease was named Rhinoerythria (red nose disease). The organism that caused the disease was soon isolated and found to be a new species in the genus *Bacillus* and the new species was named *Bacillus rudolphia.* Two general biology lab manual authors became interested in the disease and set out to find an antibiotic that could treat the disease.* After an extensive search of the literature, they found that Antibiotic X has been effective against several species of *Bacillus* and they decided to try the antibiotic on the organism that caused Rhinoerythria.

1. State the question that was being studied in this research. _____
_____
_____
_____

*Actually, the disease is fictitious — the authors, however, are not.

2. Write the research hypothesis that is being proposed for this research. _____
_____
_____
_____
_____

3. Sometimes, researchers propose several different hypotheses, in other words, all possible answers that they could come up with. Also, they often propose a **null hypothesis,** which states that there is no effect of the treatment on the outcome of the experiment. Thus, the results of the experiment will support one hypothesis or the other. Write the null hypothesis for this research.

_____
_____
_____
_____
_____

4. For this research, state the:

Independent variable: _____
_____

Dependent variable: _____
_____

5. For this research, state all of the control variables that you can think of.

_____
_____
_____
_____

6. If this experiment were actually conducted, what type of results would you expect in the control and experimental groups if the original research hypothesis was accepted?*_____
_____
_____
_____

*This experiment was never conducted since the disease is fictitious, thus the researchers are not rich and are not currently vacationing in the Bahamas).

YOU DON'T HAVE TO HAVE A WHITE LAB COAT TO USE THE SCIENTIFIC METHOD

You don't have to be a scientist in order to use the scientific method. Any time that you ask a question, propose answers, and then systematically test all of the answers, you are using scientific reasoning. Some of these questions are practical everyday life questions and some are off-the-wall problems like the one below.

You have three cups: one of two black ping pong balls; one of two white ping pong balls; and one containing one black and one white ball. The cups have lids, which were labeled for the contents of the cups: BB, WW, and BW. The lids have been rearranged so that each cup is now incorrectly labeled. You are allowed to draw one ball out of a cup at a time without looking inside. The ball does NOT have to be placed back in the cup. You need to devise a plan to determine the contents of each cup with the fewest number of draws.

1. What is your hypothesis as to the answer to the problem? The fewest number of draws that would have to be made before correctly determining the contents of all three cups is _____.

2. If you were given the cups and ping pong balls, could you demonstrate how to determine the contents of the cups with the number of draws that you stated (you are actually "testing" your hypothesis here)? How would you go about it?

_____

_____

_____

_____

_____

3. Can you repeat this demonstration over and over with the same results (make replications)?

_____

_____

4. Can any of your classmates or your instructor identify the contents with fewer draws than you hypothesized? _____

5. If so, what does this say about your hypothesis even though it seemed to be correct earlier?

_____

_____

6. Can anything like this occur in scientific investigation? Explain. _____

_____

_____

_____

_____

_____

_____

_____

_____

ANOTHER PROBLEM
(But Not Using the Scientific Method)

Below are three rows of 3 dots. Connect all 9 dots by drawing 4 straight lines without lifting your pencil from the paper (all four lines connect end to end).

•   •   •

•   •   •

•   •   •

If you are not able to solve the problem, ask your instructor. In solving this problem, most people try to draw the four lines without going outside of the rows or columns (in other words, staying within the "box"). Sometimes scientific investigation needs to go "outside the box" and propose a hypothesis that seems ridiculous at the time. It is very important for scientists to maintain objectivity and examine all new discoveries carefully and not dismiss them too soon. Throughout the history of science, many new ideas were rejected prematurely. In each case, it took years before the new knowledge became generally accepted. However, once it was accepted, that knowledge became the foundation for even more discoveries.

Ask your instructor for examples of hypotheses that were originally considered "out of the box," but were later confirmed to be correct. In a few words, write one or two of them here.

_____

_____

_____

_____

SCIENTIFIC MEASUREMENT

Scientific measurements are always made in metric system units. The **meter** is the standard unit of length, the **liter** is the standard unit of volume, and the **gram** is the standard unit of mass. Prefixes for unit values less than the standard unit include:

　centi- (one hundredth)

　milli- (one thousandth)

　micro- (one millionth)

　nano- (one billionth).

Thus a centimeter is 1/100 of a meter and a milliliter is 1/1000 of a liter. In some areas of biology, units larger than the standard units are needed. The most common is the prefix kilo- which means "one thousand" (for example, a kilogram is equal to 1000 grams).

The major advantage of this system is that conversions can be made from one unit to another by multiplying by 10 (or multiples of 10) or by dividing by 10 (or multiples of 10). For example, centimeters can be converted into millimeters by multiplying by 10 and millimeters can be converted into centimeters by dividing by 10 (or multiplying by 1/10). These conversions are easier to make and are more consistent than the American Standard System in which, for example, there are 12 inches in a foot, 3 feet in a yard, and 5280 feet in a mile.

Because of the ease of conversion, metric measurements are now the standard for almost all countries as well as for scientific measurements. In the United States, the American Standard System is used by the average citizen. However, all commercial products give metric equivalents to the American Standard units, and because of imports from other countries, some metric measurements are fairly commonplace in American life. List three places where you would encounter metric measurements of length, volume, or mass.

1. _____

2. _____

3. _____

If you stayed within the boundaries of the United States all of your life, could you get away with not learning anything about the metric system? Why or why not?_____

_____

_____

_____

If your instructor wants you to know how to convert between units of the metric system, you must also complete the assignment in Appendix A of this manual.

MEASUREMENT OF LENGTH

The standard unit of length, the meter, is equal to 39.37 inches — a little more than one yard (one inch equals 2.54 centimeters). On a metric ruler, the smallest units are **millimeters.** There are 10 millimeters in a **centimeter** and the ruler is numbered in centimeters (one meter rulers are also numbered in decimeters which include 10 centimeters). If an object's length is measured as 8.6 centimeters, then it is the same as 86 millimeters.

Use a metric ruler to measure the following items to the nearest tenth of a centimeter.

(a) the width of a 3 X 5 card

_____

(b) the width of a microscope slide

_____

(c) the height of a business card

_____

(d) the height (top to bottom) of this page

_____

MEASUREMENT OF VOLUME

Volume is usually measured in liters and milliliters (pipettes can measure down to tenths or hundredths of a milliliter). One liter equals 1.06 quarts. Figure 1.1 shows a number of devices used to measure volume. Beakers and Erlenmeyer flasks are only approximate measuring devices. Volumetric flasks are very accurate, but only for the volume noted on the flask. Graduated cylinders are also very accurate and some can measure to tenths of a milliliter. Dropper pipettes can be used to dispense small volumes but are somewhat cumbersome and inaccurate when used to measure large volumes.

Pour water into a 100 ml volumetric flask until it reaches the white circle on the neck. Add or remove water with a dropper until you get the correct volume. Without spilling any water, pour the liquid into a 100 ml graduated cylinder. If the graduated cylinder is made of glass, notice that the surface of the liquid is a thick line or, more accurately, a bow called the **meniscus.** The most accurate measurement is the lowest part of the meniscus (the bottom of the "bow"). The top part of the meniscus is merely where water is clinging to the inside of the glass.

What is the volume of the water in the cylinder?

_____

Now, pour the water into a 250 ml beaker. Does the volume read exactly 100 ml? _____

Most pipettes can measure maximums of 1 ml, 5 ml, and 10 ml. These maximum volumes are usually printed on the side of the pipette along with the smallest units that can be accurately measured. There are several methods of drawing fluid up into the pipette — ask your instructor which one you should use. Draw the colored liquid up into a one milliliter pipette to the 0 line and release 1.0 ml into a 10 ml graduated cylinder. Repeat this procedure one more time. Then measure out 0.5 ml and add it to the cylinder. Since pipettes vary in structure, your instructor will demonstrate how to obtain these volumes.

How much colored liquid *should* be in the 10 ml cylinder? _____

How much liquid do you actually measure in the cylinder? _____

If there is a discrepancy, how do you explain it?

_____

_____

_____

Dropper pipettes can approximate volumes, especially those less than one milliliter. With a dropper pipette, count the drops of colored liquid that you place in a 10 ml graduated cylinder. Stop when you reach exactly one milliliter. How many drops are there in the one milliliter volume?

_____

How many drops would there be in 1/10 of a milliliter? _____

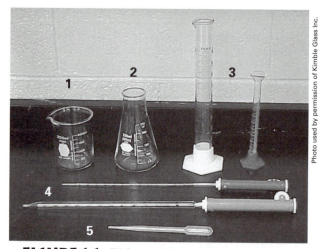

FIGURE 1.1 **Volume Measurement Devices**
1. beaker
2. Erlenmeyer flask
3. graduated cylinders
4. pipettes
5. dropper pipette

MEASUREMENT OF MASS

The standard unit of mass is the gram, but this unit is so small that the kilogram (one thousand grams) is often used. In fact, one gram equals 0.035 ounces or, somewhat easier to visualize, one ounce equals 28.3 grams. Mass and weight are often considered to be the same, but technically weight is the measurement of the pull of gravity on a mass. For instance, since the moon has a mass that is only one fifth of Earth, your weight on the moon would be one fifth of what it is on Earth. Your mass, however, would be the same on both.*

Mass can be measured on a triple beam balance (see Figure 1.2) down to tenths of a gram. If electronic balances are used mass can be measured in

_____

*So if you want to lose weight, go to the moon. If you want to lose mass, go on a diet!)

hundredths or thousandths of a gram. When measuring powder or liquid with a triple beam balance, the mass of an empty container is determined first and then the material is placed in the container until the total mass equals the mass of the container plus the desired mass of the material. To determine the mass of an unknown amount of powder or liquid, the mass of the dry empty container is measured first and then the total mass of the container plus the material is measured. The mass of the container is then subtracted from the total mass to obtain the mass of the material alone.

A triple beam balance has a different sized mass on each of the three beams. To measure the mass of an object, move the masses on the three beams to the far left. The arrow to the right of the three beams should line up with the white line on the body of the balance. Place an empty 250 ml Erlenmeyer flask on the pan of the triple beam balance. The arrow at the right end of the beam should point upward. Move the mass on the middle beam in increments of 100 grams until the arrow on the beam points downward (make sure the mass fits in the notch on the beam). Back up the mass by 100 grams (the next notch to the left). Then slide the mass on the back beam in increments of 10 grams until the arrow points downward. Move that mass back to the left by one position. Then slide the metal clip on the front beam to the right until the line on the arrow

lines up with the line on the body of the balance. Add up the mass on the three beams for the total mass.

What is the mass of the flask in grams?

_____

Pour some water into the flask. What is the mass of the flask plus the water? _____

_____

What is the mass of the water alone? _____

_____

Pour the water from the flask into a 100 ml graduated cylinder. What is the volume of the water?

_____

Under conditions of standard temperature and pressure, each milliliter of water will measure one gram. How close are your mass and volume measurements for the water?

_____

_____

_____

To gain perspective on the size of one gram, what is the mass of a penny? _____

OPTIONAL: If you have access to an electronic balance, push the bar down to turn the machine on. The digital scale should read 0.00. Place a dry 250 ml beaker on the balance. The mass appears on the digital scale. Now push the bar down one more time. Presto! The digital display reads 0.00 again. Now the mass of any amount of water that you add to the beaker will be read directly on the digital display. Subtracting the mass of the container is not needed!

Photo used by permission of Ohaus Corporation.

FIGURE 1.2   **Triple Beam Balance**

REVIEW QUESTIONS

1. An "educated guess" as to the answer to a scientific question is a _____.

2. The _____ _____ of acquiring knowledge involves carefully testing all possible answers to a question from the natural world.

3. For years, laboratory rats have been fed W&S animal food. A new type of food (Happy Rat) is now available and claims to be better than W&S. An experiment is designed to test this claim. Two identical groups of rats are set up. One group is given W&S food and the other is given Happy Rat food. The type of food given is the _____ variable. Rats are weighed daily to determine if they gain weight — this is the _____ variable.

4. The statement "there is no difference in weight gain between the two groups of rats" is called the _____.

5. When two groups are set up in an experiment, the group that receives the independent variable is the _____ _____ and the group without this variable is the _____ _____.

6. The independent variable is the only variable that is allowed to be different. All other variables that must remain constant are called the _____ _____.

7. An accepted hypothesis *(circle the correct answer)*:
   (A) will eventually become a scientific law
   (B) is considered to be scientific proof
   (C) could later be shown to be false
   (D) does not need to be verified by other scientists

8. In order to be valid, all scientific experiments *(circle the correct answer)*:
   (A) must prove a hypothesis
   (B) must be directed toward proving a scientific theory or law
   (C) should be repeated with the same result
   (D) must be the result of inductive logic

9. Are the following instruments used to measure length, volume, or mass?

   triple beam balance _____   graduated cylinder _____

   pipette _____   metric ruler _____

10. In the metric system, the standard unit of mass is the _____. For volume, the standard unit is the _____ and for length, it is the _____.

11. One nanometer is equal to _____ of a meter.

12. One microliter is equal to _____ of a liter.

13. Under conditions of standard temperature and pressure, one liter of water would have a mass of

    _____.

14. Water tends to cling to glass surfaces but not to plastic. If you place water in a plastic graduated cylinder,

    would it have a meniscus? _____

15. Let's say you love chocolate dearly! If someone offered you 5 ounces or 100 grams of chocolate, which

    one would you choose? _____

16. Let's say you ate too much chocolate over a period of time and gained so much weight that you could not
    run very far. If you were required to run either 100 yards or 100 meters, which would you choose?

    _____

The Microscope

OBJECTIVES

After completion of this exercise, the student should be able to do each of the following:

- Identify, locate, and give the functions of the main parts of the **compound microscope** and the **stereomicroscope**.

- Properly carry, care for, and put away both types of microscopes (be able to list steps in putting the microscopes away).

- Properly examine a slide under the compound microscope.

- Properly examine a plastomount or small opaque object under the stereoscope.

- Explain how or if the apparent direction of movement differs from the true direction of movement when observing a specimen under the compound microscope and under the stereoscope.

- Determine the total magnification afforded by either type of microscope in any given position.

- Explain how each of the following changes as the magnification changes on a compound microscope: field of view, working distance, and light intensity.

- Properly prepare and observe a wet mount under the compound microscope.

- Explain the main advantage(s) of each of the following over the compound light microscope: **phase contrast microscope, transmission electron microscope**, and **scanning electron microscope**.

- Compare and contrast the use of the transmission electron microscope with that of the scanning electron microscope and be able to distinguish between pictures taken by each.

- Locate a small animal or alga in pond water using the stereomicroscope. Transfer it to a slide with coverslip. Observe it under a compound microscope, estimate its size, sketch it, and identify it using such resources as the lab manual or other books provided by the instructor.

- Answer the review questions at the end of this exercise.

COMPOUND LIGHT MICROSCOPE

CARE AND USE

Much of the work in a biology lab involves the use of microscopes. With them one can open a window into new worlds. However, microscopes are quite expensive ($600–$1200) and should be handled with great care. If something goes wrong with the microscope, notify the laboratory instructor. Do not attempt to fix it or force any mechanical part.

1. Label the illustration of the compound microscope (Figure 2.1) with the following terms:

   Ocular — lens nearest the eye

   Body tube — keeps ocular and objective lenses at proper distance from each other

   Nosepiece — permits interchange of low, medium, and high power objectives

Pointer — found in ocular, can be moved by turning the black eyepiece

Arm — supports body tube and adjustment knobs

Objectives — contains lenses of various magnification

Coarse adjustment — changes the distance between the slide and the objective in order to focus an image

Fine adjustment — permits exact focusing

Base — bears the weight of microscope

Stage — supports slides

Stage clips — holds slide steady

Iris diaphragm — regulates amount of light going through specimen

Illuminator — provides light source

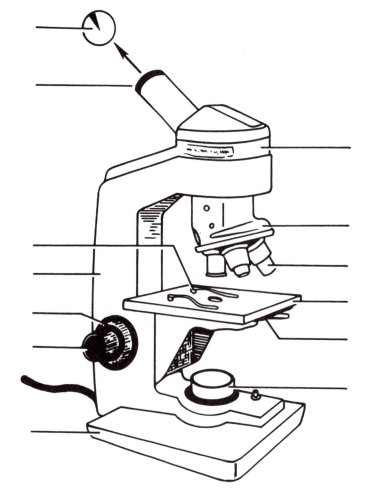

FIGURE 2.1   Compound Light Microscope

2. Obtain a compound light microscope from the cabinet. Note that the number on the microscope corresponds with the number of the cabinet cubicle. Carry the microscope in an upright position with one hand holding the arm and the other supporting the base. Place it on the lab table in front of you, approximately four inches from the edge. Now get a prepared slide designated by the instructor.

3. Clean all of the exposed lenses at the beginning of each laboratory with special **lens paper**. (*DO NOT USE PAPER TOWELING, KIMWIPES®, OR CLOTH* as this will scratch the lenses.) If the view through the microscope becomes blurred, additional cleaning with lens paper may be necessary.

4. Unwind the lamp cord and insert the plug in the nearest electrical socket.

5. Make sure that the low power objective is clicked into position under the body tube. Turn the substage switch on.

6. Place the slide (coverslip up) on the stage and center the specimen over the opening in the stage.

7. Watching from the side, use the coarse adjustment knob to move the objective and the slide as close as possible (the objective should not touch the slide). On some microscope models the coarse adjustment knob will move the nosepiece down toward the stage and on other models the knob will raise the stage toward the objective. Check with your instructor to determine which type of microscope is in your lab.

8. Next, while looking through the ocular, use the coarse adjustment knob to slowly move the objective upward or the stage downward until the material comes into focus. It should do so in less than a full turn. If it does not, check to see that the material is centered on the stage, lower the objective or raise the stage (watching from the side) and try again. (Always focus by increasing the distance between the objective and the slide, not by decreasing it.) Learn to keep both eyes open when looking through the microscope; this prevents eyestrain and headaches.

9. Obtain a sharp focus using the fine adjustment knob. (Always get the object to be studied in focus first under low power and later move to a higher power.)

10. Regulate the light intensity by adjusting the iris diaphragm. Reducing the light will usually give greater contrast.

11. To increase the magnification be sure the area you wish to examine specifically is in the center of the field; then, watching from the side to be sure that the objective clears the slide, turn the nosepiece until the next higher power objective clicks into position. The material should now be in view and should require only slight focusing with the fine adjustment. *NEVER FOCUS WITH THE COARSE ADJUSTMENT UNDER HIGH POWER.*

12. Before removing the slide, always return the microscope to low power and turn the coarse adjustment knob until the objective and stage are as close as possible.

NOTE

When using the microscope, train yourself to keep both eyes open, concentrating only on the object under observation. This will help eliminate the possibility of eyestrain and headaches.

MAGNIFICATION

Magnification tells how much larger the object appears under the scope than it actually is. Each objective is engraved with its magnifying power. The magnifying power of the ocular is 10X (times). When the ocular is used with one of the objective lenses, the **total magnification** is calculated by multiplying the power of each lens by the other. Complete the chart below. The scanning lens is usually 4X.

*Power	Magnification of Objective	Magnification of Ocular	Total Magnification
*Low	X	=	
*Med	X	=	
*High	X	=	

*not oil immersion lens

FIELD OF VIEW

The circular field you see when you look through the ocular is called the **field of view**. The field of view changes in size at different magnifications. At what magnification can you view the largest area on

the slide? _____ At what magnification

is the field of view the smallest? _____

_____

OPTIONAL: MEASUREMENT WITH THE COMPOUND MICROSCOPE

The metric system provides convenient units for expressing small measurements often used with the microscope. The most useful units are millimeters and micrometers. One micrometer (μm) = 1/1000 millimeter (mm) and 1 millimeter = approximately

1/25 inch. 1 micrometer = _____ inch.

In order to estimate sizes of microscopic objects, it is necessary to know the size of your field of view. Using a plastic ruler, in the same fashion as you would a slide, estimate the diameter of this field on low and medium power. Since the ruler is too thick and the markings too far apart to measure the diameter of the high power field, you will have to use another type of ruler. Measure an object such as a brine shrimp under medium power, and then **use the organism as your ruler.**

At low power the diameter of the field of view is _____ mm.

At medium power the diameter the field of view is _____ mm.

The brine shrimp is _____ mm long.

At high power the diameter of the field of view is _____ mm.

DEPTH OF FOCUS

The **depth of focus** is the thickness of an object which is all in sharp focus at the same time. On a microscope this is fairly thin, so, to adequately view a thick specimen, you will have to focus up and down over its body. Pick out a slide with three colored threads, which have been mounted at varying depths. Use the compound microscope to determine the order of threads. *REMEMBER:* No two threads will be in focus at the same time. Not all of the slides are mounted in the same order, so check with the instructor if you are not sure. Medium power will probably help you most, along with dim light.

Bottom Color _____

Middle Color _____

Top Color _____

ILLUMINATION AND WORKING DISTANCE

The **working distance** (space between slide and objective) and the **light intensity** change as you switch objectives. Focus on a slide, and without adjusting the diaphragm, switch from low to high power and compare these qualities.

Does the working distance increase or decrease as the magnification increases? _____

Does the light intensity increase or decrease as the magnification increases? _____

OPTIONAL: USING THE OIL IMMERSION LENS

Some of the microscopes in the lab have an oil immersion objective lens with a magnification of 93X, 95X or 100X. This lens is used mostly for observing microorganisms and the finer details of cell structure. The lens gets its name because a drop of oil is placed between the slide and the objective lens. Immersion oil has the same optical density as glass so stray light rays aren't lost or distorted. Remember, illumination decreases with higher magnifications. Also, if oil isn't used, light rays become distorted when moving through the slide, then air, and then the objective lens.

Before using this objective lens, locate, center and focus an object on a prepared slide under high power. Swing the high objective out of the way and place a drop of immersion oil on the area of the coverslip where the object is being observed. Swing the oil immersion lens into position (DO NOT MOVE THE HIGH POWER OBJECTIVE LENS THROUGH THE OIL AND GET OIL ON IT) and focus with the fine adjustment knob. When you have finished observing the slide, clean the oil immersion lens with lens paper (blot only) to remove all of the oil. If the lens is not cleaned thoroughly, it will eventually give only cloudy images. Also clean the slide completely before putting it back in the slide box.

TYPES OF SLIDES

1. **Prepared slides** have been made in advance by time-consuming techniques. Living material is first killed and cleared by solvents, embedded in paraffin wax, and often sliced into thin sections onto a slide. Then the paraffin is removed from the tissue by solvents and the sliced sections are stained by one or more permanent dyes. Balsam

glue is placed over the tissue and a thin glass cover slip is carefully put over it.

The label often indicates the name of the organism, the particular part or structure of that organism, and the type of slide, if any. (Note these key abbreviations: **w.m. — whole mount, x.s. or c.s. — cross section,** and **l.s. — longitudinal section.**)

Prepared slides are expensive, so treat them carefully. Never stack one slide on top of another. Always note the label on the side and put the slide back in the designated box.

2. When living tissues or cells are desired for microscopic investigation, **wet mounts** are often made in the lab such as the following (see Figure 2.2):

- Pick a leaf from an *Elodea* plant and place it on a blank slide.

- Add a drop of water. The leaf must be a healthy one taken from a rosette of leaves at the growing tip of a stalk.

NOTE:

The leaf in Figure 2.2 does not represent an <u>*Elodea*</u> *leaf, but rather shows how to make a wet mount.*

- Slowly lower the coverslip from one side of the water drop so that any air bubbles present will be pushed out as the coverslip comes down. There should be enough water to reach all edges of the coverslip, but not so much that it leaks out on the side.

STEREOMICROSCOPE

The **stereomicroscope** (also known as the **dissecting scope** or **binocular scope**) has two oculars. Although magnification is much lower than with a compound light microscope, the depth of focus is similar to that of our eyes. Also, transmission of light through the object is not always necessary since there are two sources of illumination; a substage light and an above-stage light. Thus, thick, opaque objects can be viewed with this microscope.

1. Obtain a stereoscope from the cabinet using the same care as you did with the compound microscope.

2. Label the illustration of the stereomicroscope, (Figure 2.3) with the following terms: magnification knob (not found on all models), oculars,

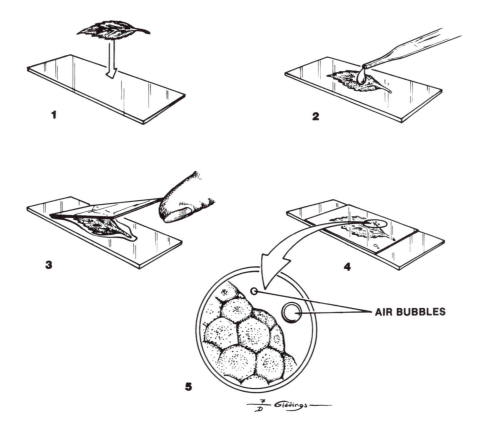

FIGURE 2.2 **Preparing A Wet Mount**

above-stage light, substage light, arm, focusing knob, light switch, base.

3. Get a fingerbowl with a small preserved animal in it. Following the directions below, examine it under the stereomicroscope.

4. Place object on stage and center the area to be examined.

5. If there is a magnification knob, turn it to the low power. The total magnification for low is 10X; high power is 20X. If there is no knob, the microscope magnifies 15X. For material that is transparent, the substage light will work best; if the object is opaque, use the overhead or above-stage light. (It is seldom advisable to use both lights at once.)

6. Focus by turning the large, black focusing knob.

7. If your scope has two powers, you may want to switch it to high power to look at the object more closely.

8. Try different kinds of objects on the stereomicroscope to see what this piece of equipment is capable of, e.g., butterfly wing, housefly, fingertip, plastomount, dish of pond water, etc.

OTHER TYPES OF MICROSCOPES

1. In each lab there may be one **teaching** or "**team**" scope. This is just like the compound scope you use, except there are two oculars, so two people can look at the same thing at the same time. There is also a green arrow pointer which can be moved by means of a knob on the body tube.

2. There may be one **phase-contrast microscope** in each lab. This scope uses regular light rays for illumination, but, by filtering out certain rays, allows the observer to see objects (particularly tiny moving objects such as cilia and flagella) more distinctly. This microscope is especially good for observing live material.

3. In most cases, material observed under electron microscopes is first killed and fixed in a condition as close to the living one as possible. Extremely thin sections of the material are made for viewing under the **transmission electron microscope**, which is able to increase magnification and resolution (clarity) by passing electrons, rather than light rays, through the specimen. Huge electromagnets are used to spread the electrons, in place of glass lenses used to spread light rays. Small organisms or groups of cells may be

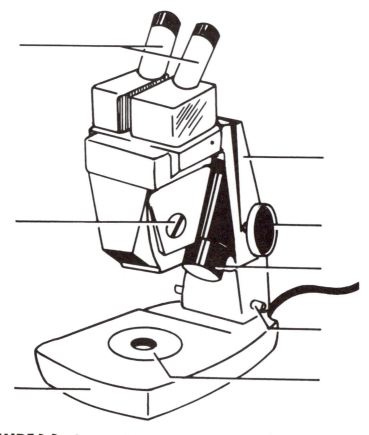

FIGURE 2.3 **Stereomicroscope (Dissecting Microscope)**

observed in the **scanning electron microscope**, which uses a moving beam of electrons to bounce the electrons off of the specimen. These are detected electronically and an image is produced on a viewing screen. The biggest advantage of the scanning electron microscope is its great depth of focus, coupled with high magnification.

4. Observe the scanning and transmission electron micrographs in Figures 2.4 and 2.5 and those available in the lab. Also look at the material on demonstration under the phase-contrast and teaching microscopes.

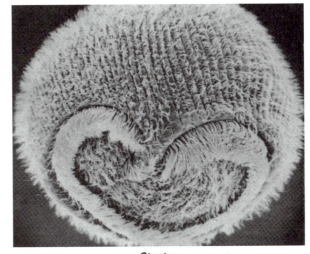

Stentor
(500X)

AN ELECTRON MICROGRAPH OF AN ERYTHROCYTE
(RED BLOOD CELL).

FIGURE 2.4   Scanning Electron Micrograph

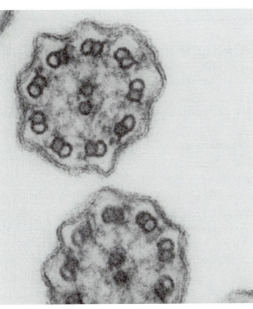

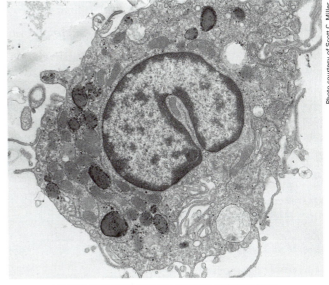

AN ELECTRON MICROGRAPH OF LYSOSOMES.

AN ELECTRON MICROGRAPH OF CILIA, SHOWING THE CHARACTERISTIC "9 + 2" ARRANGEMENT OF MICROTUBULES IN THE CROSS SECTIONS.

FIGURE 2.5   Transmission Electron Micrograph

A WINDOW INTO A NEW WORLD

If you feel confident in handling a microscope properly, obtain a small dish with pond or aquarium water. Use the stereoscope to locate any small animals, plants, rocks, etc. Then use a small medicine dropper to collect some of this water and make a wet mount with a slide and coverslip. Use the compound light microscope to view the small organisms. Dim the light. The light creates heat which can kill small animals in a short time. You will find that they move rather quickly out of the range of view. It will take a bit of practice to follow an organism with a microscope which turns all images upside down and backwards — if you want to go left, push the slide right, etc.

After you have located an interesting organism, attempt to "key it out." A **key** is an arrangement of characteristics of species which aid in their identification. While a key can never include **all** the organisms of an area, it should enable one to identify most of the commonly-encountered species.

This key, by which you will identify major groups of pond organisms, is referred to as **dichotomous.** That is, one needs only to make a single choice — whether the unknown has this particular trait or doesn't.

While observing one particular organism, follow the first couplet and make the choice. Then go to the next couplet as directed by the key. Once you believe you have identified the organism, check it with the illustrations provided in Figure 2.6.

KEY TO ORGANISMS COMMONLY FOUND IN FRESH WATER

1a.	Cells submicroscopic (usually appearing as dots, dashes, or commas under high power. May be clustered or in chains)	Bacteria
1b.	Cells not microscopic	2
2a.	Organisms green	3
2b.	Organisms not green	10
3a.	Pigment blue-green, dispersed evenly in cells; gelatinous sheath often present	Blue-green Algae
3b.	Pigment not blue-green, pigment usually found in plastids	4
4a.	Pigments brown, golden brown, or greenish yellow	5
4b.	Pigments leaf green	6
5a.	Cells non-flagellate, cell walls decorated with striations, bumps, and grooves	Diatoms
5b.	Cells non-flagellate, pigments greenish-yellow, cell walls not decorated	Yellow-green Algae
5c.	Cells flagellate, pigments golden brown, cell walls not decorated	Golden-brown Algae
6a.	Cells with flagella or cilia	7
6b.	Cells without flagella or cilia	8
7a.	Cells inflexible — always keep same shape	Flagellated Green Algae or Spores of Green Algae
7b.	Cells flexible — can shape a little as they move	Euglenoids
8a.	Cells arranged in colonies (chains, balls, sheets, etc.)	Green Algae
8b.	Cells usually not arranged in colonies	9
9a.	Cells divided into 2 equal parts by constriction	Desmids
9b.	Cells not divided into two equal parts, cell shape usually spherical or ovoid	Unicellular Green Algae
10a.	Organism unicellular or colonial (protists)	11
10b.	Organisms multicellular (organs of digestion usually visible inside animals)	12
11a.	Cilia or flagella present, pseudopodia absent	13
11b.	No cilia or flagella, pseudopodia fingerlike or pointed in radially symmetrical forms. May have shell	Sarcodines
12a.	Body shape wormlike	14
12b.	Body shape not wormlike	15
13a.	Cells with cilia	Ciliates
13b.	Cells with long whiplike flagella	Flagellates
14a.	Body segmented	Bristle Worms
14b.	Body unsegmented, movement by S-shaped slashing	Nematodes
15a.	Jointed appendages present	16
15b.	Jointed appendages absent, forked foot, cilia around mouth	Rotifers
16a.	Have two pr. of antennae and/or legs on abdomen	Crustaceans
16b.	Have one pr. antennae, no legs on abdomen	Insects

NOTE:

Organisms are not drawn to scale. Your instructor may ask you to tell how you worked through the key.

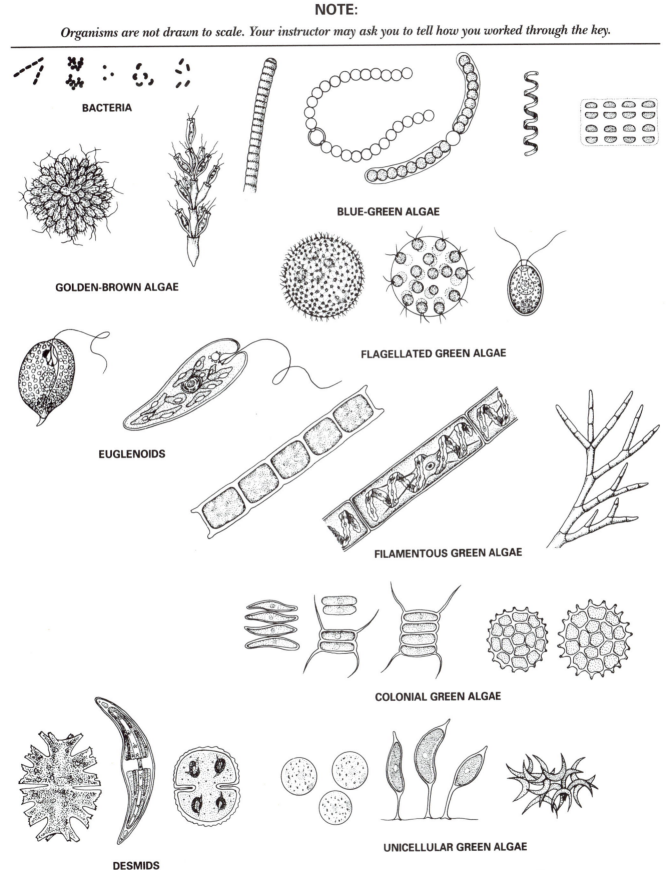

BACTERIA

BLUE-GREEN ALGAE

GOLDEN-BROWN ALGAE

FLAGELLATED GREEN ALGAE

EUGLENOIDS

FILAMENTOUS GREEN ALGAE

COLONIAL GREEN ALGAE

DESMIDS

UNICELLULAR GREEN ALGAE

FIGURE 2.6   **Microscopic Organisms Commonly Found In Pond Water**

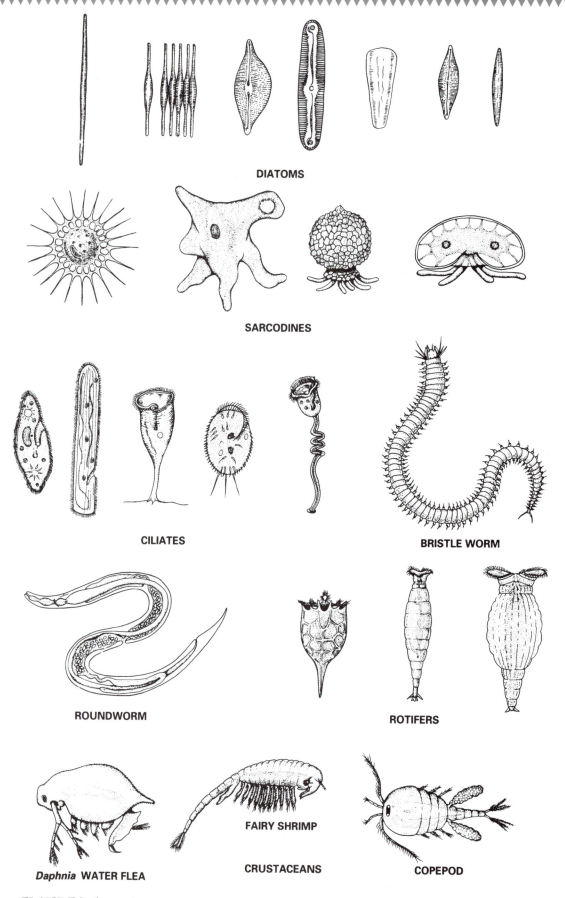

DIATOMS

SARCODINES

CILIATES

BRISTLE WORM

ROUNDWORM

ROTIFERS

Daphnia **WATER FLEA**

FAIRY SHRIMP

CRUSTACEANS

COPEPOD

FIGURE 2.6   Microscopic Organisms Commonly Found In Pond Water *(continued)*

PUTTING AWAY AND STORAGE OF MICROSCOPES (ALL TYPES)

1. Remove the slide or object being observed and be sure all parts of the microscope are dry and free of oil.

2. Set the microscope on the lowest magnification possible and move objectives close to stage.

3. Turn off the light and wind the cord **loosely** around your hand, then loop the plug through the coil and place the coil on the microscope stage. Cover the microscope with its plastic dust cover.

4. Put the microscope back in the proper cubicle in the cabinet.

REVIEW QUESTIONS

1. What is the function of the pointer on a microscope? _____

2. The magnifying power of the ocular lens is _____.

3. If the objective lens is 10X, what is the total magnification of the microscope? _____

4. What is meant by the working distance of a microscope? _____
_____

5. Which of the four types of microscopes discussed does not invert the image? _____

6. Which microscopes do not use a beam of light? _____

7. What do the following abbreviations mean?

    w.m. _____

    c.s. _____

    l.s. _____

8. What does the term "opaque" mean?_____

9. What is a wet mount? _____

10. Describe the "field of view." _____

11. How is a "prepared slide" different from a "wet mount"?

    Prepared slide _____

    Wet mount _____

12. How do the following change when you go from low power to high power magnification?

    Working distance _____

    Field of view _____

    Light intensity _____

13. What about the following microscopes gives them an advantage over the compound light microscope?

    Phase-contrast microscope _____

    Transmission electron microscope _____

    Scanning electron microscope _____

14. Matching: Match the function with the correct term.

_____   1. Ocular   a. provides light source

_____   2. Nosepiece   b   supports slides

_____   3. Arm   c. contain lenses of various magnification

_____   4. Coarse adjustment   d. lens nearest the eye

_____   5. Base   e. found in ocular, can be moved by turning the black eyepiece

_____   6. Iris diaphragm   f. hold slide steady

_____   7. Illuminator   g. changes distance between slide and objective lens by moving
                                          objectives or stage
_____   8. Body tube
                                      h. permits exact focusing
_____   9. Pointer
                                      i. supports body tube and adjustment knobs
_____   10. Objectives
                                      j. keeps ocular and objective lenses at proper distance from each
_____   11. Stage   other

_____   12. Stage clips   k. permits interchange of low, medium, and high power objectives

_____   13. Fine adjustment   l. bears the weight of microscope

                                     m. regulates amount of light going through specimen

15. Practice using the dichotomous key on page 18 in your lab manual and key out the following organisms:

Organism A
— Not submicroscopic (you can see it using the microscope).
— It has green color.
— The color is NOT blue-green.
— The pigment (color) is a leaf-green.
— The organism is not moving and no flagella can be seen.
— The cells are arranged in a chain.

What is it?_____

Organism B
— Not submicroscopic.
— Organism is not green.
— Organism is unicellular (single cell).
— A flagella is present.
— The flagella is long and whip-like.

What is it?_____

Organism C
— The cells are not submicroscopic.
— The organism is NOT green.
— It is not unicellular and appears to be multicelled.
— The body shape is worm-like.
— The body is unsegmented (no sections or divisions visible, smooth outer exterior), it moves in a whip-
   like fashion.

What is it?_____

Chemical Aspects of Life

OBJECTIVES

After completion of this exercise, the student should be able to do each of the following:

- Define **organic** and **inorganic** compounds and describe a simple test to distinguish one from the other.

- Name the elements most often found in living organisms.

- Describe the three major groups of organic compounds which compose living organisms, and give examples.

- Distinguish between the three types of **carbohydrates**, and give an example of each.

- Define: **polymer**, **monomer**, **dehydration synthesis**, and **hydrolysis**.

- Define: **reducing sugar**, **emulsion**, **emulsifier**, **fats**, **oils**, **amino acid**, **glycerol**, and **fatty acid**.

- Explain the results of each experiment using each of the following: Benedict's reagent, iodine, and Biuret reagent.

- Answer the review questions at the end of this exercise.

INTRODUCTION

All living organisms are composed of compounds which are usually divided into two groups, organic and inorganic. It was once believed that organic compounds were produced only by living organisms. However, many of the thousands of organic compounds are now being made in laboratories. Therefore, a standard definition of an **organic compound** is one whose molecules contain carbon in the form of chains or rings; thus, a compound which does not contain carbon arranged in chains or rings is considered **inorganic**.

Most cells are 70–90% water; the bulk of their dry weight consists of carbon (C), hydrogen (H), oxygen (O), nitrogen (N), and phosphorus (P), variously arranged into four major types of organic compounds: **carbohydrates**, **lipids**, **proteins**, and **nucleic acids**. The more complex members of these categories are made up of chains of smaller molecules (**monomers**) strung together more or less like beads in a necklace. Such complex molecules are called **polymers**. In living organisms, polymers are made by **dehydration synthesis**, the loss of a water molecule between each pair of monomers. Conversely, polymers can be digested (broken up into monomers) by the addition of a molecule of water between each pair of monomers. This process is known as **hydrolysis**.

TEST FOR ORGANIC AND INORGANIC COMPOUNDS

Heat the test tubes, each containing a small amount of one of the substances below, over an open flame until all reactions inside the tubes seem to stop. Note the residue in each test tube. If the deposit is black, the substance left on the test tube wall is carbon and the compound was organic. Fill in the chart below.

	Organic	Inorganic
sugar	_____	_____
meat	_____	_____
table salt	_____	_____
baking soda	_____	_____
unknown	_____	_____

CARBOHYDRATES

The approximate ratio of carbohydrates is 1 C: 2 H: 1 O. Simple sugars, or **monosaccharides**, are the simplest of the carbohydrates. There are many types of monosaccharides, but the most prevalent in living organisms is **glucose**. See Figure 3.1.

REDUCING SUGARS

Sugars which consist of two monosaccharides linked together are called **disaccharides**. When monosaccharides join, it is in the closed position (Figure 3.1). All monosaccharides and some disaccharides have the ability to add electrons to (reduce) other molecules. These sugars are called **reducing sugars**. Reducing sugars have free **carbonyl** groups ($- C \overset{O}{\lessgtr}$ or $- C \overset{O}{\underset{H}{\lessgtr}}$) in close proximity to **hydroxyl** groups ($- OH$), and this is where the electrons come from. Frequently, when two monosaccharides join to form a disaccharide, the carbonyl group gets tied up in the linkage, so it is not available to release electrons. This is the case with some disaccharides and all **polysaccharides** (long polymers of monosaccharides).

Benedict's reagent, which is used for testing for reducing sugars, contains copper ions in alkaline solution. The blue color of the reagent is characteristic of solutions containing copper ions. When Benedict's reagent is heated in the presence of a reducing sugar, the copper ions are reduced to metallic copper. This forms a **precipitate** (a substance which settles out of solution) which colors the contents of the tube green to brick red or brown, depending on how much reducing sugar is present. Before the development of paper test strips, this test was used by diabetics to test for sugar (in this case, glucose) in the urine.

1. Since all solutions in this test must be heated, prepare a hot water bath by setting a beaker about half full of tap water on a hotplate. Once steam is rising, you can turn the temperature back to medium.

2. Number test tubes 1–9 with a wax pencil and place in a test tube rack.

3. Add the test material to the first 8 tubes indicated in Table 3.1. Add 2 ml of Benedicts Reagent to each tube. Mix well.

4. Place in a hot water bath for 2 minutes, remove using a test tube holder, cool, and record the color (in your own words) on Table 3.1.

5. In tube 9 combine the sucrose and acid and heat in the water bath for 10 minutes. **Then** remove from heat and add the Benedict's to tube 9. Reheat, cool, and record the results on Table 3.1.

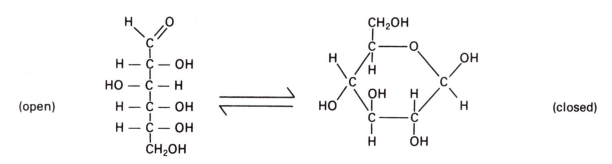

FIGURE 3.1 Two Forms of Glucose in Solution

TABLE 3.1

TUBE	TEST MATERIAL	BENEDICT'S	OBSERVATIONS	TEST RESULTS (+ or −)
1	2 ml tap water (control)	2 ml		
2	2 ml glucose	2 ml		
3	2 ml milk (lactose)	2 ml		
4	2 ml apple juice	2 ml		
5	2 ml starch	2 ml		
6	2 ml molasses (diluted)	2 ml		
7	2 ml sucrose	2 ml		
8	2 ml 10% HCl (control)	2 ml		
9	2 ml sucrose – 2 ml 10% HCl	2 ml		

NOTE

One dropperful is approximately 2 ml.

STARCH

Starch is a polysaccharide consisting of many glucose molecules linked together. Iodine (I_2-KI) is added to detect the presence of starch. The iodine molecules get stuck in the spirals of the starch molecule and cause a blue-black color to appear. A red color is an indication of the presence of certain **dextrins**, intermediate products of starch digestion whose spirals are shorter.

1. Thoroughly clean the tubes from the previous experiment.

2. Add the materials to the first seven tubes as indicated in Table 3.2. **No heating** is needed to produce a reaction.

3. Observe, and record the results on Table 3.2.

4. To the eighth tube, add 1 ml of saliva or bacterial amylase to l ml of starch solution. Let this sit for 10–15 minutes, then pour half of the contents into a clean test tube.

5. Add iodine to one of these tubes and test the other with Benedict's solution.

From the results of testing tube 8, what did the saliva do to the starch spirals? _____

What did the amylase convert the starch to? ____

_____

TABLE 3.2

TUBE	TEST MATERIAL	IODINE	OBSERVATIONS	TEST RESULTS (+ or −)
1	2 ml starch	5 drops		
2	2 ml glucose	5 drops		
3	2 ml water	5 drops		
4	2 ml sucrose	5 drops		
5	cellulose (cotton)	5 drops		
6	potato (small piece)	5 drops		
7	bread (small piece)	5 drops		
8	starch – saliva	5 drops		

LIPIDS

Lipids are oily or waxy compounds. They are generally insoluble in water, but are soluble in organic solvents, such as ether, acetone, carbon tetrachloride, and chloroform. The largest class of lipids, the **triglycerides**, are composed of two kinds of component molecules, **fatty acids**, and **glycerol** (an alcohol) (see Figure 3.2). Triglycerides which are liquid at room temperature are called **oils**, while those which are solid at room temperature are called **fats**.

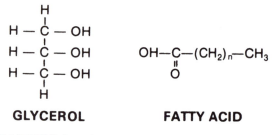

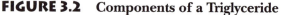

GLYCEROL   **FATTY ACID**

FIGURE 3.2   Components of a Triglyceride

Lipids are very important to living systems because they serve as concentrated sources of stored energy (nearly twice as rich in calories per gram as carbohydrates or proteins), and, in entire organisms as thermal insulation and shock absorbing pads for organs, bones, and muscles.

SUDAN STAIN TEST FOR LIPIDS

1. Add 3 drops of vegetable oil to a test tube half filled with tap water.
2. Shake thoroughly and observe the way the oil is only temporarily dispersed. This is an **emulsion**, a mixture of two liquids, each insoluble in the other.
3. Now add a small amount of lipid-specific red Sudan stain and mix again.
4. Add several droppers full of a liquid detergent to the tube and shake again. Allow to stand, noting that the two phases (oil and water) are no longer distinctly separated. Detergent is often termed an **emulsifier**. Its molecules are water-soluble on one end and lipid-soluble on the other. These surround small oil droplets, water-soluble end out, and allow the droplets to stay suspended in the water.

Which layer stained red with the Sudan? _____

How did the detergent affect the continuity of the lipid layer? _____

GREASY SPOT TEST FOR LIPIDS

Place a drop of oil and a drop of sucrose solution each on a piece of paper. After 15 minutes, hold the paper toward the light. A transparent to translucent spot is a positive test for a lipid.

PROTEINS

Proteins are polymers made up of **amino acids** linked by peptide bonds. The "R" group on the amino acid structural formula shown in Figure 3.3 can be one of 20 different attachments. Therefore, there are 20 amino acids commonly found in living systems.

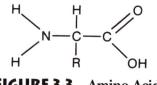

FIGURE 3.3   Amino Acid

The number of ways in which 20 different subunits can be arranged in a chain is extremely large, consequently there is a greater variety of protein molecules in living systems than any other kind of molecule. And the functions which proteins perform are just as varied.

BIURET TEST FOR PROTEIN

Biuret reagent (a blue-green solution) contains a strong solution of sodium hydroxide (NaOH) or potassium hydroxide (KOH) and a very dilute solution of copper sulfate ($CuSO_4$). If protein is present, the solution will change to violet because of a reaction with the peptide bonds that link the individual amino acids into the protein chain.

1. Clean out test tubes used previously and number test tubes 1–7 with a wax pencil.
2. Add the materials to these tubes as indicated in Table 3.3 and mix well. **No heating** is needed to produce a reaction.
3. Replace test tubes in the rack for 2 minutes, and record your results on Table 3.3. Base your conclusion only on the presence or absence of the violet color.

NOTE

One dropperful is approximately 2 ml.

TABLE 3.3

TUBE	TEST MATERIAL	BIURET REAGENT	OBSERVATIONS	TEST RESULTS (+ or −)
1	2 ml tap water	2 ml		
2	2 ml sucrose	2 ml		
3	2 ml albumin	2 ml		
4	2 ml milk	2 ml		
5	bread – 1 chunk	2 ml		
6	ground peanuts – small amount	2 ml		
7	2 ml vegetable oil	2 ml		

REVIEW QUESTIONS

1. Would Benedict's reagent give a positive reaction with all carbohydrates? _____

   Why? _____

2. Put a drop of iodine on the corner of this page. Does paper contain starch? _____

3. What chemical elements are present in starch? _____

4. What is a reducing sugar? _____

   _____

5. What are the component molecules making up triglycerides? _____

6. Dry cleaners often use carbon tetrachloride. What stains would this remove? _____

   _____

7. What are the building blocks of proteins? _____

8. How would you determine whether milk contains reducing sugars? _____

   _____

9. What could you say about a substance if it did **not** turn blue-black with iodine? _____

10. Is CO_2 (carbon dioxide) considered organic? _____ Why? _____

11. Biuret reagent is made up of _____ or _____ and _____.

12. What does "hydrolysis" mean? _____

13. The hydrolysis of sucrose resulted in the presence of what two sugars? _____

    Were they reducing sugars? _____

14. Define organic and inorganic compounds and describe a simple test to distinguish one from the other.

    _____

    _____

15. Name the elements most often found in living organisms. _____

16. Describe the 3 major groups of organic compounds which compose living organisms and give examples.

    a. _____ example _____

    b. _____ example _____

    c. _____ example _____

17. List the 3 types of carbohydrates, and give an example of each.

    a. _____ example _____

    b. _____ example _____

    c. _____ example _____

18. Define:

    Polymer _____

    Monomer _____

    Dehydration synthesis _____

    Hydrolysis_____

19. Why has water been included as one of the test materials for reducing sugars, starch, and protein?

    _____

20. CIRCLE OR FILL IN THE CORRECT ANSWER:

    a. Reducing sugars are usually (monsaccharides, disaccharides, or polysaccharides) that release (protons, neutrons, electrons). These (protons, neutrons, electrons) are added to other molecules.

    b. _____ is the reagent used to test for reducing sugars. When a reducing sugar is

       present the copper ions in the _____ reagent are reduced to

       _____. A _____ is formed and settled out of

       the solution. The color of the _____ is anything from green to brick red or brown.

    c. _____ is the chemical substance used to detect the presence of starch.

    d. The largest class of lipids is the _____.

    e. Lipids are soluble in (water, ether, acetone, chloroform).

21. Why are lipids very important to living systems?

    a _____

    b. _____

    c. _____

22. Give an example of an emulsion. _____

23. Give an example of an emulsifier. _____

24. Name the chemical substance used to indicate the presence of lipids. _____

25. What reagent is used to test for the presence of protein in foods? _____

26. The tested substance will turn _____ when mixed with this reagent, if protein is present.

Cell Structure

OBJECTIVES

After completion of this exercise, the student should be able to do each of the following:

- Distinguish between prokaryotic and eukaryotic cells. Give examples of each type.
- Identify on slides and explain the functions of the following cell parts: **cell wall, plastids** (including **chromoplasts, amyloplasts,** and **chloroplasts**), **cell membrane, nucleus,** and **nucleolus**.
- Determine whether a cell is a plant cell or an animal cell on the basis of its observable structure.
- On a stained blood slide, distinguish among red blood cells, white blood cells, and platelets.
- List the eukaryotic cell organelles and describe the structure and function of each one.
- Identify the eukaryotic cell organelles on plant and animal cell models.
- Answer the review questions at the end of this exercise.

INTRODUCTION

All living cells are enclosed by a **cell (plasma) membrane** which surrounds the **cytoplasm** which composes the volume of the cell. Some cells (most notably plant cells) have a rigid **cell wall** on the outside of the cell membrane. Cells also differ in the degree of organization and complexity in the cytoplasm. **Eukaryotic** cells have specialized structures called **organelles** ("little organs") in the cytoplasm. Many of these organelles are enclosed by membranes — the largest of these organelles being the nucleus. All plant and animal cells are eukaryotic. **Prokaryotic** cells do not have any internal membranes and no membrane-bound organelles (and thus no nucleus). Prokaryotic cells have almost the same cell chemistry and metabolism as eukaryotic cells but the reactions are not compartmentalized into discrete structures and their hereditary material (DNA) is spread out in the cytoplasm rather than enclosed in a nucleus.

EUKARYOTIC CELLS — PLANTS

Plant cells are eukaryotic, but the nucleus may not be clearly visible if the cell is unstained or if there are many other darker organelles to obstruct the view.

ELODEA CELLS

Look closely at a wet mount of *Elodea* under high power and pick out a single cell. All plant cells are surrounded by a **cell wall** on the outside of the **cell membrane**. The cell walls in unstained cells appear as glassy lines. The cell membrane is held tightly against the cell wall and is nearly impossible to detect under normal conditions. Inside the cell membrane is the complex colloid known as **protoplasm**. Sometimes it is subdivided into the **cytoplasm** (all protoplasm except inside the nucleus) and **nucleoplasm** (that inside the nucleus). The cell's inner packages float in the protoplasm.

Such packages are called **organelles**. The most prominent organelle in the *Elodea* is the **chloroplast**. Chloroplasts contain the green cholorophyll pigment, enzymes and other molecules that function in the production of food through photosynthesis. These are the green ovals which may be seen flowing around the periphery of the cell as **cyclosis** (**cytoplasmic flow**) occurs. At some point the chloroplasts will appear to pile up as if held back by a barrier. If you look closely, you may be able to see the greyish ghostlike **nucleus** in the middle of the pile.

The largest, although most transparent organelle, is the **vacuole** (also known as the **sap vacuole**) — a "bag" of water and minerals which occupies most of the center of the cell. This organelle is bound by a membrane known as the **tonoplast**. When plant cells lose water and shrink, it is the vacuole which loses most of the fluid. Explain how you can tell through the microscope that the vacuole is present, even if not visible. *HINT:* Notice the path of the moving chloroplasts.

Draw one *Elodea* cell. Label the cell wall, chloroplasts, and cytoplasm.

ONION LEAF EPIDERMIS

An onion bulb is actually a collection of food storage leaves. "Peel" the very thin membrane from the **concave** side of the layer of onion and mount in tap water. Do not cut out a chunk with a scalpel. Make sure the membrane lays very flat (not wrinkled or folded on your slide). Dim the light and find the **nucleus** — much easier to see here, since there are no chloroplasts. Close examination should reveal one or more **nucleoli** in the nucleus. Also note the movement of small particles in the nearly transparent cytoplasm. This is Brownian movement, which will be discussed in a later exercise. Methylene blue stain or iodine should be used if you have difficulty locating the nucleus or nucleoli. Draw a few onion cells with a nucleus in the space below. Label the cell wall, nucleus, nucleoli, and cytoplasm.

CARROT ROOT

Shave a thin slice of carrot onto a slide and prepare a wet mount of it. Observe the orange-yellow bodies called **chromoplasts**, another type of plastid. Chromoplasts are enriched in pigments that give flowers, fruits, and fall leaves their colors and gives carrots their orange color. Draw a few of the cells with chromoplasts in the space below. Label the cell wall, cytoplasm, and chromoplasts. *HINT:* Look along the edge where the section is thinnest.

CELLS OF THE POTATO

White potatoes are stems inflated with starch, stored within the cells in colorless plastids (**amyloplasts**) containing starch grains. Amyloplasts are also known as **leucoplasts**. Cut a wedge-shaped sliver of potato. (This ensures that one edge will be thin enough to see through.) Add a drop of iodine and a coverslip and observe under the microscope. Iodine stains starch a purple or blue-black color.

By dimming the light with the diaphragm lever, you may be able to see the faint growth rings in some of the grains and the nearly transparent cell walls, which look rather like soap bubbles. Some of the starch grains may have broken out of the cell walls when making your slice. The staining may take a few minutes, since the iodine must diffuse through both the cell membrane and the amyloplast membrane.

Draw a few of the potato cells with amyloplasts. Label the cell wall, cytoplasm, and amyloplasts.

EUKARYOTIC CELLS — ANIMALS

Although animal cells do not have cell walls or plastids, they have many other parts in common with plant cells. (Refer to your textbook for a comparison between plant and animal cells.) Animal cells, like plant cells, are eukaryotic.

HUMAN CHEEK CELLS

The cells on your skin are constantly sloughing away. In most areas they dry as they detach; such cells are not helpful in studying structure, since this breaks down as the cells dry. Such surface-coverings, composed of **epithelial cells**, are also formed on the inner linings of your cheeks. The cells here are still moist and will show structure if collected.

Rub the inner side of your cheek with a tooth pick or a tongue depressor to pick up some cells. Scrape the cloudy (cell-containing) fluid onto a slide, add a drop of dilute methylene blue, and observe.

These cells function only as shielding, and therefore have no elaborate internal structure, but you should be able to identify the **cell membrane** (no cell wall, you notice) of the pancake-shaped cells, the thin **cytoplasm, nucleus,** and **nucleolus.** Nucleoplasm will stain darkest. Draw one or two epithelial cells in the space provided below. Label the cell membrane, cytoplasm, nucleus, and nucleolus.

HUMAN BLOOD: STAINED

Wright's stain is used to make white blood cells more visible among the **red blood cells** (RBC's). Obtain a prepared slide of human blood and see if you can recognize the RBC's. There are several types of **white blood cells** (WBC's), all larger than the RBC's, with differing shaped purple-stained nuclei. See how many different types of WBC's you can find. Find at least two.

You will have to look for these, since there are only 1 or 2 for every 1000 RBC's in normal blood. A person whose body was fighting infection or who had a blood disease such as leukemia would produce more WBC's.

Among all of the cells are scraps of cells originating in the bone marrow — the **platelets**. These greyish scraps are very important in the initiation of blood clotting. They have no nuclei, being only cell fragments. (The red blood cells have no nuclei either, but theirs have disintegrated.) Draw a few red blood cells, a white blood cell, and a few platelets in the space provided on the next page.

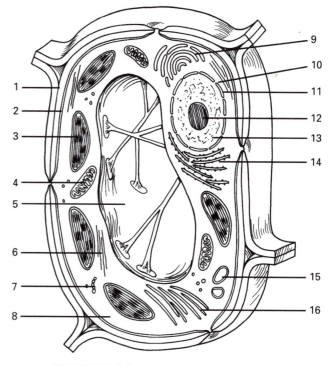

FIGURE 4.1   A Typical Plant Cell

1. Cell wall
2. Cell (plasma) membrane
3. Chloroplast
4. Mitochondrion
5. Vacuole
6. Microfilament
7. Plastid
8. Cytoplasm
9. Golgi apparatus
10. Nuclear pore
11. Nuclear membrane
12. Nucleolus
13. Chromatin
14. Rough endoplasmic reticulum
15. Vesicle
16. Smooth endoplasmic reticulum

SMALLER CELL ORGANELLES

In addition to the nucleus, plastids, and vacuoles, eukaryotic cells have other smaller organelles. These organelles, their structure and their function are summarized in Table 4.1. The appearance of these structures are shown in Figure 4.1 (typical plant cell) and Figure 4.2 (typical animal cell) although they are drawn larger than they should be in comparison to the cell (many of these organelles cannot be seen well or at all without an electron microscope).

Using models of plant and animal cells in the lab, identify the organelles shown in Figures 4.1 and 4.2.

TABLE 4.1   Structure and function of components of a eukaryotic cell.

Component	Structure	Function
Cell (plasma) membrane	Composed of protein and phospholipid molecules	Provides form to cell and controls passage of materials into and out of cell
Cell wall	Cellulose fibrils	Provides structure and rigidity to plant cell
Cytoplasm	Fluid to jelly-like substance	Suspending medium of organelles
Endoplasmic reticulum	Interconnecting hollow channels	Supporting framework of cell; cell transport
Ribosomes	Granules of nucleic acid	Protein synthesis
Mitochondria	Double-layered sacs with cristae	Production of ATP
Golgi apparatus	Flattened sacs with vacuoles	Synthesizes carbohydrates and packages molecules for secretion
Lysosomes	Membrane-surrounded sacs of enzymes	Digests foreign molecules and worn cells
Centrosome	Mass of two rod-like centrioles	Organizes spindle fibers and assists mitosis
Vacuoles	Membranous sacs	Stores and excretes substances within the cytoplasm
Fibrils and microtubules	Protein strands	Supports cytoplasm and transports materials
Cilia and flagella	Cytoplasmic extensions from cell	Movement of particles along cell surface or move cell
Nucleus	Nuclear membrane, nucleolus, and chromatin (DNA)	Direct cell activity; forms ribosomes
Chloroplast	Inner (grana) membrane within outer membrane	Photosynthesis

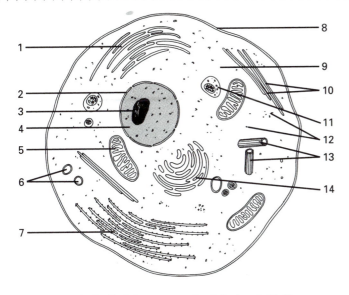

FIGURE 4.2   A Typical Animal Cell

1. Smooth endoplasmic reticulum
2. Nuclear membrane
3. Nucleolus
4. Nucleoplasm
5. Mitochondrion
6. Vesicles
7. Rough endoplasmic reticulum

8. Cell membrane
9. Cytoplasm
10. Microtubules
11. Lysosome
12. Ribosomes
13. Centrioles
14. Golgi apparatus

2. Observe the demonstration slide of *Anabaena* (a blue-green algae) under the oil immersion lens. Make a drawing in the space below.

How do the sizes of these cells compare to the eukaryotic plant and animal cells?

PROKARYOTIC CELLS

The only living cells that are prokaryotic are the bacteria and bacteria-like groups (such as the blue-green algae). Examples of these groups will be studied in detail in Exercise 13, but they are introduced here so that their size and structure can be compared to eukaryotic cells.

1. Observe the demonstration slide of a typical bacillus (rod-shaped) bacterium under the oil immersion lens. Make a drawing in the space below.

Are any internal structures visible?

___

NOTE

Remember that the plant and animal cells were observed under 430X total magnification and the total magnification with the oil immersion lens is over twice as much — 950X or 1000X.

REVIEW QUESTIONS

1. What cell parts were seen in the plant cells? _____

   _____

2. What cell parts were seen in the animal cells? _____

3. Name a human cell that does not have a nucleus. _____

4. Chromoplasts can be found in _____ cells.

5. Chloroplasts were found in the _____ plant.

6. Amyloplasts store _____ in the potato cells.

7. What is the ratio of WBC's to RBC's: _____ : _____

8. List the differences between animal cell and plant cell.

   Animal   Plant

   1. _____   _____

   2. _____   _____

   3. _____   _____

   4. _____   _____

   5. _____   _____

9. Cells that have no organized nucleus and no internal membrane structure are called _____.

10. Are most species of organisms composed of prokaryotic or eukaryotic cells? Explain. _____

    _____

11. Tiny specialized structures in the cytoplasm of cells are referred to generally as _____.

12. Which cell structure could aptly be nicknamed the "power house of the cell"? _____

    The "control center" _____

    The "transportation system"? _____

13. _____ are the site of protein synthesis in the cell.

14. _____ in the cytoplasm of cells digest foreign material and worn-out cells.

Exchange Between Cells and Their Environment

OBJECTIVES

After completion of this exercise, the student should be able to do each of the following:

- Explain what causes **Brownian movement**. Be able to recognize it and describe it.
- Define **diffusion** and tell how it was demonstrated in the lab (include every demonstration of diffusion).
- Define **osmosis** and explain how it was demonstrated in the lab.
- Use the following terms in describing solutions and their relationships to each other: **solute**, **solvent**, **hypertonic**, **hypotonic**, and **isotonic**.
- Explain the difference between **hemolysis** and **crenation** of erythrocytes and tell how each may occur.
- Distinguish between **turgor** and **plasmolysis** of plant cells and explain how each was demonstrated in the lab.
- Answer the review questions at the end of this exercise.

INTRODUCTION

For a cell to live and function, certain substances must enter the cell (e.g., food, water, and oxygen) while others must leave the cell (e.g., waste materials). Cellular organization and makeup is so complex that it cannot be easily studied, and is far from being completely understood. However, there are many properties of many living cells which can be explained in terms of physical and chemical laws.

Most physical phenomena occurring in cells are consequences of molecules being in constant oscillating motion, which increases as temperature increases. This molecular motion produces a variety of effects: Brownian movement, diffusion, osmosis, facilitated diffusion through cell membranes, and others. Most of these phenomena are extremely important to the functioning of living systems; life is as much a result of specific physical phenomena as a result of specific chemical reactions. Many of these phenomena are easily observed in living and non-living systems.

BROWNIAN MOVEMENT

Brownian movement is visible under high power of the microscope. Minute particles of solid matter, bacteria and other tiny organisms in a liquid suspension seem to move constantly in an erratic, random course. This motion is due to the bombardment of the larger particles by the molecules in the fluid. Energy which keeps particles larger than a molecule in motion is the kinetic energy of the molecules in the surrounding medium; these strike the larger particles on all sides and impart to it a movement similar to their own kinetic motion. The movement of larger particles is obviously slower than that of smaller ones.

Tap a tiny bit of powdered carmine dye from the tip of a toothpick onto a drop of water on a slide, add a coverslip, and observe the quivering motion of the particles under high power. (A mass movement of particles in any one direction is due to the microscope not being quite level, using only one stage clip, or catching the coverslip under the stage clips. This is not what you are looking for.)

DIFFUSION

Particles in solution tend to move from areas where they are highly concentrated to areas where they are less concentrated until a uniform distribution of particles is achieved. The phenomenon of particles scattering in this manner is called **diffusion.**

1. In front of the labroom the instructor will open a bottle of Top Job or similar substance. The fluid molecules will evaporate and escape from the bottle into the air. Note how long it takes your sense of smell to detect the gaseous molecules.

   What causes the liquid to change into a gas? _____

   _____

   Which students detected the gaseous molecules first? _____ Last? _____

   What might cause irregularities in the distribution of these molecules in the room? _____

2. When a solid dissolves in water, the molecules or ions scatter throughout the water until their distribution is uniform. This is easy to observe if the dissolving substance is colored.

   Fill a small test tube 2/3 full of tap water, then drop a crystal of purple $KMnO_4$ (potassium permanganate) into it. Set the tube in a place where it will not be jostled and observe the dispersal of the colored material throughout the remainder of the lab period.

   How might the dispersal be speeded up? _____

   _____

3. On a petri dish of agar gel place a crystal of potassium permanganate approximately 5 centimeters from an equal size crystal of malachite green. Note the rate of diffusion. Find the molecular weights of each compound from the jar label.

   m.w. of malachite green _____

   m.w. of potassium permanganate _____

   How does the molecular weight of the compound affect the rate of diffusion? _____

   _____

DIFFUSION THROUGH A SEMIPERMEABLE MEMBRANE

A **semipermeable**, or **selectively permeable membrane** is one that allows some particles (molecules and ions) to pass through it and restricts the movement of others. **Cell membranes** are semipermeable and certain substances can freely diffuse through them without any expenditure of energy by the cell (this is called **passive transport**). The selection of particles by the cell membrane is based on size and other factors.

Passage of water molecules is virtually impossible for a cell to stop. The diffusion of water molecules through a semipermeable membrane is called **osmosis**. If a sugar solution is separated from pure water by a semipermeable membrane, there will be a greater movement of water molecules from the pure water into the sugar solution than vice versa. The water molecules will move from an area of high concentration of water to an area of lower concentration of water. Only the water molecules will diffuse because sugar molecules are too large to pass through most semipermeable membranes (including cell membranes).

Cellophane dialysis tubing is a synthetic semipermeable membrane which permits diffusion of water and other small molecules while restricting passage of large molecules. We will create artificial "cells" with strips of this tubing. (The tubing is flattened when dry. Soak the strip in water for about a minute, then open by rolling between your thumb and fingers.)

1. For this experiment obtain a 20 cm strip of dialysis tubing and tie one end by twisting the tubing and then folding it double before securing tightly with string. Fill to about 5 cm of the top with syrup or an 80% glucose solution. Insert a 1 ml pipette, and tie the open end shut very

securely around the pipette. Support the bag in a beaker of water as shown in Figure 5.1 and note the position of the column of fluid for about 25 minutes. (If there is no change after 5 minutes, check for leaks, which appear as wiggly lines in the water.) It will take a few minutes to build up pressure in the bag.

NOTE

Water level should not go above string. This helps to view the water level changes more easily.

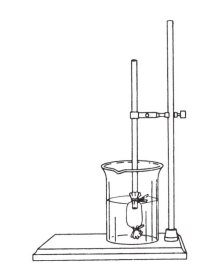

FIGURE 5.1   **Apparatus For Experiment 1**

2. Fill a 10 cm strip of tubing with a starch solution and tie off both ends tightly.

NOTE

Fold the ends of the bag before tying and remove most of the air.

Rinse with tap water, then submerge in water containing enough iodine (I_2–KI) to approximate the color of beer (see Figure 5.2). When starch molecules touch iodine, a blue or purplish color appears. Observe the bag after about 15 minutes for any color change. Based on the location of the stain, you should be able to answer the following questions.

Could iodine get through the membrane? _____

How do you know? _____

_____

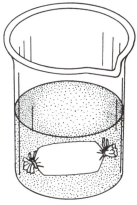

FIGURE 5.2   **Starch-Iodine Experiment 2**

Could starch get through the membrane? _____

How do you know? _____

_____

Which experiment (1 or 2) demonstrates osmosis?

_____

Why is the other experiment **not** a demonstration of osmosis? _____

_____

3. When two substances (usually one is a liquid) are mixed, the one in greatest quantity is called the **solvent** and the one in least quantity that is dissolved is the **solute**.

With a cup of coffee containing 1 teaspoon of sugar, the solvent is _____

and the solutes are _____ and _____.

When the solute concentration outside a cell is greater than the solute concentration inside a cell, the solution is said to be **hypertonic** to the cell. If the solute concentration outside the cell is less than that inside the cell, the solution is **hypotonic** to the cell, and if the solute concentration is equal in the cell and the solution, the cell is in an **isotonic** solution. Most cells are designed to function best in isotonic conditions, but some have the ability to tolerate solutions unlike their own. Most can control the movement of large solute molecules, but cannot stop the diffusion of the solvent.

In the situations pictured in Figure 5.3, assume that the solute molecule is too big to penetrate the membrane. Label each solution (hyper-, hypo-, or isotonic) and use an arrow to show the

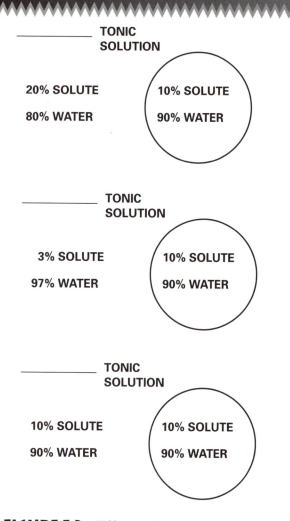

_____ TONIC
SOLUTION

20% SOLUTE   10% SOLUTE
80% WATER   90% WATER

_____ TONIC
SOLUTION

3% SOLUTE   10% SOLUTE
97% WATER   90% WATER

_____ TONIC
SOLUTION

10% SOLUTE   10% SOLUTE
90% WATER   90% WATER

FIGURE 5.3   **Effects of Osmosis on Cells**

direction of the heavier flow of solvent (water). If the flow in and out is equal, use opposite-facing arrows. The circles indicate cells.

LYSIS AND CRENATION

Red blood cells (erythrocytes) can easily be used to illustrate the osmotic reactions of a living animal cell. The plasma membrane of RBC's is permeable to water but relatively impermeable to salts. If placed in an isotonic saline solution (normal plasma is 0.85% NaCl) it retains its normal shape; if placed in a hypotonic solution, it swells and bursts (**lysis**, specifically known as **hemolysis** in red blood cells) releasing the hemoglobin and producing ghosts (empty cell membranes); when placed in a hypertonic solution, the cells lose water and shrink (**crenation**) showing wavy, irregular outlines.

1. Obtain three clean slides and, with a wax pencil, mark one 5%, one 0.85% and one 0% (distilled water). Set up three microscopes.

2. Place a drop of the appropriate saline solution (NaCl) on each slide.

3. Obtain some blood from the instructor, i.e., cow's blood.

4. Deposit a small drop of blood **beside** the drop of saline.

5. Add a coverslip to each slide immediately, raising and lowering the coverslip a couple of times to mix the blood and saline. Add more of the proper solution at the edge of the coverslip, if necessary.

6. Choose an area where the cells are thinly distributed and observe under high power. The differences are best compared if three microscopes are used at once, side by side.

7. Refer to Figure 5.3 to label the red blood cell illustrations in Figure 5.4 with the proper solution designated and the condition of the cell (hemolyzed, normal, or crenated).

TURGOR AND PLASMOLYSIS

The **protoplast** (plasma membrane plus all cytoplasm) of a plant cell is confined by a cellulose **wall**, which the plasma membrane secretes. This wall is fairly flexible, but normally plant cells contain enough water to push the cell membrane firmly against the wall, a condition called **turgor**. The rigidity of non-woody plants is due to the turgid condition of individual cells, collectively.

If a plant cell is placed in a hypotonic solution, water will flow into the cell until the wall pressure (resistance of the wall) equals the turgor pressure (osmosis causing the protoplast to swell outward). For this reason plant cells do not rupture as animal cells do in hypotonic situations. It is virtually impossible to tell whether a cell is in an isotonic or hypotonic solution by just observing the cell, but in hypertonic solutions, plant cell protoplasts lose water and shrink (**plasmolyze**). This is analogous to crenation in animal cells.

If you want to make lettuce salad crisp, would you put it in salt water or tap water? _____

Mount a leaf of _Elodea_ in water on a slide and observe. These leaves are two cells thick, so you should be able to focus up and down to see that the cells in one layer are larger then those in the other. When one layer is in focus you may be able to see the shadowy outlines of cell walls in the other layer.

As the cells warm, the cytoplasm will begin to flow faster, carrying the green egg-shaped chloroplasts and the nearly transparent nucleus around the cell. Do the chloroplasts and cytoplasm flow uniformly all over the cell or around the edges? (Don't let the other layer's wall shadows fool you.)

_____

Blot off the tap water and add salt solution (7% NaCl) by adding the salt at one side of the coverslip and drawing it under by using a paper towel to absorb the fluid from the opposite side of the coverslip. Let it sit for a minute or two and observe again. The cell contains a **vacuole** which holds most of the cell's water, similar to a water bag inside the cell. Note the clear space forming between the cell wall and the cell membrane.

Has water left or entered the cell in the salt solution? _____

Rinse the salt solution from the leaf and remount in tap water. Observe periodically for about 5 minutes.
How is the cell changing? _____
_____

Account for this change in terms of osmosis and the characteristics of the solution. _____
_____

Draw some _Elodea_ cells in tap water and some cells in salt water. Label the cell wall, cell membrane and chloroplasts.

Type of Solution

(_____% NaCl)

Condition of Cells

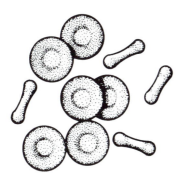

Type of Solution

(_____% NaCl)

Condition of Cells

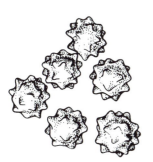

Type of Solution

(_____% NaCl)

Condition of Cells

FIGURE 5.4 **Effects of Osmosis on Human Red Blood Cells**

REVIEW QUESTIONS

1. Which is the largest molecule, starch, iodine, or water? _____

    How do you know? _____

2. Does Brownian movement occur in the cytoplasm of living cells? _____

3. What causes Brownian movement? _____

    _____

4. Would the salt in the ocean be considered the solute, or the solvent? _____

5. A red blood cell is dropped into a 2% salt solution. In relationship to the cell's contents, this solution is:

       a. hypertonic   b. hypotonic   c. isotonic

6. Do plant cells have cell membranes? _____

7. Turgor occurs in plant cells, but not in animal cells, due to the presence of a _____ which offsets osmotic pressures.

8. Crenation occurs in cells in a solution which is:

       a. hypertonic   b. hypotonic   c. isotonic

9. What chemical substance was used in the lab to demonstrate Brownian movement? _____

10. Define **Diffusion**: _____

11. How can the diffusion of molecules be accelerated? (*Hint:* Think about how you can cause tea to sweeten faster.)

    a. _____

    b. _____

    c. _____

12. What material was used in the lab to demonstrate the diffusion of vapor molecules? _____

13. Describe a selectively permeable membrane. _____

14. The diffusion of water molecules through a semipermeable membrane is termed _____.

15. What material did you use to represent an "artificial cell" in Experiments 1 and 2? _____

16. What is the solute used in Experiment "1" on page 37? _____

17. What is the solute used in Experiment "2" on page 37? _____

18. Define:

    Hemolysis _____

    Crenation _____

    Turgor _____

    Plasmolysis _____

Enzymes

OBJECTIVES

After completion of this exercise, the student should be able to do each of the following:

- Explain how changes in enzyme concentration affect enzyme activity.
- Explain the effects of extreme pH and heat on protein structure and function.
- Explain how cheese is produced. Use the following terms: **curd**, **rennilase**, **enzyme**, **whey**, **catalyst**, and **coagulation**.
- Answer the review questions at the end of this exercise.

INTRODUCTION

Enzymes are proteins which act as **catalysts** (substances which regulate the rate of chemical reactions, but which do not permanently unite with the reactants). The **substrate**, the molecule that the enzyme will alter, must physically combine with the enzyme in much the same way as a key fits into a lock. This combination somehow stresses the internal chemical bonds of the substrate and forces the reaction to occur. The region of the enzyme where contact is made between the substrate or substrates and enzyme is called the **active site**. The more molecules of enzyme that are present, the more molecules of substrate can be processed at any one time. Therefore, the rate of reaction depends in part on the concentration of the enzyme. These events also depend critically on the physical shape of the enzyme and its complementarity to the substrate. Each enzyme is specific, that is, it catalyzes only one chemical reaction at a time. The shape of an enzyme is controlled, as with all proteins, by the formation of very weak **hydrogen bonds** that hold the long chain of amino acids in a particular three dimensional shape. Heat, acids, and bases will all break these weak bonds and ruin the unique shape of a protein. The process is called **denaturation** and it destroys enzyme activity.

Most enzymes work within the confines of living cells, but many digestive enzymes, such as amylase, are secreted to the outside. In the oral cavity, amylase catalyzes the breakdown of the substrate, **starch**, into the product **maltose**. Starch is a long **polysaccharide polymer**, of the simple six carbon sugar glucose which is a **monomer**. Amylase breaks these sugars off the starch molecule two at a time to produce the **disaccharide** maltose. The enzyme is not used up in the reaction and one molecule of enzyme can act over and over again.

EFFECT OF ENZYME CONCENTRATION ON ENZYME ACTIVITY

In this experiment and the others that follow, you will test for the presence or absence of starch as an indication that amylase has acted. Iodine or Lugol stain is added to test for the presence of starch. If starch is still present, the stain will turn black or blue-black. If the enzyme has converted the starch to maltose, the stain will remain light brown.

In your lab you will find several stock bottles of prepared solutions.

1. Number test tubes 1–8 with a wax pencil and place in a test tube rack.

2. Add the materials to the first 5 tubes as indicated in the table and place a clean Pasteur pipet into the test tubes.

TEST TUBE	TEST MATERIAL
1	4ml enzyme/amylase
2	3ml pH 7 buffer
3	3ml distilled water
4	3ml starch
5	3ml iodine
6	1:2 enzyme dilution
7	1:4 enzyme dilution
8	1:8 enzyme dilution

NOTE

A buffer solution is a substance that consists of acid and base and is resistant to pH change. The buffer will add or release hydrogen ions as they are needed or remove them during a chemical reaction.

3. You will now prepare several different concentrations of the enzyme stock. Take the remaining three empty test tubes labeled 6, 7, and 8.

NOTE

Be careful not to mix your pipets and test tubes.

a. Place 20 drops of enzyme stock from test tube 1 into test tube 6, add 20 drops of distilled water, and mix. Test tube 6 contains an enzyme solution one half as concentrated as the original. Place a clean pipet into test tube 6.

b. Fill a clean pipet from test tube 6 and put 20 drops into test tube 7 replacing the unused portion into test tube 6, add 20 drops of distilled water and mix. Place a clean pipet into test tube 7. Test tube 7 is now half as concentrated as tube 6 and one quarter as concentrated as the original stock in test tube 1.

c. Fill a clean pipet from test tube 7 and put 20 drops into test tube 8, replacing the unused portion into test tube 7. Add 20 drops of distilled water and mix. Place a clean pipet into test tube 8. Test tube 8 now is a solution one eighth as concentrated as test tube 1.

4. Take your two porcelain spot plates and place them side by side to form an array of four rows and six columns. With your wax pencil, label the rows on the left edge A to D, and the columns on the top edge 1 to 6 as shown below.

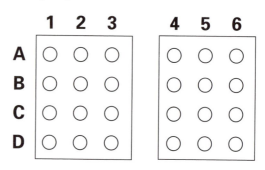

5. Place 5 drops of starch solution in each of the 24 wells in the spot plates. Then add 2 drops of pH 7 buffer to each well. **Stir the plates to mix thoroughly.**

NOTE

To mix drops in a well, hold the spot plate firmly on the table and rotate it in small circles.

6. Perform the following step as quickly as possible. Place one drop of undiluted enzyme, test tube 1, to each of the wells in Row A; one drop of test tube 6 to each well of Row B; one drop of test tube 7 to each well of Row C, and one drop of test tube 8 to each well of Row D. Try to add the enzymes as close in time as possible. When you finish, stir the plates and start timing. This is considered Time 0.

7. Immediately add 1 drop of iodine to each of the wells in column 1, stir, and record your observations for the four wells on the following table. Write + if the well stained black or blue-black, and − if the stain is unchanged.

ENZYME/ TEST TUBE	TIME 0	TIME 5	TIME 10	TIME 15	TIME 20	TIME 25
1	___	___	___	___	___	___
6	___	___	___	___	___	___
7	___	___	___	___	___	___
8	___	___	___	___	___	___

8. Five minutes after you added the iodine to column 1, add it to each of the wells in column 2, stir, and record your observations on the table. This is considered Time 5.

9. Continue to add iodine to columns 3–6 at 5 minute intervals, stir, and record your observations on the table.

10. When you have finished, carefully wash and dry the spot plates and test tubes 6, 7, and 8.

EFFECT OF pH ON ENZYME ACTIVITY

In this experiment you will keep the concentration of enzyme constant and vary the pH of the environment in which it acts.

1. Use the undiluted enzyme, test tube 1, and the pH 7 buffer, test tube 2, from the previous experiment and place in a test tube rack. The pH 7 buffer will serve as the control.

2. Write the number 6 on a clean test tube and add 3 ml of pH 2 buffer. The buffer of pH 2 will maintain acid conditions similar to those in your stomach.

3. Write the number 7 on a clean test tube and add 3 ml of pH 12 buffer which will maintain a very basic environment.

4. Use **only** one of the spot plates and rotate it so that you have three rows and four columns as shown here.

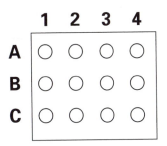

5. Place 5 drops of starch in each of the 12 wells in the spot plate.

6. Place 2 drops of pH 2 buffer in each well in Row A, 2 drops of pH 7 buffer in each well in Row B, and 2 drops of pH 12 buffer in each well in Row C. **Stir the plate to mix thoroughly.**

7. As close in time as possible, place one drop of your **unheated** enzyme, test tube 1, in each well and **stir**.

8. Immediately add 1 drop of iodine to each well of column 1, stir, and record your observations on the table. This is considered Time 0.

9. Continue to add iodine to columns 2–4 at 5 minute intervals, stir, and record your observations on the table.

10. When you have finished, carefully wash and dry the spot plate and test tubes 6 and 7.

BUFFER	TIME 0	TIME 5	TIME 10	TIME 15
pH 2	___	___	___	___
pH 7	___	___	___	___
pH 12	___	___	___	___

EFFECT OF TEMPERATURE ON ENZYME ACTIVITY

You probably realize that cooked food spoils more slowly than many raw foods. One reason for this is the denaturing effect of heat. Usually above 40°C, most enzymes active in bacteria that is found in the food become denatured. Their secondary or tertiary protein structure breaks down and the cells die because their enzymes won't work.

1. Write the number 6 on a clean test tube and add 2 ml of undiluted enzyme from test tube 1.

2. Place in a hot water bath (boiling water) for 5 minutes, remove using a test tube holder, and place in the test tube rack.

3. Use **only** one of the spot plates and rotate it so that you have three rows and four columns as shown here.

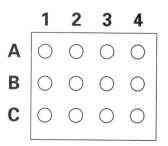

4. Place 5 drops of starch in each well of **Rows A and B only**.

5. Place 2 drops of pH 7 buffer in each well. **Stir the plate to mix thoroughly**.

6. As close in time as possible, place 1 drop of your unheated enzyme, test tube 1, in each well in Row A, and place 1 drop of your boiled enzyme, test tube 6, in each well in Row B. **Stir plates to mix thoroughly**.

7. Immediately add 1 drop of iodine to each well of column 1, stir, and record your observations on the table. This is considered Time 0.

8. Continue to add iodine to columns 2–4 at 5 minute intervals, stir, and record your observations on the table.

9. When you have finished, carefully wash and dry the spot plate and all test tubes.

ENZYME	TIME 0	TIME 5	TIME 10	TIME 15
Unheated	_____	_____	_____	_____
Heated	_____	_____	_____	_____

MAKING CHEESE

Milk (as shown by your results with the Biuret reagent, Exercise 3) contains a considerable amount of protein. The human stomach produces **rennin**, which converts this milk protein (**casein**) to insoluble **paracasein**, a shorter chainlength product of casein hydrolysis. This precipitates out of solution forming a **curd** and leaves the water and other dissolved substances in the milk behind as a watery **whey**. About three-fourths of the world population cannot produce rennin, and consequently, cannot digest milk protein as easily as those who have it.

In this experiment, cheese will be made from whole milk by the use of a similar enzyme (**rennilase**) produced by bacteria.

1. Put 1/2 cup or 125 ml whole milk in a beaker and warm to 32° C (88–90° F).

2. Add 3 drops of rennilase to the warmed milk and stir. Remove from heat and allow to stand undisturbed for 15 minutes.

3. By now a curd has been formed. Break up the curd and filter to remove the whey, using a piece of cheesecloth.

4. Salt to taste and press the curd between cheesecloth or paper towels.

5. For further processing, the cheese should be wrapped or placed in an air-tight container and left to age at 35° F to develop flavor.

REVIEW QUESTIONS

1. Define:

   Catalyst _____

   _____

   Denaturation _____

   _____

   Buffer _____

   _____

   Active site _____

   _____

   Substrate _____

   _____

2. What was Little Miss Muffet actually eating? _____

   _____

3. What were your conclusions regarding the effect of enzyme concentration on reaction rate? _____

   _____

4. What were your conclusions about the effect of acids and bases on enzyme activity? _____

   _____

5. Your mouth produces amylase and your stomach is very acidic. What do you think happens to the amylase

   when you swallow?_____

   _____

6. People preserve food by pickling (soaking in vinegar, which is acetic acid). How do you think this method

   works? _____

   _____

7. What were your conclusions about the effect of heat on enzyme activity? _____

   _____

Energy Capture, Fermentation, and Cellular Respiration

EXERCISE

7

EXERCISE · EXERCISE · EXERCISE

OBJECTIVES

After completion of this exercise, the student should be able to do each of the following:

- Explain the function of mitochondria and recognize them in specially-prepared slides.
- Write the generalized formula for respiration.
- Describe a simple test for CO_2 production in respiration.
- Explain the results of the experiment involving the respirometer.
- Describe the method used to demonstrate that the electron transport chain in respiration is operating.
- Answer the review questions at the end of this exercise.

MITOCHONDRIA

The mitochondria are the powerhouses of the cells; it is here that most of the organic fuels such as glucose are burned and ATP (an energy storage and transfer molecule) is produced. Although the chemical reactions of the process are intricate and extensive, the process may be briefly described by the following equation:

$$C_6H_{12}O_6 \text{ (glucose)} + 6O_2 \longrightarrow 6H_2O + 6CO_2 + \text{energy (to form ATP)}$$

On demonstration is a slide of tissue which was stained specifically to show mitochondria. Look for the small blue-green specks in the cytoplasm.

CARBON DIOXIDE PRODUCTION

The CO_2 you exhale from your lungs is one of the by-products from burning or respiring fuel by the cells of your body. To prove that you exhale CO_2, use a straw to blow bubbles gently into a test tube of limewater. The CO_2 combines with the $Ca(OH)_2$ or **limewater** to form a white precipitate often known as **chalk**.

$$Ca(OH)_2 + CO_2 \longrightarrow CaCO_3 \text{ (chalk)} + H_2O$$

OXYGEN CONSUMPTION

Seeds are essentially dormant plant embryos, accompanied by a carbohydrate fuel supply to help them

47

until they can make leaves and roots of their own and make their own food. When soaked in water, the embryos come out of their dormancy and begin to respire their food supply. We will use germinating seeds to demonstrate oxygen consumption by putting them in an airtight system called a **respirometer** (Figure 7.1). The consumption of O_2 will cause a weak vacuum in the system.

However, if only the seeds were enclosed, their CO_2 production would cancel out the vacuum caused by O_2 disappearance. For this reason potassium hydroxide (KOH) is added to absorb the CO_2. Thus, only oxygen volume will change.

$$2 \ KOH + CO_2 \longrightarrow K_2CO_3 + H_2O$$

We will also test seeds killed from soaking in formalin (if they are completely dead, they shouldn't respire) and aquarium gravel (non-living things cannot respire). The gravel will also allow us to make adjustments during the test period if air pressure or room temperature should change. If the tube containing gravel should show changes in readings from either of these causes, we should have to add or subtract that amount from our live-seed data to be sure exactly how much gas was used. Such a setup is called a **control**, to make sure our results were really due to the experiment performed and not to some outside factor.

Each table should set up one live-seed and one dead-seed respirometer. **One** table in the class should also prepare a gravel control.

1. Half fill the tube with seeds (or gravel), add a wad of cotton, and then 5 or 6 pellets of KOH. (The cotton keeps the KOH from touching and killing the live seeds.)

CAUTION

Do not handle KOH with bare fingers — it is lye!

---

2. Insert the stoppers with their tubing and place a clamp on the vertical tube.

3. Adjust the respirometers in the ring stand clamps so the horizontal tubes are level.

4. Let the apparatus sit for 2–3 minutes. The warmth of your hands on the tubes must dissipate, and the gas pressure must return to room pressure.

5. With a dropper add just enough dye to the horizontal tube to block the opening. As O_2 is used,

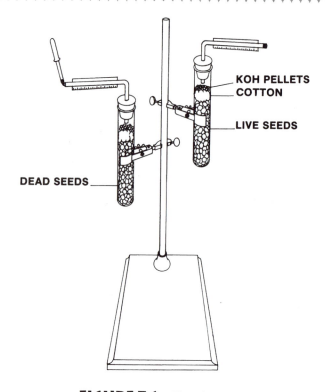

KOH PELLETS
COTTON
LIVE SEEDS
DEAD SEEDS

FIGURE 7.1 Respirometer

the drop will be pulled past the ruler and you can make readings.

6. Record your results on Table 7.1. (If the dye reaches the end of the ruler before readings are finished, unclamp the vertical tube and **gently** blow the dye back to the other end).

CAUTION

In reading the tubes, don't stand too close. Body temperature can raise the temperature of the tubes and change the position of the drop.

DEMONSTRATION OF ELECTRON TRANSPORT CHAIN

Respiration involves two chemical pathways: the **Krebs Cycle**, and the **Electron Transport Chain**. Each glucose molecule is broken down into two 3-carbon molecules (pyruvic acid) and each goes through a Krebs Cycle in which 8 high energy hydrogen electrons are released. These electrons (and 4 others released in other reactions) are passed along a chain of electron acceptor molecules (the Electron Transport Chain). As they are passed along, the electrons gradually lose most of their

TABLE 7.1

TIME (MIN)	GRAVEL (CM TRAVELED BY DROP FROM 1ST POSITION)	LIVE SEEDS	DEAD SEEDS
0			
1			
2			
3			
4			
5			
6			
7			
8			
9			
TOTAL MOVED			

energy and some of this energy is harnessed to generate ATP.

For many years, the chemical tetrazolium chloride ($C_{19}H_{15}N_4Cl$) has been used to determine if a seed embryo is alive and respiring. In the presence of this chemical, live embryos will turn a carmine red color and dead embryos will not change color. Live embryos are respiring and their electron transport chains are working. Some of the electrons that are being passed along the ETC are intercepted by tetrazolium chloride and the reduction of the chemical causes it to turn from clear to red.

Two beakers of soaked corn kernels (labeled 1 and 2) will be used for this experiment. In one beaker, the seeds have previously been placed in boiling water thus killing the embryos in the seeds. Place several kernels from each beaker in separate Petri dish halves labeled as 1 and 2. Place the seeds with flat side down and with a razor blade, cut them into right and left halves. Place several drops of 0.1% tetrazolium chloride in each dish and, using forceps, place the kernels cut side down into the drops. BE CAREFUL NOT TO GET TETRAZOLIUM CHLORIDE ON YOUR HANDS — THE CHEMICAL IS POISONOUS. After 30 minutes and after one hour, lift the dish halves and look beneath at the cut sections of the kernels.

Which numbered plate has the red stained embryos? _____

Which numbered dish has living respiring seed embryos? _____

FERMENTATION

A type of fungus called common baker's yeast produces CO_2 and ethyl alcohol (C_2H_5OH) when respiring glucose. This doesn't require oxygen and is called **fermentation.** CO_2 bubbles make the bread rise. Other yeasts produce the CO_2 bubbles in beer and some wines. The ethyl alcohol which is also produced evaporates during the baking of bread, but is retained in beer and wine.

A yeast-glucose suspension poured into a fermentation tube (Figure 7.2) (which is then tilted to remove any bubbles from the top of the tube) will show a collection of bubbles in a few minutes. Examine such a tube on demonstration, and shade in the yeast-glucose levels at the beginning of the experiment and after 10 or 20 minutes.

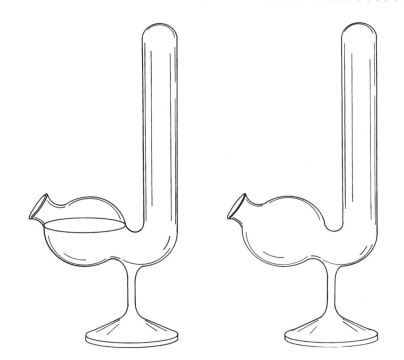

FIGURE 7.2 Yeast Fermentation

REVIEW QUESTIONS

1. Calcium carbonate ($CaCO_3$) is commonly known as _____.

2. Are bean seeds living?_____ How can you tell? _____
   _____

3. Name three products of fermentation by yeast. _____
   _____

4. What is the main product of respiration?_____

5. Describe a simple test for CO_2 production in respiration. You may use a labeled diagram for your explanation or list the steps. _____
   _____

6. In the experiment involving the respirometer, why is it important to use potassium hydroxide (KOH)? What function does it serve?_____
   _____

7. Why will we be testing seeds that have been soaked in formalin? _____
   _____

8. Describe how fermentation was demonstrated. _____

_____

9. Define the following terms:

Mitochondria (function and location) _____

Limewater _____

Respirometer _____

Potassium hydroxide (KOH) _____

Formalin _____

Control (as in the experiment set-up) _____

Fermentation _____

Ethyl alcohol_____

Photosynthesis

OBJECTIVES

After completion of this exercise, the student should be able to do each of the following:

- Explain the effect of light intensity on photosynthesis.
- Explain the experiments which involved chromatography, the relationship between duration of exposure of leaves to light and starch formation, and CO_2 consumption by a photosynthesizing plant.
- Explain the effect of light quality (wavelength) on photosynthesis.
- Explain how a spectrophotometer works, using plant pigments as an example. Explain what an absorption spectrum shows.
- Write the generalized equation for photosynthesis.
- Answer the review questions at the end of this exercise.

INTRODUCTION

The maintenance of life requires the continual use of energy. Radiant energy from the sun — this planet's only energy source — is captured by the process of photosynthesis and transformed into usable chemical fuels. All living things are directly or indirectly dependent upon this process.

The overall chemical equation for photosynthesis may be stated as follows:

$$6CO_2 + 6H_2O \xrightarrow[\text{chlorophyll}]{\text{light energy}} C_6H_{12}O_6 + 6O_2$$

This process takes place in the chloroplasts of the cell. The glucose ($C_6H_{12}O_6$) produced is usually converted into starch. This light-capturing process is affected by the duration, intensity, and quality of light and concentrations of CO_2, O_2, and water. About 90% of the photosynthesis on earth takes place in the sea, in algae. In multicellular plants, the photosynthetic organ (leaf) may vary in structure according to the habitat. The basic anatomy of the organ is designed to accommodate the needs of the process to a maximum. Most plants utilize **accessory pigments** to capture selected wavelengths of light in addition to those captured by **chlorophyll** directly.

DURATION OF LIGHT—DEMONSTRATION

Starch, the carbohydrate storage product of photosynthesis, may be stored anywhere in the leaf where photosynthesis is occurring or be shipped to a root or leaf for storage. We will use *Coleus* leaves to show that only the leaf areas which contain chlorophyll can make and store starch. This plant has vari-colored leaves; some areas contain only chlorophyll and look green, others contain the accessory pigment **anthocyanin** and look pink, still other areas contain no pigment and appear creamy white. Where chlorophyll and anthocyanin both occur, the color is a rusty brown-red.

The leaves were bleached in boiling alcohol, then stained with iodine to show the location of any starch deposits. Leaf A came from a plant which had been exposed to constant light for 72 hours prior to bleaching. Leaf B came from a plant which had been kept in the dark for 72 hours. Use colored pencils to indicate the areas of pigments and starch on the outlines in Figure 8.1.

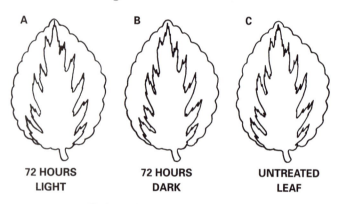

A	B	C
72 HOURS LIGHT	72 HOURS DARK	UNTREATED LEAF

FIGURE 8.1 *Coleus* **Leaves**

LIGHT INTENSITY

Since oxygen is a by-product of photosynthesis, the rate of oxygen production may be used to measure the rate of photosynthesis. Submerged aquatic plants release their oxygen in the form of bubbles, which may be counted over a period of time to determine **photosynthesis rate** indirectly. This may be done as a demonstration.

1. Select a fresh, crisp sprig of *Elodea* about 15cm in length.
2. While the plant is still submerged, cut 2–3 mm from its base.
3. Place the sprig upside down in a test tube filled with 0.25% sodium bicarbonate. This is a buffer to absorb toxic materials evolved.

4. Keeping the plant submerged, position a light source 1 ft away and adjust so that the light shines directly on the plant.
5. Place a beaker of water between the plant and the light to prevent overheating the plant. Allow the system to stand 7–10 minutes, or until bubbles begin to appear regularly.
6. Count the bubbles produced each minute for a 5-minute period and average.
7. Move the light back 2 ft from the plant, wait 5 minutes and repeat counting.
8. Move light back to 3 ft from plant and repeat the bubble counting.

Ft. from Light	Bubbles Per Minute					Average
	1	2	3	4	5	

Conclusions regarding light intensity and photosynthesis:

LIGHT QUALITY

Using the setup from the experiment above, place the lamp at the distance which produced the most bubbles per min (highest photosynthetic rate).

1. Place a colored filter between the test tube and the heat shield beaker and allow to set for 5 minutes. (How does this affect the color of light reaching the plant?)
2. Count bubbles for 5 minutes as in previous experiment.
3. Repeat steps 1 and 2 for each color filter available.

Color of Filter	Bubbles Per Minute					Average
	1	2	3	4	5	

Color of Filter	Bubbles Per Minute					Average
	1	2	3	4	5	

Conclusions regarding light quality and photosynthesis:

How would you expect a fluorescent light works?

LIGHT COMPOSITION AND COLOR

White light from the sun or a lamp is actually composed of many wavelengths. The feature of light quality called **color** actually represents different wavelengths of light and energy. The color of an object depends not on the wavelengths of light absorbed by it, but those transmitted or reflected by it. Thus, the chlorophyll pigments look green because the green wavelengths are transmitted, while the red and blue regions of the spectrum are mostly absorbed. The four main things which happen to light that strikes leaves can easily be demonstrated (Figure 8.2).

Absorbed: The results of the photosynthetic experiments demonstrated this indirectly.

Reflected: Examine the shiny surface of a leaf. The "shine" is reflected light.

Transmitted: Hold a bottle of chlorophyll extract up to a light. Transmitted light comes through.

Fluoresced: This does not normally happen in an intact leaf. In the pigment extract, the molecules are detached from their photosystems and cannot keep the energy they absorb, so they reradiate or fluoresce it. Hold the pigment extract so a strong light shines on it from over your shoulder. It should fluoresce a deep wine red.

When a beam of light passes through a prism, as in a **spectrophotometer** (Figure 8.3), it emerges as a spectrum of colors (different wavelengths) like a rainbow. (See Figure 8.4.) If a colored solution is placed between a light source and a prism, the solution will absorb some of the wavelengths and transmit others. This spectrum from which specific wavelengths have been absorbed is known as the **absorption spectrum** of that substance. Thus, the absorption spectrum of a pigment may be used to identify it, somewhat like a "fingerprint."

A plant cannot use all wavelengths of light, but by having pigments of various colors, it can capture a greater array of wavelengths than possible with chlorophyll alone. (There are several chlorophylls; your extract will contain various proportions of them.) Because of the variations in the composition of chloroplast pigments from plant to plant, the absorption spectrum in your text for just one chlorophyll, will not exactly match the one you will make. You are to determine the absorption spectrum for the available extract(s) as follows:

1. Turn the spectrophotometer on and allow it to warm up for about 15 min.

2. Fill one of the special "spec" tubes with acetone, insert into the sample holder, and close the lid. (This is your "blank" — the chlorophyll extract was made in acetone.)

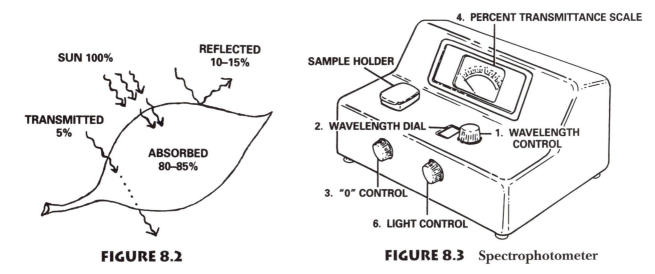

FIGURE 8.2

FIGURE 8.3 Spectrophotometer

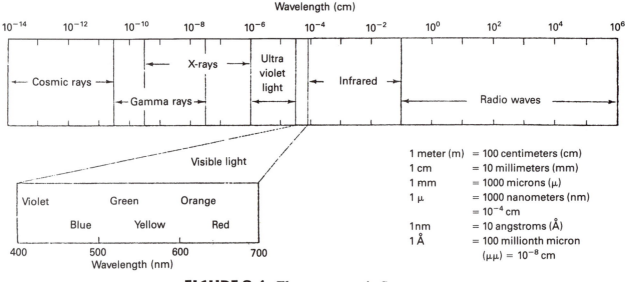

FIGURE 8.4 Electromagnetic Spectrum

3. Set the zero point on the transmittance scale with the "0" control. Read all data when you cannot see the arrow's reflection in the mirror. This eliminates **parallax** error. (You will have to repeat this zeroing for each new wavelength, since the acetone has its own absorption spectrum and you adjust the machine to "see" only chlorophyll, not acetone.)

4. Turn the wavelength control to 430 nm and set the absorbency at zero with the light control. Do not adjust the light control again.

5. Prepare a dilute sample of pigment by adding a few drops of extract to a "spec" tube half full of acetone. Adjust this sample so that it has an absorbency of about 0.7 at 430 nm. This assures you enough pigment to be detected throughout the range examined.

6. Now set the wavelength to 380 nm. Using the blank, re-zero absorbency. Insert pigment sample and read absorbency. Record your results on Table 8.1.

7. Repeat step 6, making readings every 20 nm up to 700 nm.

8. Plot points on graph paper and connect with a line to give the absorption spectrum.

If more than one plant extract is available, each team should gather data from one source and pool data on the blackboard. Graph the individual spectra so they may be distinguished from one another.

TABLE 8.1

nm	PLANT: Absorbency	PLANT: Absorbency
380		
400		
420		
440		
460		
480		
500		
520		
540		
560		
580		
600		
620		
640		
660		
680		
700		

CHROMATOGRAPHY

Chloroplast and chromoplast pigment solutions are obtained by rupturing plant cells and extracting the pigments with an organic solvent such as acetone.

CAUTION

The solvents used in this exercise are quite flammable. Keep solutions stoppered to avoid excess inhalation or accumulation of fumes in the room.

Pigment extracts will be prepared already, by homogenizing a small quantity of leaves in a blender with acetone then filtering to remove cellular debris. This extract stains intensely, so avoid spilling.

By a technique known as **paper chromatography**, mixtures of similar compounds may be separated on the basis of their different affinities for the solvent phase (acetone-ether) and the solid phase (paper) and hence different migration rates along the paper. Those with greatest affinity for the paper will travel slowest and those with greatest affinity for the solvent will travel the fastest.

1. Obtain a strip of filter paper which will fit easily into a test tube provided. Handle the paper only by the edges; oil from your fingers may interfere with the flow of the solvent.

2. Dip a micropipette into the extract. DO NOT cap the tube with your finger; the fluid will rise in the tube and remain by itself due to capillary action.

3. About 1 inch from the bottom of the paper, draw a quick line with a pencil completely across, leaving as thin a line as possible.

NOTE

Do not use ink because it disappears.

---

4. Repeat pigment applications 3 or 4 times, drying between times to prevent a wide band from forming — blow or hold under a lamp.

5. Pour about 1/2 inch of chromatography solvent into the tube **first**, then insert the paper and stopper **loosely** with a cork (ether fumes expand in a warm room and may "pop" the cork out.) The pigment line should not be submerged in the solvent.

6. Set the chromatogram aside for about 10–15 minutes to develop.

7. Remove the chromatogram before the uppermost color band runs off the top and allow to air dry. Mark the position of the solvent with a pencil.

8. Time and light cause the colors to fade out, so outline the top of each color band as soon as the chromatogram is dry. Then draw the color bands on the chromatogram in Figure 8.5.

9. You will note that an orange band of **carotenes** moves about as fast as the solvent front. Lower on the page will be one (or more) yellow **xanthophyll** bands, a bluish-green **chlorophyll a** band, a yellow-green **chlorophyll b** band, and in some chromatograms, a grey band between the carotene and xanthophyll; this is **pheophytin**, a breakdown product of chlorophyll indicating that the extract is becoming old. Red **anthocyanins** may show up at the base line as well. These are insoluble in the solvent. Label the bands on your chromatogram correctly.

10. Calculate the Rf for each of these pigments.

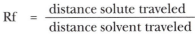

$$Rf = \frac{\text{distance solute traveled}}{\text{distance solvent traveled}}$$

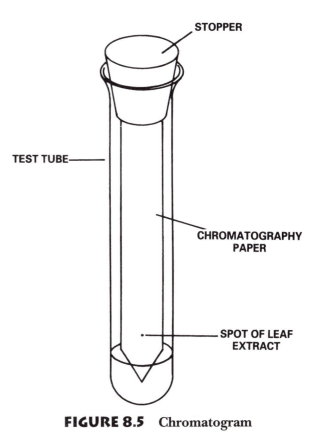

STOPPER

TEST TUBE

CHROMATOGRAPHY PAPER

SPOT OF LEAF EXTRACT

FIGURE 8.5 Chromatogram

CO₂ CONSUMPTION IN PHOTOSYNTHESIS

1. Into a test tube which is two-thirds filled with water, place a large leafy stem of *Elodea* and then 4 or 5 drops of 1% solution of phenol red. The drops of phenol red should turn the water a red to pink color. **Phenol red** is a pH-indicating compound which gives a red color in a basic or neutral solution (pH 7-14) but the color changes to yellow if the solution becomes acidic (pH 0–6.9)

2. Now insert a straw into the test tube and blow gently **only** until the water becomes orange-yellow in color. Some of the exhaled CO_2 is dissolving in the water to form carbonic acid (popularly known as carbonated water).

$$H_2O + CO_2 \longleftrightarrow H_2CO_3$$

3. Then immediately cease the blowing, remove the straw, and stopper the tube. Place the test tube in a well-lit area for photosynthesis to resume. After 10–20 minutes note any color change.

What was taken in by the *Elodea* plant during photosynthesis? _____

_____

_____

Considering the summary equation for photosynthesis, what is **the basis for** this change in color? _____

_____

_____

REVIEW QUESTIONS

1. Glucose is produced in which organelle? _____

2. Phenol red dropped into distilled water would produce what color? _____

3. What acid is produced by bubbling CO_2 into H_2O? _____

4. Name the bluish-green pigment in leaves. _____

_____

5. Can an albino corn leaf produce any food? _____ Explain. _____

_____

6. How does the intensity of light affect photosynthesis? _____

_____

7. What does a small Rf value indicate about the characteristics of the moving molecules? _____

_____

_____

8. Is it possible to have an Rf value greater than one? _____ Why? _____

_____

_____

9. According to your spectrophotometric data, which colors of light are used in photosynthesis? _____

_____

10. What is the main product of photosynthesis? _____

11. How did we demonstrate the use or consumption of CO_2 by a plant? Include the name of the chemical indicator and what will cause it to change color. _____

_____

_____

12. Define the following terms:

Chloroplast (function and location) _____

Elodea _____

Phenol red _____

pH_____

Chlorophyll _____

Carbonic acid _____

Paper chromatography _____

Carotenes_____

Xanthophyll_____

Chlorophyll a_____

Chlorophyll b _____

Pheophytin _____

Anthocyanins_____

Spectrophotometer _____

13. How did the duration of light affect the leaf's ability to form starch?_____

_____

_____

14. How was the photosynthetic role affected by changes in light intensity? _____

_____

_____

15. Which color filter yielded the highest rate of photosynthesis?_____

Lowest?_____

Mitosis and Meiosis

OBJECTIVES

After completion of this exercise, the student should be able to do each of the following:

- List the events that occur as a plant cell divides.
- List the events that occur as an animal cell divides.
- Demonstrate a procedure to stain tissue for the identification of mitotic stages in once living tissue.
- Manipulate chromosome models to demonstrate the events of meiosis I and II and crossing over.
- Describe the process of gametogenesis.
- Answer the review questions at the end of this exercise.

INTRODUCTION

The first part of the cell theory states that cells are the structural and functional units of living organisms. The second part of the cell theory states that all cells come from pre-existing cells. Cell division is the most common way of accomplishing this. Cell division consists of 2 processes: nuclear division (by **mitosis** or **meiosis**) and **cytokinesis** (division of the cytoplasm and associated parts).

Nuclear division by mitosis results in two daughter cells which are identical (have the same number of chromosomes) as the parent cell. This enabled you to grow from a fertilized egg into an adult human being; it is also the method by which cells are made to replace those killed or removed by a wound. In multicellular organisms, some cells (such as red blood cells, muscle cells, and liver cells) give

up the ability to divide mitotically in order to become specialized for some specific task. Except for such very specialized cells, the cells of any organism are potentially able to undergo mitotic division at any time.

Nuclear division by meiosis results in daughter cells in which the chromosome number is reduced to one-half of that in the parent cell. Specific reproductive cells (such as plant spores and animal eggs and sperm) in most eukaryotic organisms are capable of nuclear division by meiosis.

MITOSIS

PLANT CELL DIVISION

Since the chromosomes of plants are usually larger than those of animals and more easily seen, we will start with them, and go on to animal cells once the stages are identified here (Figures 9.1–9.7). Select a prepared, stained slide of *Allium* (onion) root tip. Refer also to Figure 9.9.

The rounded end of the root is the tip, and is covered by the cells of the **root cap**. These cells have relatively thin cytoplasm and are not dividing. Their function is simply protection of deeper tissues. The cells which will be dividing are concentrated just behind this cap in an area which is called the **meristem**. The cytoplasm of these cells is much denser and many dividing cells should be visible if you move the slide around a bit. The cells are not dividing in sequence, so consecutive stages will not be right next to each other. Shop around until you find **all** of the stages listed.

Be able to identify any of the stages if shown under the microscope. (Just looking at the pictures is not enough. The real thing never looks like a diagram, and you may be tested from the real thing and not the pictures.) Remember that these cells are dead, so they won't change stages before your eyes. Also any one stage can be subdivided into an early, middle, and late stage, each of which will look just a tiny bit different.

ANIMAL CELL DIVISION

Pick out a slide of whitefish blastula. This is an embryonic stage of fish development. At the same time the embryo was sliced, it was essentially a ball of rapidly dividing cells. These embryos contain some oil as stored food for the embryo, and you will find dark purple spheres of this oil scattered over and among the individual cells.

Since virtually all of the cells are dividing, interphase will probably be the hardest to find. If a cell looks as if it has no nucleus, that is because the slice didn't happen to hit the nuclear material. The chromosomes are very tiny compared to those of plants, with the **asters** filling almost the whole cell.

Make sure you can identify the stages diagramed (Figures 9.1–9.7) if shown the real thing on a slide, as you did with plant cells. Refer also to Figure 9.8 for stages of mitosis in animal cells.

Interphase (Figure 9.1)

1. DNA, present in a granular-appearing mass known as **chromatin**, replicates.
2. **Nucleolus(i)** is present.
3. **Nuclear membrane** is clearly visible.

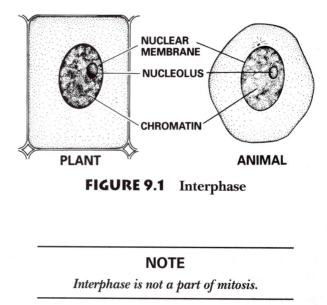

FIGURE 9.1 Interphase

NOTE

Interphase is not a part of mitosis.

Prophase (Figure 9.2–9.3)

1. DNA coils tightly to form visible **chromosomes** which appear rod-shaped. One strand of a double-stranded chromosome is known as a **chromatid**.

2. Nuclear membrane dissolves.

3. Nucleolus(i) disappears.

4. In animal cells, a pair of **centrioles** (not visible on these slides) separate and with their newly formed **asters** (visible) move to each pole.

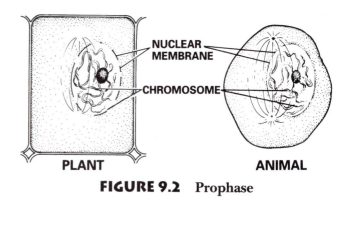

FIGURE 9.2 Prophase

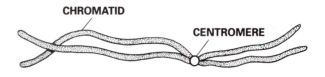

FIGURE 9.3 Double-Stranded Chromosome

Metaphase (Figure 9.4)

1. Chromosomes line up at the **equator** of the cell.

2. One **spindle fiber** of the very conspicuous spindle system attaches to the **centromere** (**kinetochore**) of each chromosome.

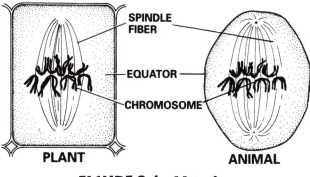

FIGURE 9.4 Metaphase

Anaphase (Figure 9.5)

1. Centromeres divide.

2. Single-stranded chromosomes (now consisting of one DNA molecule each) move to opposite poles.

FIGURE 9.5 Anaphase

Telophase (Figure 9.6)

1. Chromosomes begin to uncoil into chromatin.

2. Nucleolus (i) and nuclear membrane begin to reappear.

3. Spindle slowly dissolves.

4. In animal cells, the centrioles double in number and the aster dissolves.

5. In animal cells, cytokinesis occurs with a **cleavage furrow** dividing one cell into two cells.

6. In plant cells, cytokinesis occurs with the formation of a **cell plate**. This plate divides the one cell into two cells and will later form the **middle lamella**, a common partition between adjacent cells.

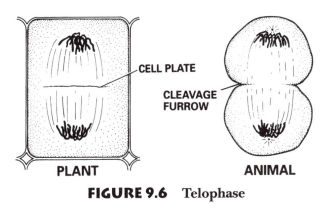

FIGURE 9.6 Telophase

Daughter Cells (Figure 9.7)

This is the same as interphase, but these cells are smaller and less mature than the original cell.

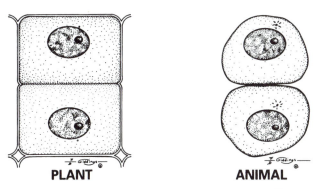

FIGURE 9.7 Interphase (Daughter Cells)

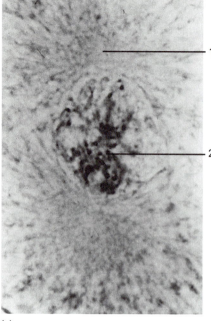

(a)

1. Centriole   2. Chromatid

Prophase
Each chromosome consists of two chromatids jointed by a centromere. Spindle fibers extend from each centriole.

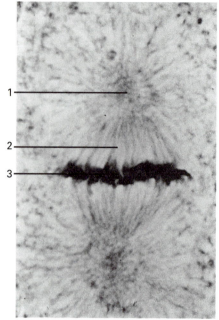

(b)

1. Centriole   3. Chromatids
2. Spindle fibers   at equator

Metaphase
The chromosomes are positioned at the equator. The spindle fibers from each centriole attach to the centromeres.

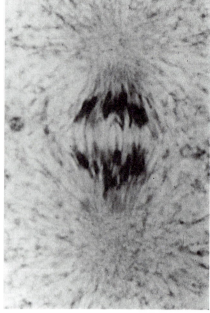

(c)

Anaphase
The centromeres split, and the sister chromatids separate as each is pulled to an opposite pole.

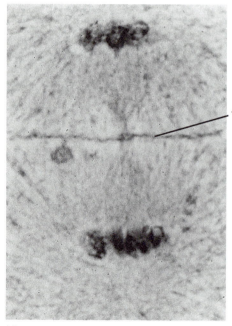

(d)

1. Cell membrane

Telophase
The chromosomes lengthen and become less distinct. The cell membrane forms between the forming daughter cells.

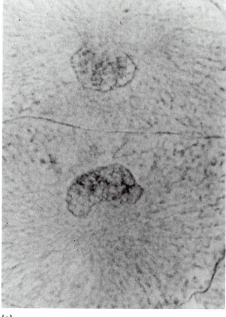

(e)

Daughter cells
Cell division is complete and the newly formed cells grow and mature.

FIGURE 9.8
Stages of Mitosis in Representative Animal Cells

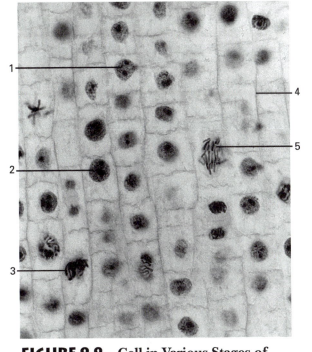

FIGURE 9.9 Cell in Various Stages of Mitosis From an Onion, *Allium*, Root Tip

1. Interphase   3. Metaphase   5. Anaphase
2. Prophase   4. Cell wall

8. Examine the slide for the stages of mitosis as well as interphase and cytokinesis.

9. Repeat the procedure if the slide is not satisfactory. **First check with your instructor.**

10. Make some diagrams below of what you observed.

PREPARING AN ONION ROOT TIP SQUASH

Your instructor has prepared some living onion root tips for you. Mitosis is usually confined to cells near the top of the root.

1. Obtain an onion bulb that shows some roots. Cut off a root and place it on a clean slide.

2. Cut off 1 mm to 2 mm of the root tip and throw the upper portion of the root away.

3. Cover the root tip with 4 drops of 1 N HCl and warm the slide over an alcohol burner (or bunsen burner) flame for 1 minute. **Do Not Boil.**

4. Blot off the excess HCl and cover the root tip with .5% aqueous toluidine blue.

5. Again, pass the slide through the alcohol burner flame for 1 minute without boiling.

6. Blot off the excess stain and add a drop of fresh stain and apply a coverslip.

7. Cover the slide with a paper towel and carefully squash the coverslip firmly with a pencil eraser or your thumb.

MEIOSIS

Meiosis reduces the chromosome number by one-half during the formation of gametes in animals and spores in plants and fungi. Meiosis involves two successive nuclear divisions that produce four daughter cells. The first division (meiosis I) is the reduction division; the second division (meiosis II) separates the chromatids. During meiosis I, chromosomes come together in pairs called **homologous chromosomes** or homologous pairs. Homologous pairs of chromosomes are chromosomes that look alike and carry genes for the same traits. Later in meiosis I, these pairs of homologous chromosomes are separated so that each gamete (or spore) contains only one of each chromosomal pair. Cells with homologous pairs of chromosomes before meiosis are said to be **diploid** or 2N (two examples of each chromosome). After meiosis, the cells have only one example of each chromosome and are said to be **haploid** or 1N. When haploid gametes unite during fertilization, the resulting zygote is diploid, having received one chromosome of each pair from each parent.

Mitotic cell division produces new cells genetically identical to the parent cell. Meiosis increases genetic variation in the population. Variation increases dramatically because, during meiosis I, each pair of homologous chromosomes come together in a process known an **synapsis**. Chromatids of homologous chromosomes may exchange parts in a process called **crossing over**.

SIMULATION OF MEIOSIS

In this exercise you will study the process of meiosis using chromosome simulation kits. Your kit contains two strands of yellow beads and two strands of red beads. One of the yellow strands represents the chromosome contribution of one parent, and one of the red strands represents the chromosome contribution of the other parent. These two strands represent homologous chromosomes. The second yellow and red strands are to be used as chromatids for each of these chromosomes.

Interphase

Place one strand of red beads and one strand of yellow beads near the center of your work area. (Recall that chromosomes at this stage would exist as diffuse chromatin and not as visible structures.) DNA synthesis occurs during interphase and each chromosome, originally composed of one strand, is now made up of two strands, or **chromatids**, joined together at the **centromere** region. Stimulate DNA replication by bringing the magnetic centromere region of the second red strand in contact with the centromere region of the first red strand. Do the same with its homolog, the yellow strand. Refer to Figure 9.10.

Prophase I

Meiosis I begins with prophase I. Homologous chromosomes come together and synapse along their entire length. The pairing of homologous chromosomes represents a difference between mitosis and meiosis. A **tetrad**, consisting of four chromatids, is formed. Entwine the two chromosomes to stimulate synapsis and then stimulate the process of crossing over. Crossing over can be simulated by popping the beads apart on a red chromatid, at the fifth bead or "gene," and doing the same with a yellow chromatid. Reconnect the red beads to the yellow chromatid and reconnect the yellow beads to the red chromatid. Proceed through prophase I of meiosis and note how crossing over results in recombination of genetic information. Refer to Figure 9.11.

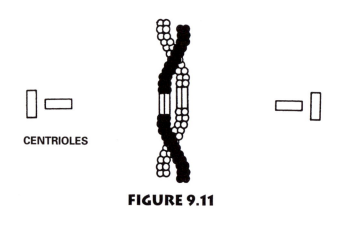

CENTRIOLES

FIGURE 9.11

Metaphase I

The crossed-over tetrads line up in the center of the cell. Position the chromosomes near the middle of the cell. Refer to Figure 9.12.

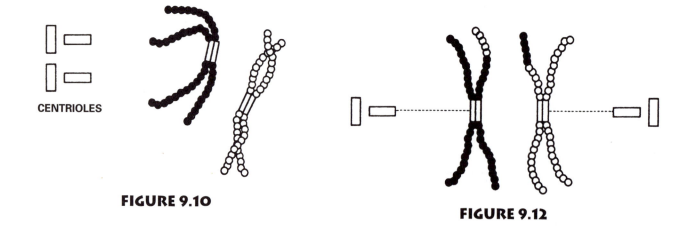

CENTRIOLES

FIGURE 9.10

FIGURE 9.12

Anaphase I

During anaphase, the homologous chromosomes separate and are "pulled" to opposite sides of the cell. Refer to Figure 9.13. This is another difference between the events of mitosis and meiosis.

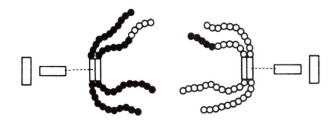

Telophase I

Place each chromosome at opposite sides of the cell. Centriole duplication takes place at the end of telophase in preparation for the next division. Formation of a nuclear envelope and division of the cytoplasm (cytokinesis) often occur at this time to produce two cells, but this is not always the case. To simulate cytokinesis, place a pencil or ruler vertically between the separated chromosome pairs. Note that each chromosome within the two daughter cells still consists of two chromatids. Refer to Figure 9.14.

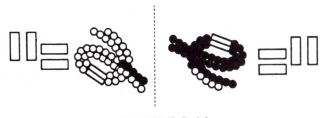

FIGURE 9.14

A second meiotic division is necessary to separate the chromatids of the chromosomes in the two daughter cells formed by the first division. This will reduce the amount of DNA to one double-helical strand per chromosome. This second division is called meiosis II. It resembles mitosis except that only one homolog from each homologous pair of chromosomes is present in each daughter cell undergoing meiosis II.

Interphase II (Interkinesis)

The amount of time spent "at rest" following telophase I depends on the type of organism, the formation of new nuclear envelopes, and the degree of chromosomal uncoiling. Because interphase II does not necessarily resemble interphase I, it is often given a different name — interkinesis. DNA replication does not occur during interkinesis. This represents another difference between mitosis and meiosis.

Prophase II

Meiosis II begins with prophase II. Replicated centrioles separate and move to opposite sides of the chromosome groups. Refer to Figure 9.15.

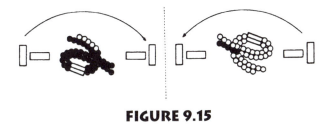

FIGURE 9.15

Metaphase II

Orient the chromosomes so they are centered in the middle of each daughter cell. Refer to Figure 9.16.

Anaphase II

The centromere regions of the chromatids now appear to be separate. Separate the chromatids of the chromosomes and pull the daughter chromosomes toward the opposite sides of each daughter cell. Since each chromatid has its own visibly separate centromere region, it can be called a chromosome. Refer to Figure 9.17.

Telophase II

Place the chromosomes at opposite sides of the dividing cell. At this time a nuclear envelope forms and the cytoplasm divides. Place a pencil or ruler between these divided cells to simulate cytokinesis. Refer to Figure 9.18.

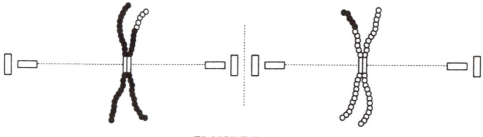

FIGURE 9.16

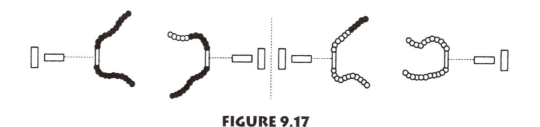

FIGURE 9.17

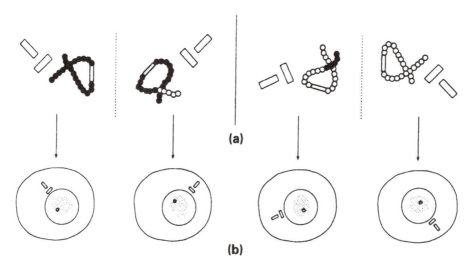

(a)

(b)

FIGURE 9.18

GAMETOGENESIS IN *Ascaris*

Gametogenesis is the formation of reproductive cells called **gametes**. Gametogenesis which occurs in the testes of males is called **spermatogenesis**, the formation of sperm. **Oogenesis**, the formation of ova or eggs, occurs in the ovaries of females. Gametogenesis will be discussed in detail in Exercise 34, but a brief description of meiosis using the parasitic roundworm *Ascaris* is described below.

In the formation of sperm, a **primary spermatocyte** undergoes meiosis I to produce two **secondary spermatocytes** which go through meiosis II each producing two haploid **spermatids**. The spermatids differentiate into **spermatozoa** or **sperm**. Therefore, one primary spermatocyte gives rise to four viable sperm.

Oogenesis is slightly different in that when meiosis occurs there is unequal division of the cytoplasm. The **primary oocyte** undergoes meiosis I to produce a large **secondary oocyte** and a small **polar body**. The secondary oocyte undergoes meiosis upon the penetration by the sperm to produce an **ootid** and another polar body. The ootid will differentiate into an **ovum** or egg. Sometimes, the first polar body will undergo meiosis II resulting in two polar bodies. Therefore, one primary oocyte may give rise to one ovum and three nonfunctional polar bodies.

1. Examine a series of slides of meiotic divisions in *Ascaris* ovaries. *Ascaris* cells contain a diploid number of four chromosomes. Refer to Figure 9.19 and locate the following: sperm penetration, primary oocyte, first polar body, secondary oocyte, second polar body, and ovum.

2. Make labeled drawings in the space provided of what you observed.

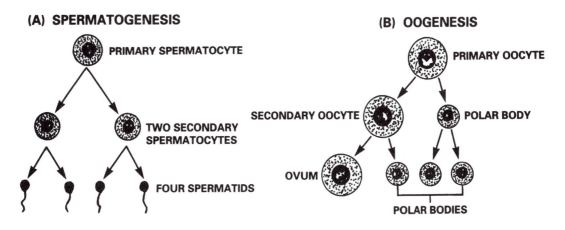

(A) SPERMATOGENESIS

PRIMARY SPERMATOCYTE

TWO SECONDARY SPERMATOCYTES

FOUR SPERMATIDS

(B) OOGENESIS

PRIMARY OOCYTE

SECONDARY OOCYTE

POLAR BODY

OVUM

POLAR BODIES

FIGURE 9.19   Meiosis in Spermatogenesis and Oogenesis

REVIEW QUESTIONS

1. What are the most obvious features of interphase in plant or animal cells? _____
   _____

2. What are the two most obvious differences in plant and animal cell division? _____
   _____

3. What is the difference between mitosis and cytokinesis? _____
   _____

4. Name the 4 major phases of mitosis in correct order. _____
   _____

5. What process is busily occurring in a meristem? _____
   _____

6. What plant was used in your study of mitosis? Why? _____
   _____

7. Define:

   Root cap _____

   Meristem _____

8. In your study of animal mitosis, what animal tissue was used? Why? _____
   _____

9. What are the major differences between mitosis and meiosis? _____
   _____

10. How can crossing over introduce additional genetic variability into the products of meiosis?_____
    _____

11. Define spermatogenesis and oogenesis. _____
    _____
    _____

Mendelian Genetics

After completion of this exercise, the student should be able to do each of the following:

- Define and apply the following terms in genetics problems: **gene**, **trait**, **genotype**, **allele**, **heterozygous**, **homozygous**, **phenotype**, **test cross**, **P generation**, **F₁ generation**, **F₂ generation**, **monohybrid cross**, **dihybrid cross**, **dominant**, **recessive**, and **codominant** **(incompletely dominant)**.

- List the different A-B-O blood types, give the possible genotype(s) responsible for each phenotype, and solve problems involving blood type.

- Use the **Punnett square** or the **forked line method** for determining the **genotype** and **phenotype** **ratios** expected from any given cross.

- Determine, by counting corn kernels, whether the kernels were produced as a result of (a) a cross between two heterozygous plants, (b) a test cross with a heterozygous plant, (c) a test cross with a homozygous dominant plant, or (d) a cross between two homozygous recessive plants.

- Explain the processes of DNA replication, transcription, and translation (protein synthesis).

- Answer any of the questions, including the review questions, in this exercise.

INTRODUCTION

In sexual reproduction of multicellular organisms, two haploid gametes unite at fertilization to form a diploid zygote, which by successive divisions will give rise to a diploid organism. Every cell of that organism contains all the genetic information present in the original zygote with one notable exception — the reproductive cells. In the formation of those reproductive cells, **meiosis** occurs and each reproductive cell thus contains only half of the genetic information present in a diploid cell. That information is contained in the chromosomes, which occur in pairs in a diploid cell. Each chromosome consists of hundreds of structural and functional subunits called **genes**. Each gene contains coded information for making a specific protein. And it is the complement of proteins (particularly enzymes) which an individual is capable of producing that determines the individual's observable physical **traits**, or characteristics, such as eye color, height, blood type, etc.

For a trait such as eye color, each individual possesses his own **phenotype** such as blue eyes. For any one trait, then, there may be a number of different possible phenotypes.

In a diploid cell, genes occur in pairs just as do the chromosomes. A trait may be controlled by only one pair of genes, or it may be controlled by more than one pair of genes. In the following discussion and activities we will consider only traits that are controlled by one pair of genes.

The members of a **homologous** (matched) pair of chromosomes are matched gene for gene, but often the two members of a gene pair may be slightly different. Different forms of a gene that occupy the same position on homologous chromosomes are called **alleles**. It is possible for more than two different alleles to exist for a particular gene location; however, one individual could only possess two of them. If an individual possesses two different alleles for a given trait, the individual is said to be **heterozygous** for that trait; however, if both alleles are alike, the individual is said to be **homozygous** for that trait. Even though the phenotype can be observed, the actual genetic makeup is not always apparent. The genetic makeup of an individual is called the **genotype**. In the case of an individual that is heterozygous for a particular trait, **both** genes may influence the resulting phenotype of that individual. When the resulting phenotype is influenced by two different alleles in the same individual, this is considered to be a type of **intermediate inheritance**, or **incomplete dominance** (codominance).

In many cases, one allele may completely mask the presence of the other one. The "stronger" allele is thus called the **dominant** gene, or allele, and the "weaker" is the **recessive**. The dominant phenotype will be present in either of the following: (1) both alleles are dominant, or (2) one allele is dominant and one is recessive. The recessive characteristic will be present only when both alleles are recessive. Inheritance of this type is said to be **dominant-recessive inheritance**.

In determining how traits are inherited, we must consider the following: Since the segregation of information during meiosis and recombination of information during fertilization is left to chance, we use laws of probability in predicting the percentage of individuals expected to exhibit certain characteristics from a given genetic cross. We need to take into consideration all possible gametes formed and all combinations of genes possible as a result of fertilization.

In this lab we will examine examples of human inheritance as well as the inheritance of certain characteristics in corn.

HUMAN INHERITANCE

Even though controlled genetic experiments are not conducted in order to determine human inheritance, it has been possible by tracing family histories to determine human inheritance for some physical features and physiological characteristics.

In this section of the lab, we will look at some examples of dominant-recessive inheritance in humans and at the inheritance of blood types.

DOMINANT-RECESSIVE INHERITANCE

Working with a lab partner, determine whether or not you possess the following characteristics. Indicate your phenotype and your possible genotype(s) and those of your lab partner. The phenotype is expressed in words describing the characteristic possessed. The genotype is expressed in a shorthand form using letters to indicate the alleles present for each trait. For a given trait, the same letter is used twice (since alleles occur in pairs). In the case of dominant-recessive inheritance, a capital letter is used to indicate the presence of the dominant allele; and a small letter to indicate the presence of a recessive allele. Use a different letter for each of the eight traits that follow.

PTC taster

The ability to taste phenyl-thio-carbamide is controlled by a dominant allele. (Get a small piece of paper saturated with the chemical and chew it; if it "tastes" just like paper, you are a non-taster.) Non-tasters have homozygous recessive genotypes.

Your phenotype: _____

Partner's phenotype: _____

Your possible genotype(s): _____

Partner's possible genotype(s): _____

If you and your partner were of the opposite sex and married, list all the possible genotypes that your

children might have: _____

_____

_____

_____

_____

If your first child was a non-taster, then your genotype would have to be _____ and your partner's would have to be _____.

U-shaped Tongue

The ability to roll the tongue into a trough-like shape when extending it from the mouth; controlled by a dominant allele.

Your phenotype: _____

Partner's phenotype: _____

Your possible genotype(s): _____

Partner's possible genotype(s): _____

If you and your partner were siblings, what possible combinations of genotypes might your parents have? (List **all possible** combinations)

_____

_____

_____

_____

_____

If your mother was a non tongue-roller then your genotype would have to be _____ and your partner's would have to be _____.

Widow's Peak

Hairline pointed over the center of the forehead; the allele for this is dominant over the allele for straight or curved hairline. See Figure 10.1.

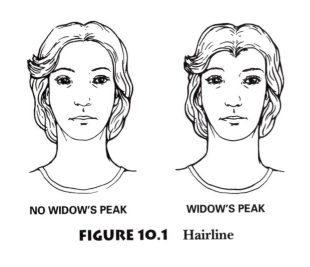

NO WIDOW'S PEAK   WIDOW'S PEAK

FIGURE 10.1 Hairline

Your phenotype: _____

Partner's phenotype: _____

Your possible genotype(s): _____

Partner's possible genotype(s): _____

Free Ear Lobe

The lobe is not attached; controlled by a dominant allele. See Figure 10.2.

Your phenotype: _____

Partner's phenotype: _____

Your possible genotype(s): _____

Partner's possible genotype(s): _____

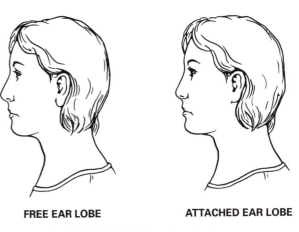

FREE EAR LOBE   ATTACHED EAR LOBE

FIGURE 10.2 Ear Lobes

Bent Little Finger

The end segment of the little finger bends toward the ring finger when placed against a straight edge; controlled by a dominant allele. If the fingers are straight and extend parallel to each other out to the very tip, this is recessive. See Figure 10.3 on the following page.

Your phenotype: _____

Partner's phenotype: _____

Your possible genotype(s): _____

Partner's possible genotype(s): _____

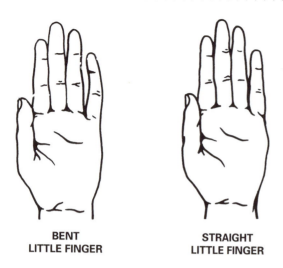

BENT
LITTLE FINGER

STRAIGHT
LITTLE FINGER

FIGURE 10.3 **Curvature of Little Finger**

Finger Hair

Presence of hair on the middle segment of the finger, dorsal surface, is controlled by a dominant allele.

Your phenotype: _____

Partner's phenotype: _____

Your possible genotype(s): _____

Partner's possible genotype(s): _____

Hitchhiker's Thumb

The ability to bend the thumb back (without force from the other hand) so that at one point it may appear slightly parallel with the extended arm; controlled by a recessive allele. See Figure 10.4.

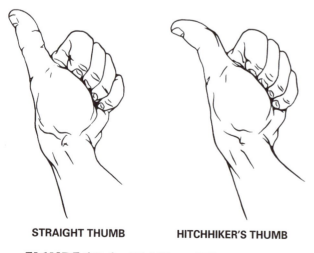

STRAIGHT THUMB   HITCHHIKER'S THUMB

FIGURE 10.4 **Mobility of Thumb Joints**

Your phenotype: _____

Partner's phenotype: _____

Your possible genotype(s): _____

Partner's possible genotype(s): _____

Interlocking Fingers

When clasping the hands together such that the fingers are interlocked, note which thumb ends up on the top naturally. If the left thumb folds over the right, this is controlled by a dominant allele.

Your phenotype: _____

Partner's phenotype: _____

Your possible genotype(s): _____

Partner's possible genotype(s): _____

SOLVING GENETICS PROBLEMS

In the following exercises you should keep in mind some basic facts and follow an orderly sequence of steps to prevent confusion while trying to find logical solutions.

1. The first step is to establish which symbols you will use, and then use them throughout the problem. Example: B = allele for brown eyes, b = allele for blue eyes.

2. Remember that each individual is diploid, and therefore must be shown to have two genes for each trait to be studied. Example: BB, Bb, or bb.

3. Remember that egg cells and sperm cells are produced by meiosis, so each gamete can have only one gene for each trait. Example: B or b.

4. The percentage, or fraction, of different gametes which can be expected to be produced by an individual is determined by the individual's genotype. If he is homozygous, all his gametes will have the same kind of gene. Example: BB individuals will produce only B gametes, bb individuals will produce only b gametes.

    If the individual is heterozygous (Bb), there are two possibilities, just as there are two possibilities when flipping a coin. The chances are 1/2 that a coin will come up "heads" on a toss. The probability that a Bb individual will produce a B gamete is also 1/2.

5. Gametes recombine at random. A B sperm does not seek out B eggs. It will fertilize any egg it meets. Thus if two Bb individuals mate, random recombination of gametes would be expected to produce offspring with the genotypes below at the frequencies given, because the probability of any egg and sperm meeting is the **product** of their probabilities of being produced in the first place.

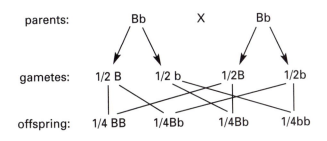

parents:	Bb	X	Bb
gametes:	1/2 B	1/2 b	1/2B	1/2b
offspring:	1/4 BB	1/4Bb	1/4Bb	1/4bb

Note that there are two ways in which Bb offspring can result from this cross, so they will be expected to occur twice as often as the other genotypes.

6. Always list the alleles in the same order throughout the problem so that you will be able to recognize similar genotypes. Example: Bb is the same as bB. List the dominant allele first in problems involving dominant and recessive alleles.

7. When dealing with genes for two or more traits at the same time, always group alleles for the same trait together. Example: (right) – RrTt, (wrong) – RTrt.

8. When dealing with genes for two or more traits at the same time, each gamete must be shown to contain **one of each kind of allele**. Example: an RrTt individual can produce RT, Rt, rT, and rt gametes; never Rr or Tt, or R or t gametes.

ABO BLOOD TYPES

Different blood types result from differences in specific proteins in the blood plasma and on the surface of the red blood cells. The proteins in the plasma of the blood are called **antibodies**. These substances act upon certain proteins present on the red blood cells called **antigens**, causing the cells to clump together. (In a few weeks you will be doing a lab on circulation in which you will determine your blood type.) Antibody A acts upon antigen A and so the two are never present together; the same is true for antibody B and antigen B.

The blood types are transmitted from parent to offspring by a series of three allelic genes. These are: an allele for the A antigen, designated I^A; an allele for the B antigen, I^B; and an allele for the absence of antigen, i. I^A and I^B are dominant over i, but are codominant with respect to each other. Thus, a person carrying $I^A I^B$ alleles has both antigens and is of the blood type AB. The possible genotypes for each blood type are given in Table 10.1.

1. If a woman with type O blood married a man with type AB blood, represent the following:

   Wife's genotype: _____

   Genetic make-up of egg cells: _____

   Husband's genotype: _____

   Genetic make-up of sperm cells:_____

   Possible combinations (i.e., possible genotypes of their offspring): _____

   _____

   _____

TABLE 10.1

BLOOD TYPE (Phenotype)	GENOTYPE	ANTIGENS ON THE RED BLOOD CELLS	ANTIBODIES IN THE PLASMA
O	ii	none	A and B
A	$I^A i$, $I^A I^A$	A	B
B	$I^B i$, $I^B I^B$	B	A
AB	$I^A I^B$	A and B	none

2. Assume that a woman with type AB blood marries a man with type AB blood and represent the following:

Wife's genotype: _____

Genetic make-up of egg cells: _____

Husband's genotype: _____

Genetic make-up of sperm cells: _____

Use a Punnett square to show possible combinations:

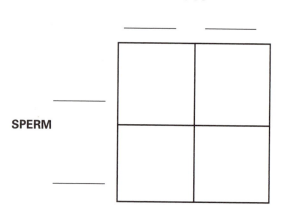

EGGS

SPERM

What are the chances (express in fraction form) that their first child will have type AB

blood? _____   type A? _____

type B? _____   type O? _____

INHERITANCE IN MAIZE (INDIAN CORN)

MONOHYBRID CROSS

A strain of corn producing pure purple kernels (PP) is crossed with a strain producing pure yellow kernels (pp). Purple is dominant with the resulting F_1 ears all bearing purple kernels. When the F_1 is self-pollinated, the resulting F_2 ears bear both purple and yellow kernels. The dominant allele for kernel

color is _____ and the recessive allele for kernel color is _____.

Count the number of purple and yellow kernels on one row of the F_2 ear without removing the kernels. Determine the ratio of purple to yellow. Now tabulate below the numbers obtained by each student (or row of students) and add these figures to get a total. Using the total numbers, determine a ratio of purple to yellow.

PURPLE	YELLOW	RATIO OF PURPLE TO YELLOW
_____	_____	_____ : _____
_____	_____	_____ : _____
_____	_____	_____ : _____
_____	_____	_____ : _____
_____	_____	_____ : _____
_____	_____	_____ : _____
TOTAL _____	_____	_____ : _____

In order to determine the expected ratio of such a cross, it is necessary to first indicate the parent (P) generation:

$$PP \quad \times \quad pp$$

Gametes   _____   _____

Next, indicate the genetic make-up of the gametes that the parent may produce. Do this in the spaces above just beneath the parents genotype. If all the eggs come from one parent and all the sperm from the other in order to produce the F_1 generation, then every plant in the F_1 generation would have what genotype? _____

Next, show a cross between two individuals in the F_1 generation. Indicate the genetic make-up of possible gametes from each in the spaces below.

F_1 generation   _____ × _____

Possible Gametes: _____   _____

Fill in the Punnett square to show the possible genotypes obtained by this cross:

What is the expected phenotypic ratio of the F_2 generation? _____

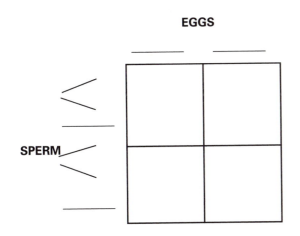

EGGS

SPERM

TEST CROSS OR BACK CROSS

To determine the genotype of an organism which exhibits the dominant feature (phenotype), a **test cross** is used. A test cross always involves a homozygous recessive and the unknown genotype. A strain producing yellow kernels (pp) is crossed with a strain of unknown genotype but with obviously purple phenotype.

Count the number of purple and yellow kernels on one row and determine the ratios and total, using numbers obtained from the other students or rows.

PURPLE	YELLOW	RATIO OF PURPLE TO YELLOW
_____	_____	_____ : _____
_____	_____	_____ : _____
_____	_____	_____ : _____
_____	_____	_____ : _____
_____	_____	_____ : _____
_____	_____	_____ : _____
TOTAL _____	_____	_____ : _____

Show the two possible outcomes of a test cross with a purple plant:

P generation _____ × _____ or _____ × _____

Possible Gametes _____ _____ _____ _____

THE DICHOTOMOUS OR FORKED-LINE METHOD

The Punnett square becomes especially time and space consuming when one is dealing with dihybrids or greater. So the **dichotomous** or **forked-line method** is usually utilized to find the possible results of crosses.

When the answer calls for only phenotypic results, then simply write the expected phenotypic ratio of the cross of one trait and multiply each of these probabilities by the expected phenotopic ratio of the other trait(s). In other words, a dihybrid is nothing more than 2 monohybrids being considered at the same time.

For example, the F_1 heterozygous purple, starchy corn (PpSs) are crossed with each other. The results will probably be: 3/4 purple, 1/4 yellow, 3/4 starch, and 1/4 sweet. Using the forked-line method, one should write:

3/4 purple
- 3/4 starchy = 9/16 purple starchy
- 1/4 sweet = 3/16 purple sweet

1/4 yellow
- 3/4 starchy = 3/16 yellow starchy
- 1/4 sweet = 1/16 yellow sweet

Thus the 9:3:3:1 ratio is derived without the use of the laborious Punnett square. Also one has a better concept of Mendelian genetics — that is, two genes governing two traits usually assort independently from each other during meiosis.

If genotypic results are needed, then write the expected genotypic ratio of one trait's cross and multiply each of these probabilities by the genotypic ratio of the other trait's cross.

For example, the F_1 heterozygous purple, starchy corn (PpSs) are self-pollinated.

1/4 PP
- 1/4 SS = 1/16 PPSS
- 2/4 Ss = 2/16 PPSs
- 1/4 ss = 1/16 PPss

2/4 Pp
- 1/4 SS = 2/16 PpSS
- 2/4 Ss = 4/16 PpSs
- 1/4 ss = 2/16 Ppss

1/4 pp
- 1/4 SS = 1/16 ppSS
- 2/4 Ss = 2/16 ppSs
- 1/4 ss = 1/16 ppss

It is doubtful that this 1:2:1:2:4:2:1:2:1 ratio is important for most problems. Some questions may ask for one or more of these probabilities such as what is the chance of an F_2 organism being PpSs?

Also with this method, trihybrids, back-crosses, and incomplete dominance are easily incorporated into the results.

For example, what is the probability that a woman (homozygous for PTC non-tasting, heterozygous for free-ear lobe, and blood type AB) and a man (heterozygous for PTC tasting, heterozygous for free-ear lobe, and blood type O) will have a baby boy who tastes PTC, has a free-ear lobe, and has A blood type?

Answer: Take each trait separately.

Woman tt (non-taster) $\times$ man Tt (taster) =
1/2 Tt, 1/2 tt

Ee (free-ear lobe) $\times$ Ee (free-ear lobe) =
1/4 EE, 1/2 Ee, 1/4 ee

∴ 3/4 free-ear lobe, 1/4 attached

$I^A I^B$ (AB blood) $\times$ ii (O blood) =
= 1/2 I^Ai, 1/2 I^Bi

Chromosomes XX $\times$ XY =
(female)   (male)
1/2 female, 1/2 male

Thus, the probability of a boy who tastes PTC, has a free-ear lobe, and A blood type is

1/2 $\times$ 3/4 $\times$ 1/2 $\times$ 1/2 = 3/32

PURPLE & STARCHY	PURPLE & SWEET	YELLOW & STARCHY	YELLOW & SWEET	RATIO
___	___	___	___	___
___	___	___	___	___
___	___	___	___	___
___	___	___	___	___
___	___	___	___	___
___	___	___	___	___

TOTAL:

___   ___   ___   ___   ___

How many phenotypes are in the F_2 generation? ___
_____

Which two phenotypes are new combinations? _____
_____

Indicate each of the following:

P generation   _____ $\times$ _____
Possible gametes: _____   _____
F_1 generation   _____ $\times$ _____
Possible gametes: _____   _____

Find the genotypic and phenotypic ratios expected in the offspring of this cross using the dichotomous or forked line method. Place your findings below.

DIHYBRID CROSS

A pure strain of corn producing purple-starchy kernels (PPSS) is crossed with a pure strain producing yellow-sweet (ppss). The resulting F_1 ears all bear purple-starchy kernels. When the F_1 is self-pollinated, the resulting F_2 generation contains various combinations.

Carefully count the number of kernels of each phenotype appearing on a row of F_2 ear. Tabulate the results as done in preceding exercises.

NOTE:

Starchy kernels are smooth and sweet kernels are wrinkled.

REVIEW QUESTIONS

Questions 3, 9, 10, and 12 are not covered in Exercise 10.

1. If the father has type A blood, the mother has type B blood, and the child has type O blood; indicate each of the following:

   Father's genotype: _____   Mother's genotype:_____

   Child's genotype: _____

2. Explain why the homozygous recessive would always be used in a test cross. _____

   _____

3. If a human male and female produce children, what proportion of their offspring would be males? What proportion would be females? Illustrate, using a Punnett square.

4. In rabbits the allele for white fat (W) is dominant over the allele for yellow fat (w). Suppose that a pure white-fatted male rabbit is mated to a pure yellow-fatted female rabbit.

   a. What color would all the first generation (F$_1$) offspring be? Why? _____

   b. What is the genotype of the female parent? The male parent? _____

   c. What is the genotype of the offspring?_____

   d. What proportion of the eggs would contain genes for white-fattedness in the female parent? Why? ____

      _____

   e. What proportion of the gametes of the F$_1$ generation will have alleles for: _____

      Yellow-fattedness only? _____

      White-fattedness only? _____

      White and yellow-fattedness in equal proportions? _____

5. Long wings in *Drosophila* are due to a dominant allele (V) and vestigial wings to its recessive allele (v). A vestigial winged male and a heterozygous long-winged female are mated. What types of gametes can be produced by the vestigial winged male? What type of gametes can be produced by the heterozygous long-winged female? What types of offspring can be produced and in what proportion? Use a Punnett square to explain.

6. A blue-eyed man whose parents were brown-eyed marries a brown-eyed woman whose father was brown-eyed and whose mother was blue-eyed. If they have children, what will the probability of their being blue-eyed? brown-eyed? _____

7. In guinea pigs short hair is dependent upon the dominant allele (S) while long hair is due to the allele (s). Rough coat is due to the dominant allele (R) and smooth coat to its recessive (r). Black is dependent upon the dominant allele (B) while white is dependent upon the allele (b). A guinea pig of genotype SSRRBB was crossed with one of the genotype ssrrbb. What was the appearance of the F_1 generation? ___

_____

Two F_1 individuals were crossed many times (ie, many offspring), what phenotypic ratio should be

expected in the F_2 generation? _____

8. Black Langshan chickens have feathered shanks while Buff Rocks have unfeathered shanks. When these two breeds are crossed the F_1 have feathered shanks. The feathered condition is due to the presence of two pairs of genes, F,f and S,s where F is dominant to both f and s. The same is true for S which is also dominant to f and s. The feathered condition is produced by the presence of one or more dominant genes. The homozygous recessive condition of f and s produces the unfeathered shanks. What would be the expected phenotypic ratio in the F_2 if the two F_1 individuals from the Langshan and Rock parents are

crossed? _____

_____

9. Normal vision (C) in man is dominant to colorblindness (c) and is sex-linked. A normal visioned man, whose father was colorblind marries a colorblind woman. What are the chances that a son will be color-blind? A daughter? Explain. _____

_____

_____

_____

10. The determiner for brown-eyes (B) is dominant to blue-eyes (b) and is not sex linked. A colorblind man with brown eyes, whose mother was blue-eyed, marries a normal visioned blue-eyed woman, whose father was colorblind. Show the expected phenotype ratio of their children involving eye color and color-blindness. _____

_____

_____

11. A man that has blood type A negative, whose father was B positive and whose mother was AB positive; married a woman that was B positive. The woman's mother was A negative and her father was AB positive.

   a. Give the genotypes of all the above mentioned individuals. _____

   _____

   _____

b. Give the genotypes and phenotypes of the F₁ generation. _____

_____

_____

12. A hemophiliac man with blue eyes married a woman that was a carrier of hemophilia and had brown eyes. The woman's mother had blue eyes. _____

_____

a. Give the genotypes of all the above mentioned individuals. _____

_____

b. Give the genotypes and phenotypes of the F₁ generation. _____

_____

c. What are this couples chances of having a normal blue-eyed son? _____

_____

13. The proteins in the plasma of the blood are called _____.

14. The proteins on the surface of the red blood cells are called _____.

15. Define:

Gene _____

Trait _____

Genotype _____

Allele _____

Heterozygous_____

Homozygous _____

Phenotype _____

Test cross _____

P generation _____

F₁ generation _____

F₂ generation _____

Monohybrid cross _____

Dihybrid cross _____

Dominant_____

Recessive _____

Codominant _____

Molecular Biology

OBJECTIVES

After completion of this exercise, the student should be able to do each of the following:

- Explain the processes of DNA replication, transcription, and translation (protein synthesis).
- Explain the mechanism of action of restriction endonucleases.
- Describe the process of producing DNA fragments.
- Describe the procedure for isolating DNA from onion cells. Describe how you would test this preparation to see if DNA is actually present.
- Describe how gel electrophoresis can separate fragments of DNA.
- State the relationship between DNA fragment size and distance traveled. Describe how this relationship can be used in determining the size of a DNA fragment.
- Discuss the principles of bacterial transformation.
- Describe how to prepare competent *E. coli* cells.
- Discuss the mechanisms of gene transfer using plasmid vectors.
- Discuss the transfer of antibiotic resistance genes and tell how to select positively for transformed cells that are antibiotic resistant.

MECHANISM OF GENE ACTION

DNA AND THE GENETIC CODE

A model of **DNA (deoxyribonucleic acid)** was produced in 1953 by Francis Crick and James Watson. Their model of a DNA molecule appeared as a spiralling ladder. The sides were composed of alternating **deoxyribose sugar** and **phosphate** molecules and the rungs were made of 2 of 4 **nitrogen bases.**

Actually, the DNA molecule is a large polymer, composed of smaller monomeric units known as nucleotides. Each nucleotide consists of a deoxyribose sugar, a phosphate, and either a **purine (adenine** or **guanine)** or a **pyrimidine (thymine** or **cytosine).** By dehydration synthesis the nucleotides are linked together to form strands. These strands run in opposite directions, or are **anti-parallel.** The strands are connected by 2 **hydrogen bonds** between adenine and thymine or 3 hydrogen bonds between guanine and cytosine.

1. Cut out the models in Figure 11.1. Tape a phosphate, a sugar, and a nitrogen base together to

form a nucleotide. Then fit them together to form a section of a DNA molecule. Do not tape together. Note (l) the hydrogen bonds, (2) the 5' carbon atom and 3' carbon of the sugar connect the phosphates and give the strand direction, (3) 1' carbon of the sugar attaches to the nitrogen base.

2. Arrange the DNA section so one strand's nitrogen bases "read" G, A, T, and G from a 3' to 5' direction.

DNA REPLICATION

Prior to cell division (mitosis or meiosis), DNA replicates to form two identical molecules. After the separation of the strands in a section of the DNA, complementary nucleotides are added to each strand by an enzyme known as **DNA Polymerase III.** Eventually, all sections of the molecule are replicated. Thus, each strand serves as a template to form a molecule identical to the original.

3. Together with a partner, recreate the DNA section in the above exercise. Now separate the strands. Then, add complementary nucleotides one at a time until two complete molecules are produced.

RNA TRANSCRIPTION

RNA (ribonucleic acid) is produced by using the template message of DNA. This process is similar to replication; however, RNA is produced and only 1 strand of DNA serves as the template. This strand is often known as the **sense** strand.

A section of DNA uncoils as **RNA polymerase** connects itself to the sense strand. Proceeding in a 5' to 3' direction of that DNA strand, the enzyme adds complementary nucleotides. These nucleotides differ from those of DNA. Thymine is absent but **uracil** is used instead as a complementary base to adenine. Also the pentose is **ribose,** not deoxyribose.

The finished product is a single-stranded RNA molecule. This molecule will be modified by other enzymes to become **transfer RNA, messenger RNA,** or **ribosomal RNA.**

4. Cut out the models of ribose, phosphate, and nitrogen bases located in Figure 11.2. Arrange these into nucleotides and tape. With a lab partner use the models from the previous exercises to create a section of DNA molecule in which one strand reads in a 5' to 3' direction the following sequence:

T A C C G A G T

Now separate the two strands of DNA to approximately a foot apart. Transcription has begun!

Match up complementary RNA nucleotides on the strand which reads T A C C G A G T in a 5' to 3' direction. Remove this RNA strand from its DNA complement and reattach the DNA "sense" strand to the other DNA strand. This RNA chain represents what will become a transfer, messenger, or ribosomal RNA.

TRANSLATION
OR PROTEIN SYNTHESIS

The process by which RNA codes for the assembly of protein is known as **translation.** All three types of RNA are involved. Two **ribosomal subunits, large** and **small,** are each composed of protein and ribosomal RNA.

The synthesis of protein begins with the attachment of the small ribosomal subunit to a particular area of a messenger RNA (mRNA). This area consists of an **initiating codon,** three nucleotides which have the nitrogen bases adenine, uracil, and guanine in a 5' to 3' direction. The small ribosomal subunit also extends past to one additional codon or 3 nucleotide set. Then the large ribosomal subunit joins the mRNA and the small ribosomal subunit complex.

A transfer RNA (tRNA), charged with one of the 20 different amino acids, unites by hydrogen bonding to the initiating codon. Another amino acid-charged tRNA bonds to the adjacent codon. The 2 amino acids are now close enough spatially to be bonded together by an enzyme. This resulting covalent bond is a peptide bond and the reaction releases water (dehydration synthesis).

The tRNA on the 5' side of the ribosome now is released and the ribosomal unit moves 3 nucleotides or 1 codon toward the 3' of the mRNA.

Another charged tRNA brings its amino acid to the ribosome and unites its anticodon to the next codon of the mRNA. Again dehydration synthesis occurs to bond the next amino acid to the growing peptide chain.

This process continues until one or more **terminating codons** are encountered. For example, UAG on the mRNA codes for a tRNA which possesses no amino acid. Thus, the chain cannot add any more amino acid and the polypeptide is complete.

Protein results when one or more polypeptides are somewhat modified by other enzymes and then united together.

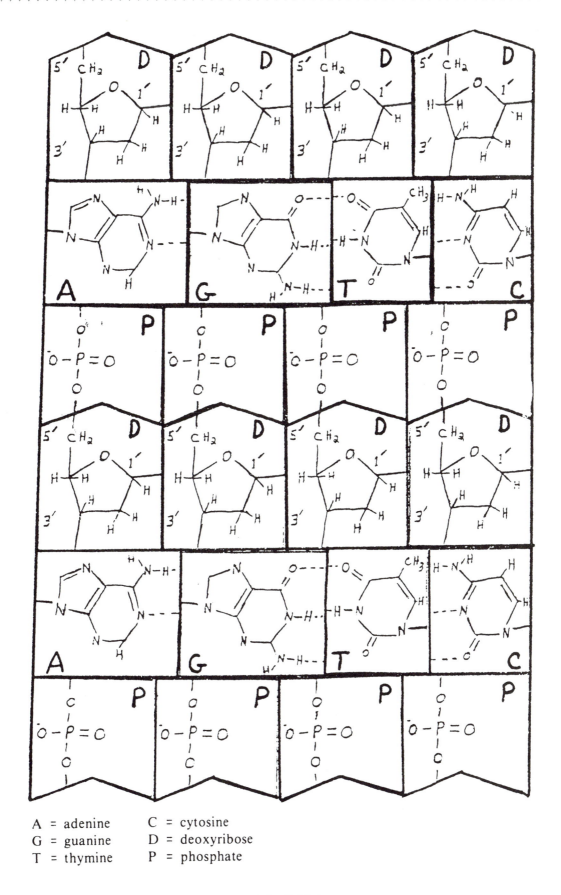

A = adenine
G = guanine
T = thymine

C = cytosine
D = deoxyribose
P = phosphate

FIGURE 11.1

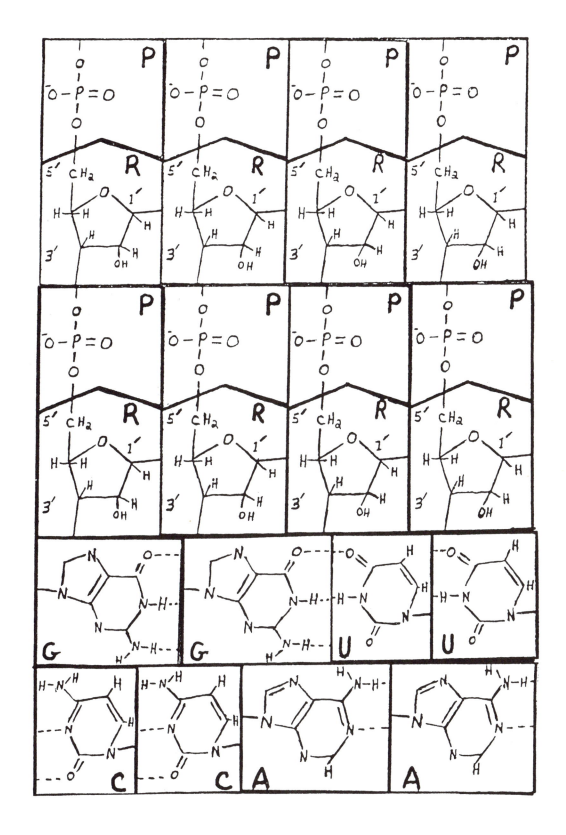

FIGURE 11.2

ISOLATION OF DNA

Isolation of DNA from living tissue is a fairly simple procedure but the directions must be followed precisely or you will get very poor results. Students will work in table groups and extract DNA from diced onions.

PROCEDURE

1. Dice a medium sized onion into cubes no larger than 3 mm.

2. Weigh out 50 g of diced onion and place in a 250 ml beaker.

3. Add 100 ml of homogenizing medium to the diced onion and place the beaker in a 60 degrees Celsius water bath for exactly 15 minutes. This step softens the onion tissue so that the homogenizing medium can penetrate. Also, this step denatures the cells' proteins and helps to dissolve the plasma membranes.

4. Quickly cool your preparation to 15 to 20 degrees Celsius in an ice bath.

5. Pour the material into a blender and fasten the lid. Turn on the blender for one minute. This step will break down cell walls, plasma membranes, and nuclear envelopes. The DNA is now mixed up in the homogenizing medium with the rest of the cell's molecules.

6. Pour the material from the blender into a 1000 ml beaker. Let the onion material stand in an ice bath for 15 to 20 minutes.

7. The material is then filtered through 4 layers of cheesecloth into a 500 ml beaker. Do not squeeze the cheesecloth but merely let the fluid trickle out of the cloth leaving the foam and other onion tissue behind.

8. Ice cold ethanol is then slowly poured down the side of the beaker until a white stringy precipitate appears in the beaker. Since DNA is not soluble in ethanol, all other molecules stay in solution and the DNA molecules "precipitate," or come together, as the white stringy material in the bottom of the beaker.

9. Place a glass stirring-rod into the beaker and move it around in large circles. As you do, rotate the rod in one direction. Notice how the stringy DNA molecules will "spool" or wind up on the rod. Perform this step **immediately** after seeing the white precipitate in Step 8. If you wait, the DNA might break down and will not spool onto the rod.

Testing Your Preparation for the Presence of DNA

How do you know that the stringy material is DNA? There is a chemical reagent named diphenylamine that will turn blue in the presence of DNA. Perform the following test to confirm that your material actually contains DNA.

1. Label 2 screw-top test tubes — one with S (DNA standard solution) and one with O (onion material). Place 3 ml of the DNA standard solution into the S tube and 3 ml of 4% sodium chloride solution in the O tube. Put the tip of the glass stirring-rod from step 9 above into the O tube and shake to remove the DNA from the rod. Shake the tube vigorously to mix the DNA with the salt solution.

2. Add 3 ml of diphenylamine into each test tube (CAUTION: BE VERY CAREFUL SINCE DIPHENYLAMINE CONTAINS CONCENTRATED ACIDS — DO NOT LET IT COME IN CONTACT WITH YOUR SKIN.)

3. Screw the caps onto the tubes, but loosely. Place the tubes in a beaker of boiling water under a fume hood. If the contents of a tube turn blue, then this is a positive test for the presence of DNA (this could take up to 15 minutes to occur).

4. Record the results in the blanks below:

   DNA standard (tube S) color? _____

   positive for DNA? _____

   Onion material (tube O) color? _____

   positive for DNA? _____

GEL ELECTROPHORESIS OF DNA FRAGMENTS*

It is not unusual these days to read in the newspaper or hear on radio/TV about the latest criminal trial in which DNA evidence is given in an attempt to convict the accused. This evidence is commonly called "DNA fingerprinting." Not only does everyone have unique fingerprints, they also have a unique set of genes possessed by no other person on

_____
* Used by permission of Carolina Biological Supply Company.

earth (except possibly for identical twins). This unique set of genes also means that the nitrogen base arrangement of the DNA of these genes are unique. The matching of the DNA from a crime scene to a suspect is, however, much more complicated than inking their fingertips!

The DNA testing is performed using a technique called gel electrophoresis. First, the DNA is split into a number of fragments by molecules called restriction enzymes or endonucleases. These fragments are then separated based on their size (essentially, molecular weight). Thus electrophoresis operates similarly to the chromatography section in Exercise 8. In electrophoresis, the DNA is placed at one end of a block of agarose gel (which is similar in consistency to firm Jello) and an electric current is run through the gel after it is placed in a buffer solution (see apparatus in Figure 11.3). The DNA fragments will migrate through the gel block along the current from the negative electrode toward the positive electrode. This migration will form bands according to the size of the fragments. A general principle is that the smaller the fragment, the farther it will travel through the gel.

Human DNA is so incredibly long that other techniques must first be used to isolate small pieces of DNA before cutting them into fragments. These techniques are beyond the scope of this exercise. Electrophoresis of DNA can be more easily demonstrated using virus DNA which is fairly short in length. In this experiment, three samples of DNA from a bacterial virus lambda (48,502 base pairs in length) will be separated using gel electrophoresis. In one sample, the DNA has been cut with the restriction enzyme EcoRI. A second sample has been cut with the restriction enzyme HindIII, and a third sample of DNA has not been cut by any enzyme and serves as a control. EcoRI and HindIII are both restriction enzymes but they cut the DNA at different places resulting in different sizes of fragments. Therefore, the lambda virus DNA will separate into different patterns of bands depending upon which enzyme was used to cut it into fragments.

PROCEDURE FOR GEL ELECTROPHORESIS OF DNA FRAGMENTS

1. Each lab group will be given a tray containing a block of agarose gel. Remove the tape from the ends of the tray.

2. Place the tray in the electrophoresis chamber being careful that the agarose gel block does not slip out of the ends of the tray. Make certain that the end of the tray that has the comb in it is pointed toward the black (or negative) end of the box.

3. Fill the chamber box with tris-borate-EDTA buffer so that it just barely covers the gel block.

4. Carefully remove the comb to expose the wells — a series of rectangular depressions in the gel block where the teeth of the comb were located. Make certain that the wells are completely filled with the buffer. Any dimples in or around the wells will disappear if enough buffer is slowly added.

5. Now it is time to load the wells with the DNA samples. One well receives the lambda virus DNA cut with EcoRI, another receives the virus DNA cut with HindIII, and a third well receives the uncut virus DNA. In the space below, draw a diagram indicating which well was loaded with which DNA sample. Use a transfer pipette to draw up one of the DNA samples. Squeeze the bulb gently to make sure that all of the air is removed from the tip of the pipette. USING BOTH HANDS, carefully place the tip of the pipette over the well and lower it until the tip is just below the surface of the buffer (but make certain that the tip does not punch into the gel block). Slowly squeeze the pipette bulb until the DNA sample fills the well. (If your instructor has extra pipettes, blocks with wells, and dye solution, you might want to practice a few times before actually loading the DNA samples.) Place the other two DNA samples in their respective wells using the same procedure but with clean pipettes. The DNA samples contain heavy molecules that causes it to sink to the bottom of the well and a dye so that the location of the DNA can be determined.

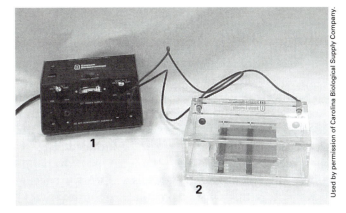

FIGURE 11.3 **Electrophoresis Apparatus**
1. power source
2. electrophoresis chamber with agarose gel block

6. Close the top of the electrophoresis chamber and connect the electrical leads (black to black and red to red). Also, connect the electrodes in the same manner to the power supply (both electrodes to the same channel if the power supply serves more than one electrophoresis chamber).

7. Turn on the power supply and set the voltage to the value chosen by your instructor. Soon the blue dye in the DNA sample can be seen moving through the gel toward the positive side of the chamber. The dye will move a little faster than the smallest DNA fragment, thus becoming a "leading edge."

8. When the dye almost reaches the end of the gel (usually around 1½ to 2 hours) turn off the power supply, disconnect the electrodes, and remove the top of the electrophoresis chamber. Carefully remove the tray containing the gel block and slide the block into a shallow plastic tray for staining.

OBSERVING THE RESULTS

Usually the lab time is over before you have the opportunity to stain your DNA samples. If time does permit, your instructor will give you directions on beginning the staining. If not, the instructor will stain your gel block for you and let you see the results during the next lab period.

While waiting for your gel block to electrophorese, the instructor will give you a block that has been run and stained in a previous lab. In this way, you can see a result while waiting for your samples to run.

Place your stained gel block on a white sheet of paper, or better yet, an overhead projector or blood typing light box. Slip a clear plastic metric ruler beneath the plastic tray containing the block. Place the zero mark of the ruler at a well and measure the distance from the well to each stained band. For the DNA cut with HindIII, there will be six visible bands (actually there are nine bands but two are too small to be seen and the band nearest the well is two combined bands since the fragments are very similar in size) and the DNA cut with EcoRI gives five bands. Enter the distance from the well to each of the bands for the HindIII sample in the following table beginning with the band closest to the well. The size of these bands is given as number of base pairs. The number of base pairs for the first band is the average of the two fused bands.

DNA sample cut by HindIII

size of band in # base pairs	distance from well to band
25,310	
9,416	
6,682	
4,361	
2,322	
2,027	

Obtain a piece of semilog paper from your instructor. This graph paper has the log number of base pairs on the y (vertical) axis and the distance from well to band in centimeters on the x (horizontal) axis. Plot the size and distance for each band from the table above onto the graph. Connect the points on the graph with a line. This line shows the relationship between band size and distance from the well. This line will hold true for any size of fragment and for any measured distance. The DNA cut with EcoRI will show five bands of different distances than the six bands of the DNA cut with HindIII. However, the relationship between the size and distance still remains the same (the line on the graph). Since you know the distance of the bands from the wells in the EcoRI sample, look up these distances on the x-axis, trace this point up until it intersects the line on the graph, and then read across to the corresponding number of base pairs on the y-axis. Enter these values on the table that appears on the following page.

DNA sample cut by EcoRI

distance from well to band	estimated number of base pairs

Your instructor will give you the values for the actual number of base pairs for the DNA fragments cut by EcoRI.

How close were your estimates to the actual values for each of the five bands? _____

_____

BACTERIAL TRANSFORMATION

The bacterium *Escherichia coli* (or *E. coli*) is an ideal organism for the molecular geneticist to manipulate and has been used extensively in recombinant DNA research. It is a common inhabitant of the human colon and can easily be grown in suspension culture in a nutrient medium such as Luria broth, or in a Petri dish on Luria broth mixed with agar (LB agar).

The single circular chromosome of *E. coli* contains about 5 million DNA base pairs, only 1/600th the haploid amount of DNA in a human cell. In addition, the *E. coli* cell may contain small circular DNA molecules (1,000 to 200,000 base pairs) called plasmids, which also carry genetic information. The plasmids are extrachromosomal; they exist separately from the chromosome. Some plasmids replicate only when the bacterial chromosome replicates and usually exist only as single copies within the bacterial cell. Others replicate autonomously and often occur in as many as 10 to 200 copies within a single bacterial cell. Certain plasmids, called R plasmids, carry genes for resistance to antibiotics such as ampicillin, kanamycin, or tetracycline.

In nature, genes can be transferred between bacteria in three ways: conjugation, transduction, or transformation. **Conjugation** is a mating process during which genetic material is transferred from one bacterium to another "sexually" different type. **Transduction** requires the presence of a virus to act as a vector (carrier) to transfer small pieces of DNA from one bacterium to another. **Bacterial transformation** involves transfer of genetic information into a cell by direct absorption of the DNA from a donor cell.

During bacterial transformation, the uptake and expression of foreign DNA by a recipient bacterium can result in conferring a particular trait to a recipient lacking that trait. Transformation can occur naturally but the incidence is extremely low and is limited to relatively few bacterial strains. These bacteria can take up DNA only during the period at the end of logarithmic growth. At this time the cells are said to be competent. Competence can be induced in *E. coli* with carefully controlled chemical growth conditions. Once competent, the cells are ready to accept DNA that is introduced from another source.

Plasmids can transfer genes (such as those for antibiotic resistance) which are a naturally-occurring part of a plasmid, or plasmids can act as carriers (vectors) for introducing foreign DNA from other bacteria, plasmids, or even eukaryotes into recipient bacterial cells. Restriction endonucleases are used to cut and insert pieces of foreign DNA into the plasmid vectors (Figure 11.4).

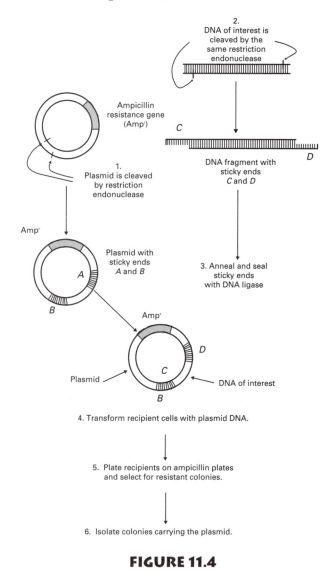

FIGURE 11.4

Restriction endonucleases do not cut the plasmid and foreign DNA straight across but, rather, makes staggered cuts leaving unpaired bases at the cut edges called "sticky ends." Since a given restriction endonuclease makes the cut between the same bases each time, all of the sticky ends are identical. The sticky ends of foreign DNA are complementary to the sticky ends of the plasmid DNA, so the foreign DNA can be inserted between the ends of the plasmid DNA, thus forming a new loop.

If these plasmid vectors also carry genes for antibiotic resistance, transformed cells containing plasmids that carry the foreign DNA of interest in addition to the antibiotic resistance gene can be selected for.

PROCEDURE FOR BACTERIAL TRANSFORMATION

AMPICILLIN RESISTANCE*

In this exercise, we will induce competent *E. coli* cells to take up the plasmid pAMP which contains a gene for ampicillin resistance. Normally, *E. coli* cells are destroyed by the antibiotic ampicillin, but *E. coli* cells that have been transformed will be able to grow on agar plates containing ampicillin.

1. Use a sterile micropipette to add 250 µL of ice cold 0.05 M CaCl$_2$ to two microtest tubes. (It is recommended that 15 ml (17 × 100 mm) polypropylene tubes can be used to improve the transformation efficiency.)

2. Sterilize an inoculating loop by flaming it and then cool it by sticking it into an LB agar plate containing *E. coli* colonies in an area where no bacteria are growing. Use the sterile inoculating loop to transfer a large (3 mm) colony of *E. coli* to one of the tubes. Be careful not to transfer any agar.

3. Vigorously tap the loop against the wall of the tube to dislodge the cell mass.

4. Suspend the cells immediately by vigorously pipetting using the sterile micropipette with a sterile tip.

5. Mark this first tube "+" and return it to the ice.

6. Repeat steps 2 to 4 for the second tube. Mark the tube "−."

7. Add 10 µL of pAMP solution (0.005 µg/µL) directly into the cell suspension in tube "+." Mix by tapping the tube with your finger.

8. Keep both tubes on ice for 15 min.

9. While the tubes are on ice, obtain two LB agar and two LB/Amp agar (LB agar containing ampicillin) plates. Label one LB agar plate "LB+" and the other "LB−." Label one LB/Amp plate "LB/Amp+" and the other "LB/Amp−." Write your name on the plate bottoms.

10. A brief pulse of heat facilitates entry of foreign DNA into the *E. coli* cells. Heat-shock cells in both the "+" and "−" tubes by holding the tubes in a 42° C water bath for 90 seconds. It is essential that cells be given a sharp and distinct shock, so take the tubes directly from the ice to the water bath.

11. Immediately return cells to ice for 2 min.

12. Use a sterile micropipette to add 250 µL of Luria broth to each tube. Mix by tapping with your finger and set at room temperature. The transformed cells are now resistant to ampicillin; they possess the gene whose product renders the antibiotic ineffective.

13. Place 100 µL of "+" cells onto one the "LB+" plates and onto one of the "LB/Amp" plates. Place 100 µL of "−" cells onto the other "LB−" plate and onto the remaining "LB/Amp−" plate.

14. Immediately spread the cells using a sterile spreading rod. (Remove the spreading rod from ethanol and briefly pass it through a flame. Cool by touching it to the agar on a part of the dish away from the bacteria. Spread the cells and once again immerse the rod in alcohol and flame it.) Repeat the procedure for each plate.

15. If time allows, allow plates to set for several minutes. Tape your plates together and inoculate *inverted* overnight at 37° C.

---

*Adapted with permission from *DNA Science: A First Course in Recombinant-DNA Technology* by David A. Micklos, DNA Learning Center of Cold Spring Harbor Laboratory, and Greg A. Freyer, Columbia University College of Physicians and Surgeons, Copyright 1990 Cold Spring Harbor Laboratory Press and Carolina Biological Supply Company. It is based on a protocol published by Douglas Hanahan, University of California, San Francisco.

ANALYSIS OF RESULTS

1. Observe the colonies through the bottom of the culture plate. Count the number of individual

colonies by using a permanent marker to mark each colony as it is counted. If cell growth is too dense to count individual colonies, record "lawn." Compare results on positive controls with another lab group.

LB + (Positive Control) _____

LB − (Positive Control) _____

LB/Amp + (Experimental) _____

LB/Amp − (Negative Control) _____

2. Compare and contrast the number of colonies on each of the following pairs of plates. What does each pair of results tell you about the experiment?

  a. LB + and LB − _____

  _____

  b. LB/Amp − and LB/Amp + _____

  _____

  c. LB/Amp + and LB + _____

  _____

REVIEW QUESTIONS

1. What molecules make up a nucleotide?

  a. _____

  b. _____

  c. _____

2. How is transcription different from translation? _____

_____

_____

_____

_____

3. On what molecule would we find the code? _____

Codon? _____

Anticodon? _____

4. What are restriction enzymes (endonucleases) and how do they work? _____

_____

_____

_____

5. Which of the following is NOT used in the isolation of DNA?

    (A) concentrated hydrochloric acid
    (B) ice cold ethanol
    (C) homogenizing medium
    (D) cheesecloth

6. Describe the function of electricity and the agarose gel in electrophoresis.

    _____

    _____

    _____

    _____

    _____

    _____

7. In gel electrophoresis, smaller DNA fragments will _____ in the agarose block than larger fragments.

8. DNA fragments move through an agarose block from the _____ end to the _____ end.

9. Diphenylamine will turn blue in the presence of _____.

10. In this experiment, DNA from _____ is cut by 2 restriction enzymes _____ and _____.

11. If you had been given DNA cut by a different restriction enzyme to run along with your other 3 samples, and 2 bands had 11,000 and 1600 base pairs, how far would these bands be from the well?_____

12. What precautions did you have to take when loading a well with a DNA sample? _____

    _____

    _____

    _____

    _____

13. What is a plasmid? How are plasmids used in genetic engineering? _____

_____

_____

_____

_____

_____

14. What does it mean when the DNA "precipitates" at the end of the DNA isolation procedure?

_____

_____

_____

_____

15. If a restriction enzyme digest resulted in DNA fragments of the following sizes: 400 base pairs, 2500 base pairs, 2000 base pairs, and 4000 base pairs, sketch the resulting separation by electrophoresis. Show starting point, positive and negative electrodes, and the resulting bands.

Life Classifications and Reproduction

OBJECTIVES

After completion of this exercise, the student should be able to do each of the following:

- Name the various **taxa** used in clarifying organisms.
- List the five **kingdoms** of life.
- Explain the difference between **asexual** and **sexual reproduction**.
- Define the types of sexual reproduction.
- Explain **alternation of generations** using appropriate terminology.
- Answer the review questions at the end of this exercise.

CLASSIFICATIONS

In the next 13 exercises, we will be looking at the diversity of life at various levels. The kingdom level of classification is the most inclusive with organisms within a kingdom sharing only a few basic characteristics. In the 5-kingdom approach of this manual, Kingdom Monera includes bacteria-like organisms whose cells have no organized nucleus (*prokaryotic*) as opposed to all other forms of life whose cells have a nucleus (*eukaryotic*). Kingdom Protista mostly includes single-celled plant-like and animal-like organisms. Kingdom Fungi includes non-photosynthetic organisms usually with non-cellulose cell walls. The final two kingdoms are Kingdom Plantae (photosynthetic, cellulose cell walls, chlorophylls a and b) and Kingdom Animalia (usually motile, multicellular, consume food).

The kingdom level is progressively subdivided into lower categories in which organisms share even more characteristics. These categories (or taxa) are:

- Phylum (Divisions in fungi and plants)
  - Class
    - Order
      - Family
        - Genus
          - Species

Such a classification system is essentially an outline, much like one you might use to help you study any subject. Such systems could be based on almost any characteristics of the organisms involved, but scientists have chosen to use apparent common ancestry as their basis. Even so, many scientists differ in their analyses of common ancestry, so you will find that there are places in which differences appear

97

when comparing different listings. For instance, some authorities feel that there are two distinct groups of bacteria-like organisms (see Exercise 13 for a discussion of Archaebacteria and Eubacteria). These classification systems consider six kingdoms instead of five or call Monera a superkingdom or kingdom and Archaebacteria and Eubacteria as kingdoms or subkingdoms respectively. Others who consider this bacterial difference to be even more important, propose three kingdoms — Archaebacteria, Eubacteria, and Eukaryotes (including the Protista, Fungi, Plantae, and Animalia as subkingdoms).

The list on the next few pages gives the basic categories that we will study. This is by no means a complete list of all the categories that exist; it includes only those groups we will examine or discuss as representative.

KINGDOM MONERA (moe-nair'-ah) or PROKARYOTAE (pro'-kar-e-o'-te)

Division Archaebacteria (ark'-e-back-ter'-e-ah) — ancient bacteria

Division Eubacteria (u'-back-ter'-e-ah) — true bacteria. ex.: cyanobacteria

KINGDOM PROTISTA (pro-tiss'-tah)

Phylum Sarcodina (sar'-cog-die'-nah) or **Rhizopoda** (rise-op'-oh-da) ex.: *Amoeba*, radiolaria, and foraminifera

Phylum Ciliophora (sill'-ee-ah'-for-ah) — ciliates. ex.: *Paramecium*

Phylum Zoomastigina (zoe'-mas-tih-ji'-nah) or **Mastigophora** (mas-e-gof'-er-ah) — flagellates. ex.: *Trypanosoma*

Phylum Sporozoa (spore'-oh-zoe'-ah) **Apicomplexa** (ap-e-com-plex'-ah) ex.: *Plasmodium* sp.

Phylum Pyrrophyta (pye-rof'-fi-tah) or **Dinoflagellata** (de'-no-flaj-e-lah'-ta) — dinoflagellates. ex.: *Gymnodinium, Gonyaulax,* and *Noctiluca*

Phylum Chrysophyta (kriss-ah'-fi-tah) — golden algae. ex.: diatoms

Phylum Euglenophyta (you'-glen-ah'-fi-tah) ex.: *Euglena*

Phylum Chlorophyta (klor-ah'-fi-tah) — green algae. ex.: *Chlamydomonas, Spirogyra, Volvox,* and *Ulva*

Phylum Phaeophyta (fay-ah'-fi-tah) — brown algae. ex.: *Laminaria, Sargassum,* and *Fucus*

Phylum Rhodophyta (row-doff'-fi-tah) — red algae

Phylum Myxomycota (my-ko-my'-co-tah) — plasmodial slime molds

Phylum Oomycota (oh-oh-my'-co-tah) — water molds

Phylum Acrasiomycota (a-kras'-e-o-my-co-tah) — cellular slime molds

KINGDOM FUNGI (fun'-jie or fun'-guy)

Division Zygomycota (zie'-go-my-co-ta) ex.: black bread mold

Division Ascomycota (ass'-koe-my-co-ta) — sac fungi. ex.: yeasts, morels, and *Aspergillus*

Division Basidiomycota (buh-sid'-ee-oh-my-co-ta) — club fungi. ex.: mushrooms and shelf fungi

Division Deuteromycota (doo'-ter-oh-my-co-ta) — imperfect fungi. ex.: *Penicillium*

KINGDOM PLANTAE (plan'-tee)

Division Bryophyta (bry-ah'-fi-tah)
    Class Hepaticae (he-pat'-ih-see) — liverworts
    Class Musci (moo'-see) — mosses

Division Psilophyta (si-ah'-fi-tah) — whisk ferns

Division Lycophyta (ly-cof'-i-tah) — club mosses

Division Sphenophyta (sphe-nof'-i-tah) — horsetails

Division Pterophyta (ter-ah'-fi-tah) — ferns

Division Coniferophyta (kon-if'-er-ah'-fi-tah) ex.: *Pinus*

Division Cycadophyta (si-kad-of'-i-tah) — cycads

Division Ginkgophyta (genko-fi'–tah) — ginkgo
Division Gnetophyta (ne-tof'-i-tah) — gnetophytes
Division Anthophyta (an-thah'-fi-tah) — flowering plants
    Class Monocotyledonae (mon'-oh-cotty-lee'-dun-ee)   ex.: grasses
    Class Dicotyledonae (di'-cotty-lee'-dun-ee)   ex.: rosebush

KINGDOM ANIMALIA (an-ih-male'-ee-ah)

Phylum Porifera (pore-if'-er-ah) — sponges
    Class Calcarea (kal-care'-ee-ah) — chalky sponges. ex.: *Grantia*
    Class Hexactinellida (hex'-act-in-el'-ih-dah) — glass sponges. ex.: Venus flower basket
    Class Demospongiae (dem'-oh-sponge'-ee-ee) — commercial, or bath sponges. ex.: *Spongia*

Phylum Cnidaria (nid-air'-ee-ah)
    Class Hydrozoa (hi'-droh-zoe'-ah) — hydroids. ex.: *Hydra* and *Obelia*
    Class Scyphozoa (skife'-oh-zoe'-ah) — jellyfish. ex.: *Aurelia*
    Class Anthozoa (an'-thow-zoe'-ah)   ex.: sea anemones and corals

Phylum Ctenophora (ten-ah'-for-ah)   ex.: comb jellies and sea walnuts

Phylum Platyhelminthes (plat'-ee-hell-min'-thees) — flatworms
    Class Turbellaria (ter'-bel-air'-ee-ah) — free-living flatworms. ex.: planaria
    Class Trematoda (treh'-mah-tode'-ah) — flukes. ex.: *Clonorchis* and sheep liver fluke
    Class Cestoda (sess-tode'-ah) — tapeworms. ex.: *Taenia* sp. and *Dipylidium caninum*

Phylum Nematoda (nee'-mah-tode'-ah) — roundworms. ex.: *Ascaris*, hookworms, and *Trichinella* sp.

Phylum Rotifera (row-tiff'-fer-ah) — rotifers, or "wheel animals"

Phylum Annelida (an-nell'-ih-dah) — segmented worms
    Class Archianellida (ark'-ee-an-nell'-ih-dah) — primitive marine worms
    Class Polychaeta (pahl'-ee-keet'-ah)   ex.: *Nereis*, bristleworms
    Class Hirudinea (ji'-roo-din'-ee-ah) — leeches
    Class Oligochaeta (oh'-lig-oh-keet'-ah)   ex.: earthworm

Phylum Arthropoda (arr-throp'-oh-dah) — joint-footed animals

  Subphylum Trilobita (try'-low-bite'-ah) — trilobites (extinct)

  Subphylum Chelicerata (kell-liss'-er-ah'-tah)
    Class Merostomata (mare'-oh-stow-mah'-tah)   ex.: *Limulus* (horseshoe crab)
    Class Arachnida (uh-rack'-nid-ah)   ex.: spiders, mites, ticks, scorpions and harvestmen

  Subphylum Mandibulata (man-dib'-you-lah'-tah)
    Class Crustacea (kruss-tay'-she-ah)   ex.: shrimp, crabs, lobsters, crayfish, and barnacles
    Class Insecta (in-sect'-ah) — insects. ex.: bees, flies, beetles, butterflies, and bugs
    Class Chilopoda (ky-lop'-oh-dah) — centipedes
    Class Diplopoda (dop-lop'-oh-day) — millipedes

Phylum Mollusca (moll-lusk'-kah) — soft-bodied animals
    Class Polyplacophora (pahl'-plak-ah'-for-ah) — chitons
    Class Monoplacophora (mon'-oh-plak-ah'-for-ah)   ex.: *Neopalina*
    Class Bivalvia (bi-val'-vee-ah) — bivalves. ex.: clams, oysters, mussels, and scallops
    Class Gastropoda (gas-trop'-poh-dah)   ex.: snails, slugs, whelks, and sea hares
    Class Cephalopoda (seff'-fal-lop'-oh-dah)   ex.: squids, octopus, and chambered nautilus
    Class Scaphopoda (skaff-op'-oh-dah)   ex.: tuskshells

Phylum Echinodermata (ee-ky'-no-derm-ah'-tah) — spiny skinned animals
    Class Echinoidea (ek'-in-oyd'-ee-ah) — sea urchins and sand dollars
    Class Holothuroidea (hole'-oh-thure-oyd'-ee-ah) — sea cucumbers
    Class Crinoidea (crin-oyd'-ee-ah) — sea lillies and feather stars
    Class Ophiuroidea (oh'-fee-you-roy'-dee-ah) — brittle stars and serpent stars
    Class Asteroidea (ass-ter-roy'-dee-ah) — starfish

Phylum Hemichordata (hem'-ee-kord-ah'-tah) — acorn worms. ex.: *Dolichoglossus* and *Balanoglossus*

Phylum Chordata (kord-ah'-tah)

Subphylum Urochordata (you'-row-kord-ah'-tah) — sea squirt, tunicates. ex: *Molgula*

Subphylum Cephalochordata (seff'-fal-lo-kord-ah'-tah) — amphioxus

Subphylum Vertebrata (vur'-tih-brah'-tah) — vertebrates

Class Agnatha (ag-nath'-ah) — jawless fish.   ex.: lampreys, hagfishes

Class Placodermi (plak'-oh-der'-mee) — extinct armored fish with jaws

Class Chondrichthyes (kon'-dra-ick'-thees) — cartilaginous fish.   ex.: sharks, skates, and rays

Class Osteichthyes (oss'-tee-ick'-thees) — bony fish.   ex.: perch, seahorses, mackerel, pike

Class Amphibia (am-fib'-ee-ah)   ex.: salamanders, frogs, and toads

Class Reptilia (rep-till'-ee-ah)   ex.: turtles, snakes, lizards, and crocodiles

Class Aves (ah'-vees) — birds.   ex.: chicken, sparrow, eagle

Class Mammalia (mam-male'ee-ah)   ex.: duckbilled platypus, kangaroo, bat, rat, apes, man

MAJOR TAXA: DEMONSTRATIONS

During this portion of the exercise, students will learn to associate major groups of organisms with kingdom, become familiar with taxa (kingdom, phylum, class, etc.), and enjoy the array of interesting and unique forms of life. Specimens (embedded in plastic, preserved in jars, on slides, or dried) are in the lab by kingdoms and in some cases phyla, and occasionally by class.

Whatever notes you take will be up to you; no formal notes are required to be taken in your lab manual. Please do **not** try to copy down all the names of the organisms; this is not what you are trying to get out of this lab.

Look carefully at the purposes of this lab. Simply observe the organisms in a particular kingdom and become familiar with them by groups. For example, you should learn that the kingdom Protista includes the algae (seaweeds as well as microscopic forms) and the protozoa (Paramecium and Euglena). Similarly, you need to learn the various types of organisms in the other kingdoms.

In the plant and animal kingdoms, you will see a breakdown into phylum and class (in plants, phyla are called divisions).

SPECIAL SAFETY CONSIDERATIONS

Be careful in handling the specimens so they don't get broken. Keep the specimens over the table, especially the bottled ones, to prevent them from dropping to the hard floor and breaking. If you pick up a specimen, be sure to return it to the same, exact position.

REPRODUCTION

Asexual reproduction is any method of producing new individuals without the fusion of **gametes**, or sex cells, and in which only one organism is involved. This is done routinely in very primitive plants, producing offspring identical to the parent. Often very elaborate structures are produced for this purpose. When no deliberate preparations are made by the parent for the production of new individuals, the process may be called **vegetative reproduction**. In some cases the center of a branched body dies and the branches left continue growth. Fragmentation may occur accidentally, of course, as when a grazing animal bites off a chunk and leaves scraps behind. Early land plants lost the ability to deliberately reproduce asexually as they became more advanced, but virtually all plants can "recreate" themselves if the scrap is large enough.

Unlike the above, **sexual reproduction** produces offspring potentially different from the parents. It involves the fusion or fertilization of two cells (or **gametes**) to produce a new cell called a zygote. Gametes may look identical (a condition known as **isogamy**) or be visibly different (**heterogamy**). The most common type of heterogamy is known as **oogamy**, in which large non-motile **eggs** are fertilized by tiny, motile **sperm**.

Most organisms have either a haploid (1N) or diploid (2N) number of chromosomes but during reproduction, intermediate cells or resulting products may have different chromosome numbers. This change of chromosome number is created by nuclear fusion of gametes or by meiosis. The reproductive history of how an organism produces another organism similar to itself is known as its **life cycle**. While there are thousands of life cycles found in nature, most organisms follow basically one of the following three schemes.

The life cycle diagrammed in Figure 12.1 is found predominantly in the Protista and Fungi Kingdoms. The major and most conspicuous organisms are haploid. They produce gametes mitotically or else

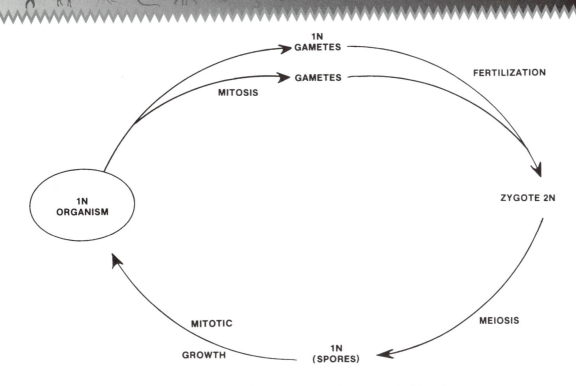

FIGURE 12.1 Life Cycle Common in Fungi

their haploid nuclei act as gametes. Fertilization, or nuclear fusion, produces a **zygote** which is usually short-lived. The zygote is diploid, and it undergoes meiosis to produce 1N spores or just young haploid organisms.

The life cycle diagrammed in Figure 12.2 is found in all animals, some fungi and some algae.

The conspicuous diploid organisms produces gametes meiotically. These fuse to form a zygote which is diploid. The zygote then grows mitotically to become a multicellular diploid organism. The gametes are the only haploid stage in this type of life cycle.

The life cycle diagrammed in Figure 12.3 is found exclusively in the Kingdom Plantae. Any one species

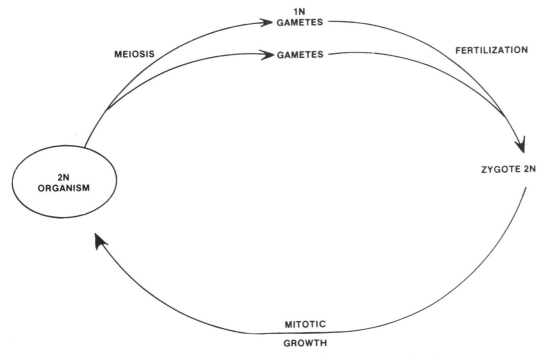

FIGURE 12.2 Life Cycle Common to Animals

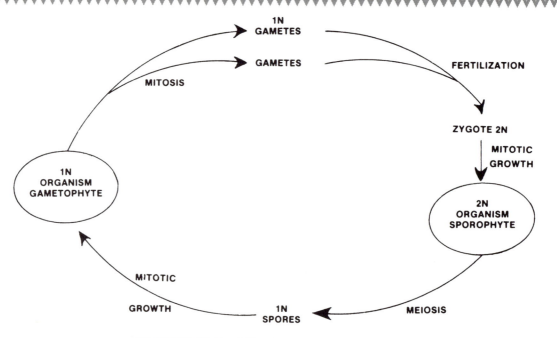

FIGURE 12.3   Life Cycle Common to Plants

exists in two multicellular forms. A multicellular haploid organism (**gametophyte**) produces gametes mitotically. These gametes fuse to form a diploid zygote. The zygote then grows mitotically to form a multicellular diploid organism (**sporophyte**). The sporophyte then produces spores by meiosis, and the spores grow mitotically into multicellular gametophytes. This cycle is often referred to as **alternation of generations**.

REVIEW QUESTIONS

1. Which is the higher taxon: Class or order? _____

    Family or class? _____

    Genus or phylum?_____

2. Is isogamy a form of asexual, or sexual reproduction? _____

3. Egg and sperm are involved in: a. asexual reproduction   b. isogamy   c. oogamy

4. The sporophyte plant has how many sets of chromosomes, one or two? _____

5. Which generation in plants produces spores? _____

6. Plant eggs and sperm, unlike animals, are produced by which process — mitosis or meiosis? _____

7. What is the haploid (1N) number of chromosomes when the diploid (2N) number is 12? _____

    26?_____44? _____

8. Fertilization produces a _____ which is _____ in chromosome number.

Kingdom Monera

OBJECTIVES

After completion of this exercise, the student should be able to do each of the following:

- Draw and identify **coccus, bacillus,** and **spirillum** bacterial cell shapes.
- Make a Gram-stained smear of bacteria and identify their Gram reactions.
- List the identifying traits of *Oscillatoria* and *Nostoc.*
- Give the meaning of **nitrogen fixation, trichome, sheath, heterocyst, colony,** and **fission.**
- Explain the difference between a colonial and a multicellular organism.
- Answer the review questions at the end of this exercise.

INTRODUCTION

Members of the Kingdom Monera show a simpler type of cellular organization than do other living organisms. Their cells are **prokaryotic,** that is, they lack the complex membrane-bound organelles found in cells such as those you observed in Exercise 4. The cells you observed in that exercise were **eukaryotic.** They possessed mitochondria, chloroplasts, Golgi bodies, etc. which prokaryotic cells lack. The few non membrane-bound organelles which prokaryotic cells do have are different in size and/or structure from those of eukaryotic cells (ex. ribosomes, flagella, etc.).

Prokaryotic unicellular organisms are divided into 2 groups — the archaebacteria and the eubacteria (true bacteria). Some texts consider these groups as kingdoms or as subkingdoms and others consider them as divisions of Kingdom Monera.

DIVISION ARCHAEBACTERIA — Ancient Bacteria

The archaebacteria, as their name implies, were once considered to be more primitive than eubacteria. Archaebacteria resemble eubacteria in appearance but differ in cell wall structure, membrane lipid structure, and protein synthesis. Some archaebacteria are anaerobic and release methane as a by-product of respiration. Others can live in extreme habitats such as high salt, high temperature, and low pH.

DIVISION EUBACTERIA — True Bacteria and Cyanobacteria

Bacteria can be **heterotrophic,** deriving their energy from organic molecules made by other living organisms, or **autotrophic,** deriving the energy they need from photosynthesis or from the oxidation of inorganic molecules. Most are heterotrophs, but those which are capable of photosynthesis may contain chlorophyll a (cyanobacteria) or may lack chlorophyll a.

Eubacteria are mostly unicellular or form chainlike or clustered groupings of cells. Their reproduction is primarily by a simple asexual process of cell division known as **fission**, in which a cell pinches into two without the complex movement of chromosomes seen in mitosis. Several bacterial species also engage in a type of sexual reproduction known as **conjugation**, in which all or part of the DNA of one bacterium is transferred to a second bacterium through a specialized cell projection known as an F pilus.

Eubacteria usually have cell walls that contain at least one layer of peptidoglycan — a unique polysaccharide in which the glucose molecules contain nitrogen. It is the cell wall which gives bacteria their characteristic shapes. Among the Eubacteria, there are only three basic cell shapes (see Figure 13.1):

spiral — **spirillum**

spherical — **coccus**

rod — **bacillus**

Spirilla are always single cells, but cocci and bacilli may show a variety of cell arrangements because their cells tend to stick together after fission. Both bacilli and cocci may be paired or in chains (**filaments**), and cocci may also be clustered in various ways. Cell arrangement is characteristic for each species of bacteria. These loose arrangements of cells which are basically independent functional units are called **colonies**. Colonial cell arrangements differ from the multicellular state seen in higher organisms in that the cells making up multicellular organisms are dependent for survival on one another to varying degrees.

Demonstration microscopes may be set up to show a variety of bacteria of each of the cell shapes and arrangements mentioned. Make sketches of bacteria showing each of the main cell shapes.

Included in the Eubacteria are six other bacterial groups:

Rickettsias and **Mycoplasmas** are so small that they challenge the resolving power of the compound light microscope.

Myxobacteria lack cell walls and move about like tiny amoebas.

Actinomycetes grow in branching chains and produce many of the antibiotics in use today.

Spirochetes, which form long, thin, flexible spirals, are best known as the group to which *Treponema pallidum*, the cause of syphilis, belongs.

Cyanobacteria, the blue-green algae, are photosynthetic, oxygen-generating prokaryotes.

THE GRAM STAIN

One of the most important procedures in identifying bacteria of medical importance is the **Gram staining technique**. Because of differences in cell wall composition, bacteria respond to this staining technique in one of two different ways, either retaining the initial violet stain when rinsed with alcohol, or bleaching out. Those that retain the stain are said to be **Gram positive**. Those that bleach out are said to be **Gram negative**.

This difference in staining properties corresponds to differences in susceptibility to different antibiotics, so it is often possible to begin treatment of an infection before the exact identity of the bacteria causing it is known, a process which requires several days at least.

Few bacteria cause infection. There are hundreds of species of bacteria living in or on the human body, including the oral cavity, which are nonpathogenic (not disease-causing).

In this activity, you will prepare a smear from scrapings taken with a toothpick from your teeth and perform the Gram staining technique on it.

1. Use the wide end of a toothpick to scrape your teeth near the gum line, then mix the scrapings thoroughly in a small drop of water at one end of a clean slide.

2. Allow the smear to dry completely, then heat it slightly by holding it briefly over your microscope's light. This makes the bacteria stick to the slide. Place the slide on a staining rack and cover the smear with crystal violet stain.

3. Leave the dye on the slide for about one minute, then rinse gently with tap water.

4. Place the slide on the staining rack again and cover the smear with iodine. This forms a complex with the crystal violet. Allow the iodine to stay on the smear for about one minute, then rinse again. At this point, all of the bacteria on the slide are stained purple.

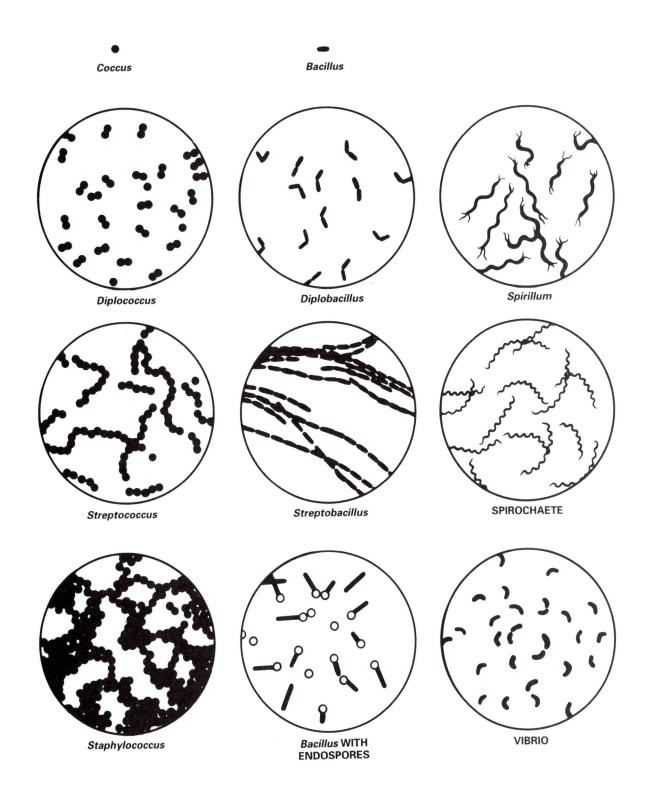

Coccus

Bacillus

Diplococcus

Diplobacillus

Spirillum

Streptococcus

Streptobacillus

SPIROCHAETE

Staphylococcus

Bacillus WITH ENDOSPORES

VIBRIO

FIGURE 13.1   Bacterial Shapes and Arrangements

5. Now hold the slide in a slanted position and drop 95% alcohol on the smear with an eye-dropper until no more purple stain shows in the alcohol coming off of the slide. At this point, the Gram positive bacteria are still purple, because they retained the stain. The Gram negative bacteria, however, are colorless, so you have to use a contrasting stain to make them show up.

6. Return the slide to the staining rack and cover the smear with safranin dye, which is red. Allow this to sit for about one minute, then rinse with tap water and blot (don't rub) dry with paper toweling.

7. Observe the slide under high power of your microscope or have the instructor focus it under oil immersion for you.

8. Try to find at least two of the three cell shapes and bacteria showing both Gram positive and Gram negative staining properties.

CYANOBACTERIA

The **cyanobacteria**, or blue-green algae, are similar in cell structure and physiology to their prokaryotic relatives, the bacteria. They also occur in unicellular and colonial forms. A characteristic of most blue-green algae is the production of a jelly-like **sheath** which surrounds both the individual cell and the colony. Reproduction is by fission or fragmentation.

Blue-green algae are found in almost any environment — lakes, streams, soil, snow, oceans, rocks, ice, deserts, and even the near-boiling waters of hot springs. They may be unicellular, filamentous, or form sheets of cells.

1. Obtain a prepared slide of *Oscillatoria*. Note the long chains of cells called **hormogonia** which are short fragments between dead cells where fragmentation occurs. This genus of blue-green algae gets its name from its oscillating movement through the water. (See Figure 13.2)

2. Prepare a slide of preserved *Nostoc*. *Nostoc* which was once referred to as "witches' butter" or "starjelly," forms large, grape-like macroscopic "colonies." The macroscopic colonies contain three types of cells (see Figure 13.3). The smaller are **vegetative cells**; the larger are known as **heterocysts**, in which **nitrogen fixation** is believed to occur. In addition, there are **akinetes** which are special spore-like reproductive cells resistant to adverse conditions of the environment. Nitrogen fixation is the incorporation of atmospheric nitrogen into inorganic compounds available to plants. It is a process that can be carried out only by some soil bacteria, cyanobacteria, and some symbiotic bacteria in association with legumes.

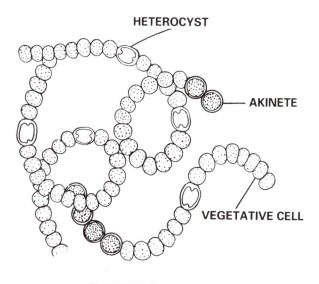

FIGURE 13.3 *Nostoc*

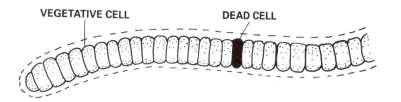

FIGURE 13.2 *Oscillatoria*

REVIEW QUESTIONS

1. Which organism is autotrophic: *Nostoc* or a spirochete? _____

2. What is meant by the term "gram positive"? _____

3. Why does *Oscillatoria* deserve its name? _____

4. What are the three basic types of bacterial shapes? _____

   _____

5. What is nitrogen fixation? _____

   _____

6. What important characteristics do bacteria and blue-green algae have in common? _____

   _____

   _____

7. What is the main difference between blue-green algae and bacteria? _____

   _____

   _____

8. Define:

   Prokaryotic _____

   Eukaryotic _____

   Heterotrophic _____

   Autotrophic _____

   Fission _____

   Conjugation _____

9. Indicate whether the following are bacteria or blue-green algae.

   Oscillatoria _____

   Diplococcus _____

   Spirochaete _____

   Rickettsia _____

   Nostoc _____

   Staphylococcus _____

10. Nitrogen fixation is believed to take place within which *Nostoc* cell? _____

    _____

11. How are *Nostoc* cells different from the *Streptococcus* bacterial cells? _____

    _____

Kingdom Protista

OBJECTIVES

After completion of this exercise, the student should be able to do each of the following:

- Give the distinguishing characteristics of the Kingdom Protista.
- Explain how protozoans differ from other protists, and list the phyla included as protozoans.
- Identify the following organisms from prepared slides and give the phylum to which each belongs:

Trypanosoma gambiense	dinoflagellates	*Euglena* sp.
Amoeba proteus	*Paramecium caudatum*	types of algae
foraminifera	*Plasmodium malariae*	fungus like protists
diatoms	radiolaria	

- List the distinguishing characteristics of the following protist phyla.

Phylum Sarcodina	Phylum Pyrrophyta	Phylum Phaeophyta
Phylum Ciliophora	Phylum Chrysophyta	Phylum Rhodophyta
Phylum Zoomastigina	Phylum Euglenophyta	Phylum Myxomycota
Phylum Sporozoa	Phylum Chlorophyta	Phylum Oomycota

- Identify and give the functions of each of the following structures:

flagella	**raphe**	**macronucleus**	**paramylum granule**
cilia	**pellicle**	**gullet**	**undulating membrane**
pseudopodia	**eyespot**	**trichocysts**	**food vacuole**
micronucleus	**oral groove**	**contractile vacuole**	**anal pore**

- Identify in prepared slides and explain the processes of **conjugation** and **transverse** or **binary fission** in *Paramecium*.
- Compare feeding and nutrient processing in *Paramecium* and *Amoeba*.
- Discuss the basic characteristics of the green algae.
- Describe the structures and/or reproductive mechanisms of *Protococcus, Spirogyra, Volvox,* and *Ulva*.
- Discuss briefly the habits and morphology of the red and brown algae.
- Explain the sexual reproduction of *Spirogyra*.
- Describe the structure of a typical plasmodial slime mold, *Physarum*.
- Answer the review questions at the end of this exercise.

INTRODUCTION

The Kingdom Protista (sometimes called Protoctista) is indeed a very diverse group of organisms. The only unifying characteristic in this group is that all its members are eukaryotic. The first four phyla are all unicellular heterotrophs and are often grouped together as **protozoans** (= first animals) because of their animal-like characteristics. The next group of six phyla are referred to as **algae** and are unicellular and multicellular plant-like organisms. The last two phyla are protists that resemble **fungi**.

PHYLUM SARCODINA

The members of this phylum move by protoplasmic extensions called **pseudopodia** (false feet). The most common of these organisms is the *Amoeba*, which we will study in detail. The amoeba is a "blob of protoplasm" with no definite shape. Many other members of this phylum produce cells of varied shapes and compositions according to their species. All members of the phylum are heterotrophic.

1. Obtain a prepared, stained slide of *Amoeba proteus*. The organism is surrounded by a **plasma membrane**. Immediately inside this membrane is a thin, clear, non-granular layer of protoplasm called **ectoplasm**. Beneath this layer is the main body mass of granular **endoplasm**. Within the endoplasm is a large disclike structure, the **nucleus**. **Food vacuoles** may also be evident, containing partially digested granules of food taken in by the cell. Observe another vacuole toward the posterior end of the cell. This is the **contractile vacuole**, which functions to pump out excess water by alternately appearing as it fills up and disappearing as it empties.

2. Prepare a wet mount from the *Amoeba* culture. For the best results, reduce the amount of light passing through the specimen. Try to identify the anatomical structures of the *Amoeba*. Observe how the *Amoeba* moves by sending out its pseudopodia. Pseudopodia are also important in food-getting. They surround the food and enclose it in a food vacuole in the process known as **phagocytosis**. Also try to observe the pumping action of the contractile vacuole. Label the indicated structures in Figure 14.1.

3. Examine a prepared slide of the class Foraminifera. The organisms belonging to this class are mostly marine and have encompassing shells of various shapes. Each shell is composed of a **calcareous** material. See Figure 14.2.

4. Examine a prepared slide of the class Radiolaria. The radiolarians have a shell that is siliceous (composed of or derived from **silica**) in nature. Refer to Figure 14.3.

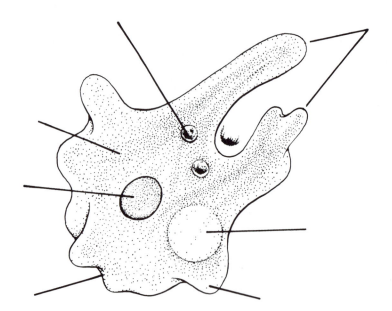

FIGURE 14.1 *Amoeba proteus*

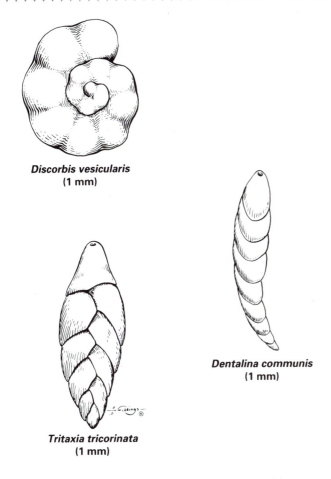

Discorbis vesicularis
(1 mm)

Dentalina communis
(1 mm)

Tritaxia tricorinata
(1 mm)

FIGURE 14.2   **Representative Foraminiferans**

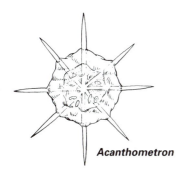

Acanthometron

Podocyrtis schomburgk

Sphaerostylus ostracion

FIGURE 14.3   **Representative Radiolarians**

PHYLUM CILIOPHORA — CILIATES

The organisms in this phylum are characterized by having hairlike projections called **cilia**, which they use for locomotion and food-getting. Cilia have the same basic structure as flagella, but are generally shorter and more numerous than flagella. Members of the phylum Ciliophora differ from other organisms in that they possess two different types of nuclei; a large **macronucleus**, and one or more smaller **micronuclei**. The macronucleus is concerned with the general control of cell activities, while the micronucleus is involved in sexual reproduction.

Like the flagellates, the ciliates are heterotrophic, a few being parasitic.

1. Using the properly-labeled pipette in the bottle containing live specimens, prepare a wet mount of *Paramecium caudatum*. In order to study the paramecium properly, you may have to slow it down. Add a drop of Protoslo to a drop of water containing some paramecia. Cover the preparation with a coverslip.

NOTE

This organism is microscopic (invisible to the naked eye), but you can be relatively sure of getting a specimen if you will siphon water into the pipette from the material in the very bottom of the jar.

The cilia cover the entire cell and are relatively uniform in size. You may not be able to see the cilia with your microscope, but they should be visible in the demonstration slide under the phase contrast microscope. The basic shape of the animal resembles a slipper and its shape is maintained by its stiff outer covering, the **pellicle**. Note that one end is rounded and the other is pointed. In observing the living specimens, note that the same end is always directed anteriorly. Which end is it? Locate the **oral groove**, the depression into which food is swept by beating cilia. This leads to a tubular **gullet** where **food vacuoles** form.

In some species undigestible wastes are released from the **anal pore**, which can only be seen when particles are being discharged through it. Two **contractile vacuoles** are present, one at each end of the cell. They contract alternately and function in expelling excess water. Just beneath the pellicle are located numerous **trichocysts** capable of being discharged like small darts under the right conditions. Their function appears to differ in different ciliates, in some, functioning in defense, and in others in immobilizing prey. In paramecia, they appear to aid in adherence to surfaces while the organisms feed.

Apply a drop of weak acetic acid at the edge of the coverslip. As it proceeds under the coverslip and the paramecia come into contact with it, the trichocysts will discharge, and you should be able to see them.

Label the indicated structures of the paramecium on Figure 14.4 using the bold-faced terms above.

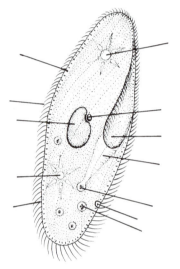

FIGURE 14.4 *Paramecium caudatum*

2. Occasionally you may observe the living paramecia reproducing. Ciliates may reproduce asexually by **binary fission** or **transverse fission**, dividing into two equal daughter cells. Since it is very difficult to see the nuclei in the living specimens, it is best to study a prepared slide. Examine prepared slides showing binary fission in *Paramecium* and sketch the various stages below.

3. In addition, paramecia may reproduce sexually, allowing for greater genetic variability, through a process called **conjugation**. During this process, two individuals come to lie with their ventral sides together, and a protoplasmic bridge forms between them. The macronucleus disintegrates and gradually disappears. The micronuclei divide several times and finally the paramecia exchange some of their nuclear material. This is a type of fertilization. Refer to your textbook for a detailed discussion of conjugation.

Examine a prepared slide showing *Paramecium* in conjugation and sketch several stages.

CONJUGATION IN *Paramecium*

4. Prepare a wet mount of pond water. Attempt to locate and identify some examples of Ciliophora. Refer to Figure 14.5 for identification and draw what you find below.

POND WATER CILIOPHORA

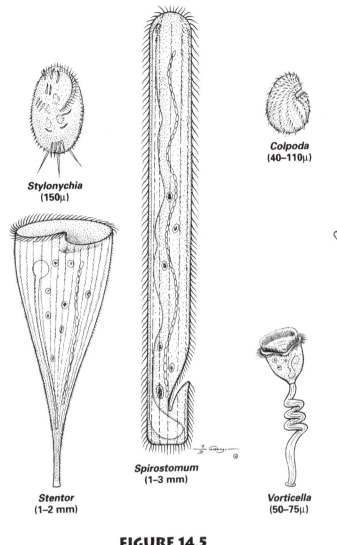

Stylonychia
(150μ)

Colpoda
(40–110μ)

Spirostomum
(1–3 mm)

Stentor
(1–2 mm)

Vorticella
(50–75μ)

FIGURE 14.5
Representatives of the Phylum Ciliophora

Obtain a slide of *Trypanosoma gambiense* and refer to Figure 14.6. This organism, carried by an intermediate invertebrate host, the **TseTse fly**, causes a neurological disease known as **African Sleeping Sickness** in man. Two morphological forms may be found on these slides: one with only an undulating membrane and one with an undulating membrane and flagellum.

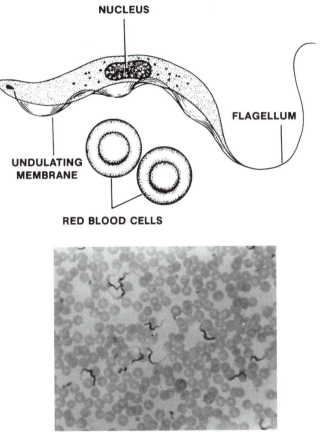

NUCLEUS

FLAGELLUM

UNDULATING MEMBRANE

RED BLOOD CELLS

Photomicrograph of *Trypanosoma gambreinse* **(430x)**
Courtesy of Carolina Biological Supply Company

FIGURE 14.6 **Trypanosome in Blood**

PHYLUM ZOOMASTIGINA —
Flagellates

All of the members of this phylum are propelled by a long whip-like projection called a **flagellum**. Almost all of the zoomastiginans have one, two, or more flagella, but in some members, a flagellum is lacking. All flagellates are heterotrophic, and most are free-living, but some are parasitic. It is believed that the first primitive eukaryotic cells were flagellates. Because of genetic variation and natural selection, some of these primitive flagellates may have given rise to plantlike organisms while others may have evolved into more highly adapted and specialized animal types.

PHYLUM SPOROZOA

Some sporozoans have pseudopods and some have flagellated gametes. The most widely known member of the sporozoan phylum is *Plasmodium*, the blood parasite which causes malaria in man. Like *Trypanosoma*, *Plasmodium* is carried by a secondary host, the female *Anopheles* mosquito, which transmits malaria to man. All of the sporozoans are highly specialized in particular parasitic ways of life. They typically have a life cycle in which a sexual stage alternates with a spore-producing asexual stage. Spores develop as the result of multiple fission of a "mother cell." Some are amoeboid and others are encapsulated.

Examine a slide of *Plasmodium malariae* and sketch your observation below.

Plasmodium malariae

PHYLUM PYRROPHYTA —
Dinoflagellates

The members of the Phylum Pyrrophyta (= fire plants) are characterized by their possession of a cell wall composed of **cellulose plates** (when they have a cell wall) and by their possession of two flagella — one directed posteriorly, and the other laterally. Many members of this group luminesce when disturbed, causing the water to "sparkle" at night, hence the name "fire plants." They also are called **dinoflagellates**.

Two members of the group, *Gymnodinium* sp. and *Gonyaulax* sp. are responsible for "**red tides**" in Florida and off the coast of California which color the water red and cause massive fish kills because of the toxic waste products they dump into the water during rapid growth periods, or "**blooms**."

Examine the slide of a **dinoflagellate** and locate the **transverse** and **longitudinal grooves** in which the two flagella lie. Also note the cellulose plates making up the cell wall. See Figure 14.7.

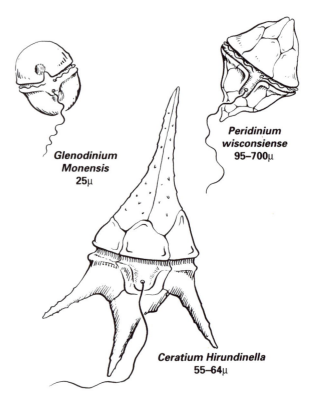

Glenodinium Monensis 25μ

Peridinium wisconsiense 95–700μ

Ceratium Hirundinella 55–64μ

FIGURE 14.7 Representative Dinoflagellates

PHYLUM CHRYSOPHYTA —
Golden Algae

Diatoms, by far the most numerous chrysophytes, are unicellular or colonial organisms of exceeding abundance in fresh and salt water, where they serve as food for small animals. Most of the plant portion of the **plankton** (floating life) are small algae like these, and by virtue of their vast numbers, their annual growth in terms of dry weight may equal all of the land plants together. They store oil instead of starch, and it is thought that much of our present undersea oil reserves may represent diatoms accumulated over millions of years, deposited with their shells on the ocean floor.

These organisms form cell walls composed largely of silica (the substance making up sand). The walls form two **valves** which fit together like the halves of a Petri dish. Diatoms which are oblong in shape are capable of smooth gliding movements on surfaces because of the presence of a **raphe**, or longitudinal slit, in their valves. Movement of the cell within the valves is thought to set up countercurrent water movement at the raphe, producing the gliding motion. Radially symmetrical forms lack raphes and thus are incapable of this type of movement.

1. Examine a prepared slide of diatoms. Note the various shapes of cells and colonies. Figure 14.8 shows some of the most commonly found diatoms.

2. Make a fresh mount of pond or bog water. Notice the many types of diatoms. Some diatoms are colonial, while others are solitary forms.

3. If time allows, make a slide of **diatomaceous earth**. This substance is an accumulation of diatom shells over a period of millions of years. Today, this solid is mined and used in silver polish, toothpaste, and water filters.

PHYLUM EUGLENOPHYTA — Euglenoids

These protists occur mostly in fresh, but often polluted, waters. Their pigments are similar to those of higher plants but they neither store starch nor produce firm cellulose cell walls. Euglenoids possess a flexible protein **pellicle** (inside the cell membrane) which allows the body to change shape while moving. Motility is achieved by an anterior whip-like **flagellum**.

1. Examine a prepared slide or make a fresh slide of *Euglena* and other euglenoids.

2. Locate as many of the labeled structures in Figure 14.9 as possible.

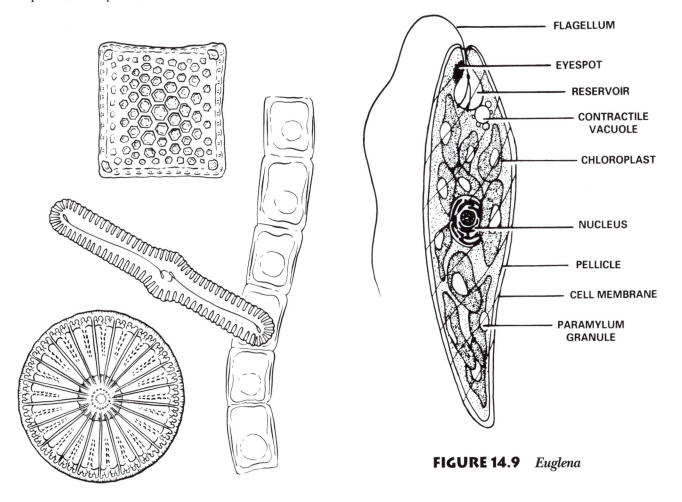

FIGURE 14.8 **Representative Diatoms**

FIGURE 14.9 *Euglena*

PHYLUM CHLOROPHYTA—
Green Algae

The green algae are the most diverse of all the algal groups. They range from single cells to colonies to multicellular organisms.

The cells of most green algae have a transparent cellulose wall and a prominent vacuole. The chlorophyll and carotene pigments are in approximately the same proportions as in most "higher plants." **Pyrenoids**, centers of starch formation, often occur on the chloroplasts.

Most green algae reproduce both asexually and sexually. Asexual reproduction occurs usually as mitotic cell division in unicellular and colonial forms, but multicellular organisms produce asexual spores and also fragment. Sexual reproduction is represented by both isogamy and heterogamy.

UNICELLULAR GREEN ALGAE

Protococcus is one of the most common, and yet simplest, of the green algae (Figure 14.10). It can be found on tree bark, fence posts, flower pots, brick walls, damp soil and rocks. *Protococcus* reproduces only by cell division. Its spherical cell wall surrounds a large cup-shaped **chloroplast**.

1. Prepare a fresh mount of *Protococcus* and examine under high power.

2. Notice the example habitat for *Protococcus* such as tree bark or flower pots. Just like many mosses, *Protococcus* often grows more heavily on the moist, northern sides of tree trunks. See Figure 14.10.

COLONIAL GREEN ALGAE

1. One common form of colonial algae is **filamentous** in which cell division occurs in only one plane. The cylindrical cells of *Spirogyra* contain spirally-shaped **chloroplasts** with conspicuous **pyrenoids**. See Figures 14.11, 14.12, and 14.13.

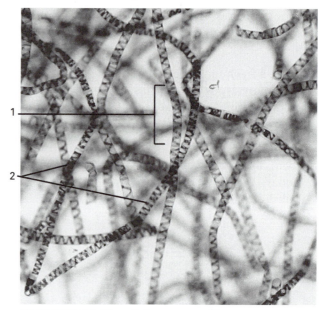

FIGURE 14.11 *Spirogyra* **are filamentous green algae commonly found in green masses on the surfaces of ponds and streams. Their chloroplasts are arranged as a spiral within the cell. Several cells comprise a filament.**

1. Single cell   2. Filaments

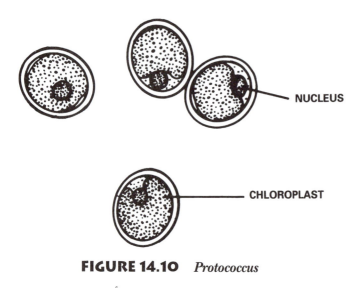

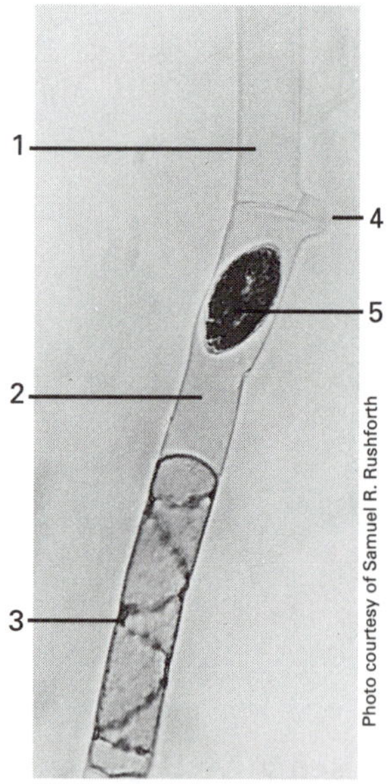

NUCLEUS

CHLOROPLAST

FIGURE 14.10 *Protococcus*

1 Upper cell
2. Lower cell
3. Chloroplasts
4. Conjugation tube
5. Zygote

Photo courtesy of Samuel R. Rushforth

FIGURE 14.12 **A self-fertile species of** *Spirogyra.* **A gamete has migrated from the upper cell to form a zygote in the lower cell.**

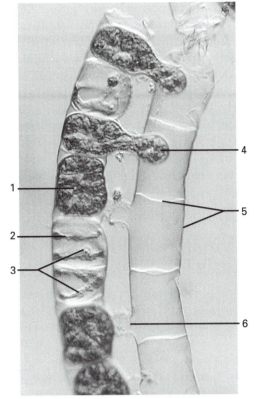

FIGURE 14.13

Spirogyra **Undergoing Conjugation**

1. Zygote (zygospore)   4. Male gamete
2. Chloroplast   5. Cell wall
3. Pyrenoid   6. Conjugation tube

Obtain a prepared slide of *Spirogyra*. Note the filaments with protrusions. Each is called a **conjugation tube**, through which the nucleus and the rest of the protoplasm of one cell act as an amoeboid gamete and pass into the adjoining cell of the other filament. A **zygote** is produced which soon forms a thick wall and undergoes a resting

stage (**zygospore**). Later, meiosis occurs to produce four haploid nuclei; three die. The remaining nucleus is incorporated into a new haploid cell from which a new filament develops by simple cell division. Label the Figure 14.14 with these terms: chloroplast, conjugation tube, nucleus, pyrenoid, and zygote.

2. (OPTIONAL) Prepare a fresh mount of *Volvox*. *Volvox* are found in cool, fresh, calm waters. *Volvox* colonies vary in size from several hundred to as many as 40,000 cells arranged in the wall of a hollow sphere and interconnected by **plasmodesmata**. Asexual reproduction occurs through the formation of **daughter colonies** (see Figure 14.15) from single vegetative cells of the parent. These detach to the interior of the parent colony, and when the parent colony disintegrates from age, the daughter colonies swim away. Sexual reproduction is an advanced type of **heterogamy** (differing gametes) called **oogamy**.

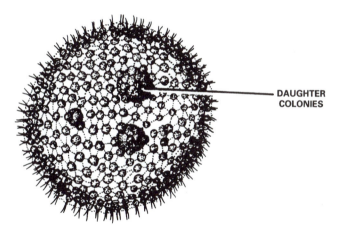

DAUGHTER COLONIES

FIGURE 14.15   *Volvox*

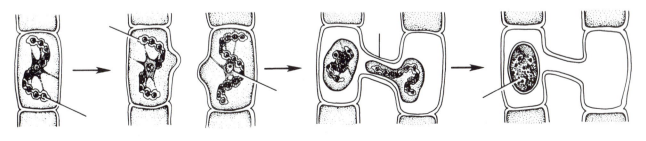

FIGURE 14.14   *Spirogyra* **Sexual Reproduction**

MULTICELLULAR GREEN ALGAE

Ulva (sea lettuce) is one of the most commonly found green algae (Figure 14.16). *Ulva* clings to submerged rocks, oyster shells, and other substrates in salt or brackish waters. It undergoes alternation of generations in which the 2N sporophyte and 1N gametophyte are similar in size and appearance. *Ulva* also reproduces asexually by fragmentation.

PHYLUM PHAEOPHYTA—
Brown Algae

The brown algae are multicellular and almost exclusively marine. They include the largest of the seaweeds, the kelps, which may grow 300 feet in length. A brown algae attaches to a solid surface by means of an adhesive **holdfast**, from which extends a stem-like **stipe**, bearing a flattened leaf-like **blade**. Deep species possess **air bladders** located along the blades, which float the plant's photosynthetic areas up into the lighted areas.

If available, observe various preserved specimens of brown algae (kelp) such as **Sargassum** and **Laminaria**. Note the holdfasts, stipes, blades, and air bladders. Label algae in Figure 14.17 with the previous terms.

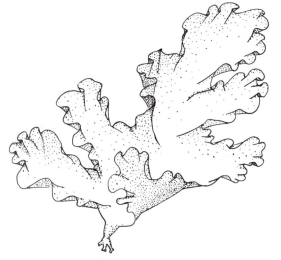

FIGURE 14.16   *Ulva*

PHYLUM RHODOPHYTA—
Red Algae

In addition to chlorophyll, the red algae contain two accessory photosynthetic pigments, **phycoerythrin** (red) and **phycocyanin** (blue). For this reason, they often are distinctly red in color. With these pigments, red algae can capture the rays which penetrate deepest into water, therefore can live deeper than most algae. This deep habitat is often below the levels of major currents, in nearly still water, enabling them to have very fragile bodies. These delicate and beautiful forms make them the darlings of seaweed collectors and their flavors the favorites of European and Oriental gourmets. Examine herbarium specimens of red algae, noting the body form, and refer to Figure 14.18.

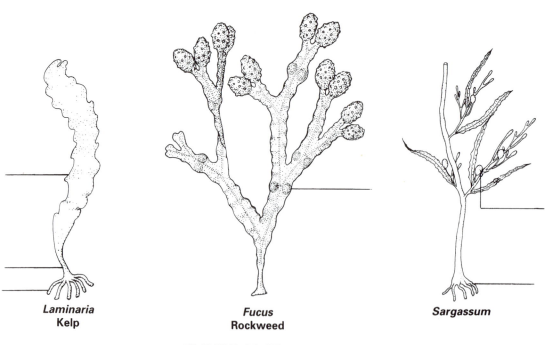

Laminaria
Kelp

Fucus
Rockweed

Sargassum

FIGURE 14.17   Brown Algae

(a) *Porphyra*

(b) *Ceramium*

THALLUS

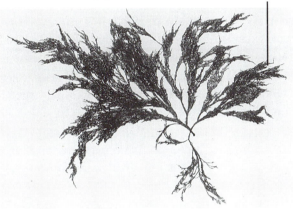

(c) *Polysiphonia.*

FIGURE 14.18
Examples of Common Marine Red Algae

amoeba's movement and process of food intake by phagocytosis. When conditions become unfavorable due to a lack of food or water, the plasmodium will produce stalks known as **sporangia** where meiosis takes place producing 1N spores.

1. Examine a culture of *Physarum* and refer to Figure 14.19. Observe the plasmodium with the high power objective lens. **Cytoplasmic streaming** may be occurring. Can you notice any food vacuoles?_____.

NOTE

A stereomicroscope may be used in part 1.

---

2. Observe any sporangia that may be present in the culture.

3. Make a drawing on the next page and label any structures you can identify.

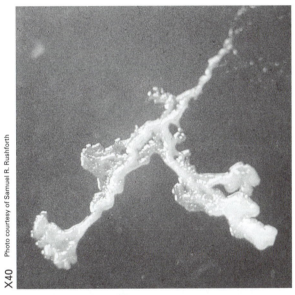

X40   Photo courtesy of Samuel R. Rushforth

FIGURE 14.19 Slime mold, *Physarum*, Growing on an Agar Culture Medium

PHYLUM MYXOMYCOTA —
Plasmodial Slime Molds

The plasmodial slime molds are not photosynthetic and they lack a cell wall around the plasma membrane. Its feeding stage is referred to as the **plasmodium** which is a single cell consisting of many nuclei. This multinucleate mass is also called a **coenocytic** mass. The plasmodium is similar to an

PHYLUM ACRASIOMYCOTA —
Cellular Slime Molds

In the cellular slime molds, individual amoeba-like cells congregate together in a single mass. This mass moves around for a while, then settles on one spot and produces a rounded sporangium. The spores from the sporangium grow into new amoeba-like cells to continue the cycle.

Physarum

PHYLUM OOMYCOTA — Water Molds

Oomycota means "egg fungi" because of their type of sexual reproduction during which large non-motile **eggs** or **oospheres** are produced within an **oogonium**. Refer to Figure 14.20. The oomycetes are commonly known as water molds, and downy mildews. One important characteristic of the oomycetes is that the cell walls are composed of cellulose instead of chitin which is a distinguishing characteristic of the Kingdom Fungi. Another feature is that the asexual **zoospores** are flagellated.

Water molds are usually saprophytes growing as a mass on dead algae and animals. One example of a parasitic water mold which attacks dying or diseased fish causes a disease called ick in ponds and aquariums. Downy mildews are usually classified as parasitic land organisms. *Phytophthora infestans* caused in the Irish potato blight in the 1850s and *Plasmopara sp.* was responsible for the downy mildew disease in grapes in France during the 1870s.

1. Examine a prepared slide of *Saprolegnia*, a common water mold. Identify the oogonia, eggs and fertilization tubes.

2. Examine a wet mount slide of a water mold mycelium (a mass of filaments) prepared from oomycetes growing on dead seeds. Identify the zoospores.

3. Make drawings and label your observations.

Photo courtesy of Samuel R. Rushforth

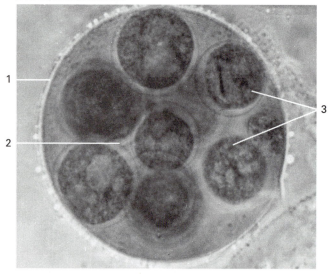

FIGURE 14.20 **A Mature Oogonium** of the Water Mold, *Saprolegnia*

1. Oogonium   2. Fertilization tube   3. Eggs

Saprolegnia

REVIEW QUESTIONS

1. *Trypanosoma gambiense* causes _____ in man.

2. By what means does *Paramecium* move? _____

3. _____ move throughout the cytoplasm in a regular pattern and finally discharge all undigested material through the anal pore of the paramecium.

4. Paramecia may reproduce asexually by means of _____ and sexually by means of _____.

5. The amoeba moves by _____ which means "_____."

6. Why would you expect a freshwater amoeba to be more likely to have contractile vacuoles than a marine amoeba? _____

7. *Plasmodium malariae* causes _____ in _____.

8. Which protists are heterotrophic: euglenoids, chrysophytes, or protozoans? _____

9. The cell wall of diatoms is largely composed of what durable material? _____

10. What is the storage product of diatoms? _____

11. How do diatoms differ from euglenoids in their movement? _____
_____

12. How does a pellicle differ from a cell wall? _____

13. Planktonic organisms are _____

14. What is a "red tide"? _____

15. What is the one characteristic that the phyla Zoomastigina, Ciliophora, Sarcodina, and Sporozoa have in common? _____

16. What is the function of a pyrenoid? _____

17. Why does *Protococcus* usually grow more heavily on the northern sides of trees? _____
_____

18. What is the function of air bladders? _____ Which phylum are they characteristic of? _____

19. What wavelengths of light would a red alga be able to capture which other green plants could not capture? _____ Why? _____

20. MATCHING: Match the phylum with the correct organism.

Organism	Phylum
_____ 1. *Trypanosoma*	a. Pyrrophyta
_____ 2. Diatoms	b. Zoomastigina
_____ 3. *Amoeba*	c. Sarcodina
_____ 4. *Paramecium*	d. Chrysophyta
_____ 5. Radiolaria	e. Ciliophora
_____ 6. *Euglena*	f. Euglenophyta
_____ 7. *Plasmodium*	g. Sporozoa
_____ 8. Dinoflagellates	
_____ 9. Foraminifera	

21. Explain these terms:

Zygospore_____

Daughter colony _____

Stipe _____

Holdfast _____

Filament_____

Conjugation tube _____

22. List the methods of asexual reproduction used by the green algae.

1. _____

2. _____

3. _____

23. Sexual reproduction in *Spirogyra* occurs through the formation of a _____.

24. *Spirogyra* is a _____ form of colonial algae.

25. Compare the structures and the reproductive mechanisms of *Protococcus*, *Spirogyra*, *Volvox*, and *Ulva*.

Type	Structure	Reproductive mechanisms
Protococcus		
Spirogyra		
Volvox		
Ulva		

Kingdom Fungi

OBJECTIVES

After completion of this exercise, the student should be able to do each of the following:

- List the major characteristics of fungi.
- Describe the typical body of a fungus, or **mycelium**.
- Identify *Rhizopus* with its four types of hyphae (**rhizoids**, **stolons**, **sporangiophores**, and **gametangia**) and explain its life cycle.
- Explain the differences between the Zygomycota, Ascomycota, Basidiomycota, and Deuteromycota, and distinguish fresh and preserved specimens of each group.
- Distinguish between pore, cup, gill, and tooth fungi.
- Explain **symbiosis**, list the three major types, and give an example of each.
- Explain what a **lichen** consists of and how it reproduces. Identify the three basic forms of lichens.
- Answer the review questions at the end of this exercise.

CHARACTERISTICS OF THE FUNGI

The fungi are a large, diverse group of **eukaryotic** organisms (see Figures 15.1–15.3). They possess cell walls but these cell walls are usually made of **chitin** instead of cellulose. Fungi lack chlorophyll so they feed themselves by being **saprophytes** (which live off dead organisms) or **parasites** (which live off living organisms). Thus, the fungi are valuable agents of decay for recycling simple compounds and elements needed for new life.

The **mycelium** or body of a fungus is usually composed of a mass of branching filaments called **hyphae**. Remember, the hyphae of a fungus are usually haploid in terms of chromosome number (diploid stages after fertilization are usually quickly converted into haploid cells by meiosis). For the most part, the mycelium is seldom seen since it grows deep inside its host. The reproductive structures, which produce spores either by mitosis or meiosis, are usually visible and characteristic of the species.

Although both asexual and sexual reproduction are possible for most fungi, there are no morphological differences which could be designated male and female; therefore, the term "mating strains" is used and the types designated as plus (+) and minus (−). A plus strain can only "mate" with a minus strain.

123

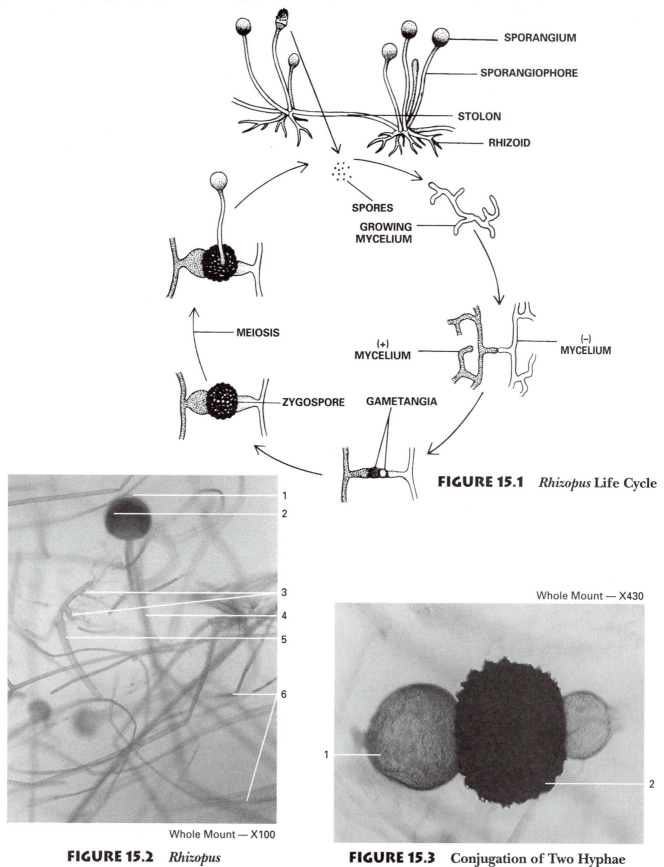

FIGURE 15.1 *Rhizopus* Life Cycle

SPORANGIUM

SPORANGIOPHORE

STOLON

RHIZOID

SPORES

GROWING MYCELIUM

MEIOSIS

(+) MYCELIUM

(−) MYCELIUM

ZYGOSPORE   GAMETANGIA

Whole Mount — X100

FIGURE 15.2 *Rhizopus*

1. Sporangium   4. Sporangiophore
2. Spores   5. Stolon
3. Rhizoids   6. Hyphae

Whole Mount — X430

FIGURE 15.3 Conjugation of Two Hyphae
in the Common Bread Mold, *Rhizopus*

1. Gametangia   2. Zygospore

DIVISION ZYGOMYCOTA —
Bread Molds

This is the most primitive of the four fungal divisions we will examine. Zygomycota hyphae have no **cross walls** (**septa**) between the nuclei. Their mycelium is therefore a mass of communal cytoplasm, making separate compartments for nuclei only during reproduction.

Rhizopus has several basic types of hyphae: **rhizoids** for food absorption, **stolons** to enable the fungus to spread quickly over the substrate surface, and **sporangiophores** with **sporangia** for asexual reproduction.

In sexual reproduction, short lateral hyphae form **gametangia** which contain either $(-)$ or $(+)$ strain nuclei as the gametes. The gametangia fuse to form a **zygospore**, or **zygosporangium**. The zygospore contains numerous diploid (2N) nuclei and develops a dark rough wall. Inside this zygospore, one diploid nucleus undergoes meiosis to produce one viable 1N nucleus. This haploid organism grows and produces an asexual sporangiophore.

1. Obtain a prepared slide of *Rhizopus nigricans*, the common black bread mold. Look for rhizoids, stolons, sporangiophores, gametangia, sporangia and zygospores.

2. Observe *Rhizopus* growing on various substrates such as bread, beans, agar, fruit, etc. Notice the sporangiophores, stolons, and rhizoids.

DIVISION ASCOMYCOTA —
Sac Fungi

Ascomycota are usually characterized by **septate** hyphae and by sexual reproductive structures called **asci** which contain eight **ascospores**. The asci are often found in large numbers within structures called **ascocarps** such as morels and cup fungi. Ascomycota also reproduce asexually, producing spores called **conidia**.

1. Make a fresh slide of common baker's yeast. Yeast have evolved to exist without hyphae and their main means of reproduction is asexual **budding**.

2. Examine various types of common sac fungi such as *Sordaria*, morels, truffles, and cup fungi. The yeast in Figure 15.4 show asexual structures. Note the illustrations are not drawn to scale.

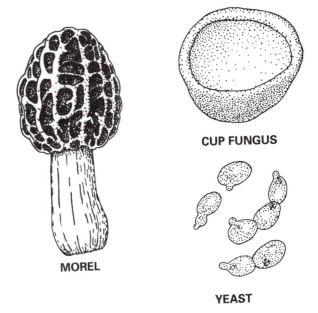

MOREL

CUP FUNGUS

YEAST

FIGURE 15.4   Various Ascomycota

DIVISION BASIDIOMYCOTA —
Club Fungi

This and the ascomycota division have a unique stage in their life cycles found in no other groups — the **dikaryon** (see Figure 15.5). At gamete fusion in other forms, both the cytoplasms and nuclei fuse immediately. In the dikaryon, the cytoplasm of the gametes fuses but the nuclei remain separate and divide in tandem for a varying period of time. This produces a multicellular body which is not really 1N (since there are two sets of chromosomes in one cell) but not actually 2N either (since the two sets are not within one nuclear membrane). This stage is designated as $N + N$, or **dikaryotic**. In the ascus or basidium, just before meiosis, the two nuclei fuse to become a genuine 2N nucleus.

Basidiomycota have septate hyphae and sexually produce spores externally on cells called **basidia**. Thousands of basidia occur in a fruiting body called a **basidiocarp**. The basidiocarp may be a mushroom, a puffball, shelf fungi, a rust or smut, or a slimy mass of gelatinous hyphae.

1. *Coprinus* — **mushroom**

   Examine a prepared slide of a x.s. of *Coprinus* basidiocarp. Note the basidia, the **sterigmata** which hold the spores to the basidia, the gills, and spores.

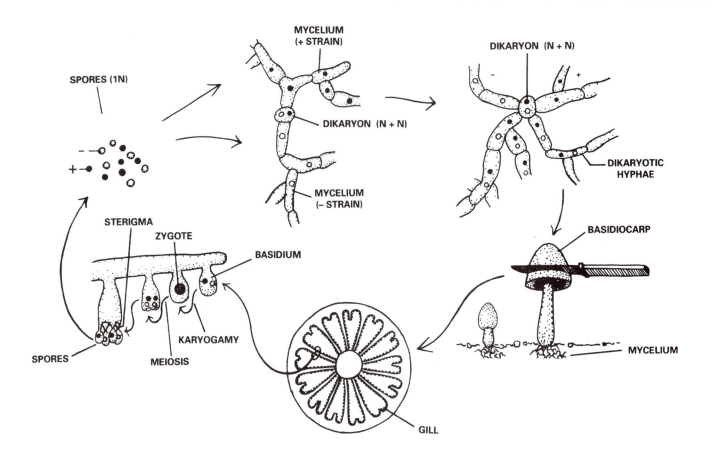

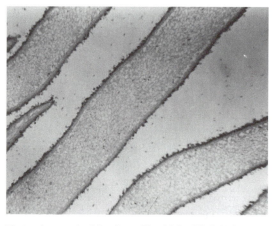

Photomicrograph of *Coprinus* gills with basidia (100x)
Courtesy of Turtox, Inc.

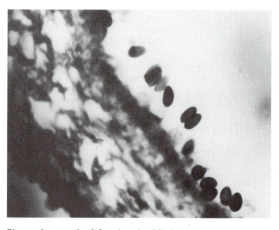

Photomicrograph of *Coprinus* basidia (430x)
Courtesy of Turtox, Inc.

FIGURE 15.5   **Basidiomycetes Life Cycle**

2. While some basidiocarps form basidia on long gills underneath the cap, others form basidia in deep narrow pores or on projecting teeth. Examine specimens of mushrooms and other fungi and note which have gills, pores, or teeth. Some fungi live in trees and produce woody basidiocarps. These are often called **shelf fungi**. Refer to Figure 15.6.

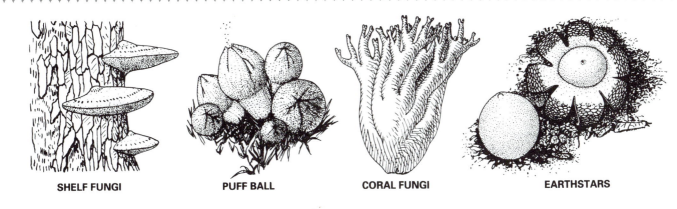

SHELF FUNGI   PUFF BALL   CORAL FUNGI   EARTHSTARS

FIGURE 15.6 Various Basidiomycota

DIVISION DEUTEROMYCOTA — Imperfect Fungi

Deuteromycetes are called the imperfect fungi because they appear to lack a sexual stage. Reproduction is asexual by **conidia**. Examples of this division include *Penicillium*, the source of the antibiotic penicillin, *Trichophyton*, which causes athlete's foot, and *Aspergillus*. (See Figure 15.7)

1. Obtain a prepared slide of *Penicillium* (Figure 15.8) or *Aspergillus* (Figure 15.9) and locate the stalked, **conidiophores**, branching up from the hyphae, on the tips of the conidiophores are chains of **conidia** or asexual spores. The erect conidiophores of *Penicillium* branch in a broom-like fashion.

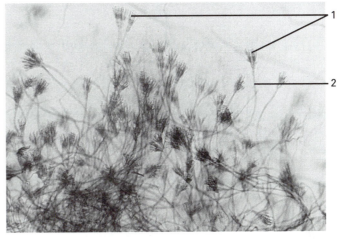

FIGURE 15.8 The Fungus *Penicillium* X100

1. Conidia (spores)   2. Conidiophore

Penicillium

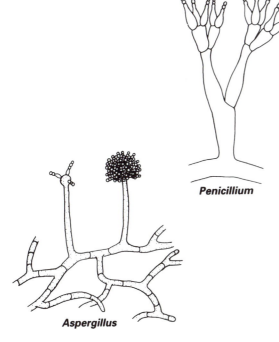

Aspergillus

FIGURE 15.7 Various Deuteromycota

FIGURE 15.9 *Aspergillus.* **The Conidia,** X430
or Spores, of this Genus are Exposed in a Radiate Pattern

1. Conidia   2. Conidiophore   3. Developing conidiophore

LICHENS

Symbiosis means "living together." There are three major types of symbiosis: **Mutualism** in which both partners benefit, **commensalism** in which one partner benefits while the other is not affected, and **parasitism** in which one partner is harmed while the other benefits.

   Lichens are examples of the first type of symbiosis, a combination of an algae (contributing food from photosynthesis) and a fungus (contributing anchorage and retaining water). The algal components of lichens (usually blue-green or green algae) can be grown by themselves. However, the fungal component (usually an ascomycota) apparently cannot survive in the absence of the proper alga. For this reason, the lichens are taxonomically classified according to the fungal partner.

1. We will not examine any specific taxonomic groups, but will look at three basic body shapes found among the lichens. **Crustose** lichens are hard, flat forms looking like stains on rocks or wood. **Foliose** lichens are more leaflike and resemble peeling paint. **Fruticose** lichens are usually erect, branching structures. Examine examples of these three types as available in the lab. (Refer to Figure 15.10)

2. Most of the algae and fungi of the lichens can reproduce independently, but the symbiotic unit reproduces by pinching off chunks of itself (**soredia**) containing a few algal cells and a few hyphae. On a prepared slide of lichen, note the algal cells near the surface, fungal hyphae, and soredia, if present.

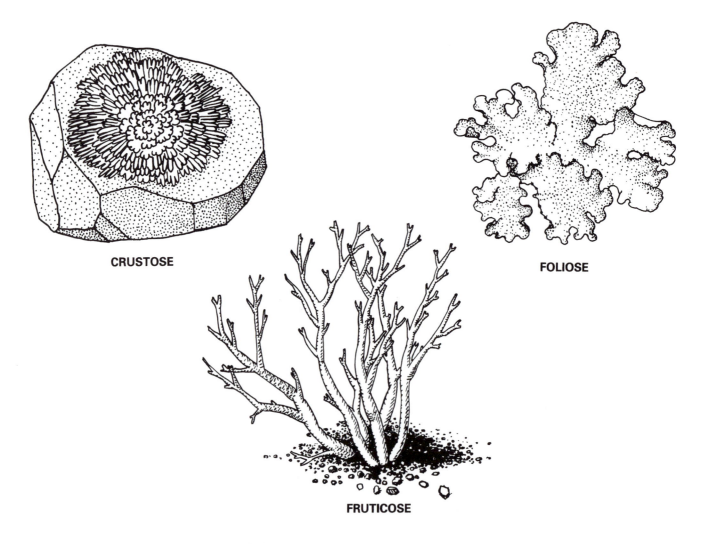

CRUSTOSE

FOLIOSE

FRUTICOSE

FIGURE 15.10   *Forms of Lichens*

REVIEW QUESTIONS

1. A shelf fungus produces which structure: basidiocarp, ascocarp, or lichen? _____

2. Saprophytes are: a. autotrophic   b. heterotrophic   c. parasitic

3. *Rhizopus* reproduces sexually by _____ (isogamy; heterogamy).

4. What economic importance do some ascomycetes have? _____

5. If you pick a mushroom, do you kill the entire fungal body? _____

    Why? _____

    _____

6. Budding is a form of _____ (sexual; asexual) reproduction.

7. What nuclear arrangement is unique to the ascomycota and basidiomycota? _____

    _____

8. Describe the typical body of a fungus, or mycelium. _____

    _____

    _____

9. The spore producing structure in *Rhizopus* is a _____ .

    For sexual reproduction the *Rhizopus* depends on the _____ which contains either

    (−) or (+) strain nuclei as gametes.

10. The zygospore gives rise to what structure? _____

11. Matching: Match the following characteristics with the specific division of fungi.

    _____ 1. The most primitive of the fungal divisions   a. Ascomycota

    _____ 2. Dikaryon stage in life cycle   b. Basidiomycota

    _____ 3. Septate hyphae   c. Zygomycota

    _____ 4. No cross walls between nuclei   d. Deuteromycota

    _____ 5. Zygospore resulting from union of (+) and (−) gametangia

    _____ 6. Sexual reproductive structures are asci

    _____ 7. Sexual reproductive structures called basidia

    _____ 8. Imperfect fungi

12. Define mutualism, parasitism, and commensalism and give an example of each.

    _____

    _____

    _____

13. What type of algae are commonly found in lichens? What is the fungal component? _____

_____

_____

14. Why are the lichens classified with the fungi? _____

_____

_____

15. Which of the 3 lichen body shapes resembles leaves? _____

16. Are lichens saprophytic like the fungi? _____

    Why? _____

_____

17. Explain why the Deuteromycota are considered to be "imperfect" fungi. _____

_____

Kingdom Plantae
Division Bryophyta

OBJECTIVES

After completion of this exercise, the student should be able to do each of the following:

- Discuss the characteristics of the bryophytes.
- Describe alternation of generations in liverworts such as *Marchantia*, and identify the associated structures.
- Describe alternation of generations in a typical moss.
- Answer the review questions at the end of this exercise.

CHARACTERISTICS OF BRYOPHYTES

The division Bryophyta contains small, autotrophic organisms which have become adapted to a terrestrial existence. The members of this division exhibit the following characteristics:

1. They are green, have **rhizoids** (rootlike structures) and may have stemlike and leaflike parts (mosses).

2. They lack true **vascular** tissue for water and food distribution and are consequently restricted to moist habitats.

3. The plant body (**thallus**) may be **dorsoventrally flattened** and **bilaterally symmetrical** (liverworts) or erect and **radially symmetrical** (mosses).

4. Like the algae, they require water for fertilization, but their reproductive organs, unlike algae, are multicellular — the **archegonium** (female) and the **antheridium** (male).

5. They have a pronounced alternation of generation cycle, in which the **gametophyte** is the more conspicuous and self-sustaining plant. The **sporophyte** is small, often ephemeral, and dependent on the gametophyte for its food and water. You may wish to review this life cycle in Figure 12.3.

CLASS HEPATICAE — Liverworts

Most liverworts live in moist shady locations; a few are essentially aquatic. *Marchantia*, the example we will study, is a flat, green, ribbonlike, dichotomously branched thallus, anchored to the substrate by rhizoids arising from the lower surface. Asexual reproduction occurs commonly by **fragmentation**; as the thallus elongates the older central portions die, leaving the tips isolated as individual new plants. Some liverworts (*Marchantia* is one) produce asexual bodies known as **gemmae**. These are pieces of thallus which have pinched themselves off and lie in a cup-like structure (**cupule** or **gemmae cup**) on the upper

131

surface of the thallus. The gemma is a miniature thallus and, in a favorable environment (having been rinsed from the cupule by rain) it will produce a thallus exactly like the one which gave rise to it, all 1N.

Male and female plants are separate in *Marchantia*. Male plants produce **antheridial receptacles**, stalks with discs on top. Buried in these discs are multicellular globes of cells, the inner ones of which will metamorphose into *sperms* and rupture through the surface of the disc via pores when rain or dew splashes onto the mature disc. These **antheridia** are 1N already, so the sperm are formed by simple mitosis and subsequent modification for swimming.

The female plant produces an **archegonial receptacle** which is topped with fingerlike projections. The **archegonia**, flask-shaped structures housing the **egg**, are hung from the undersides of these projections and are shielded by fringelike **involucres**. The archegonium is also 1N and the egg is formed by modification. At first the flask is solid, but then the bottom bulges out freeing the egg. The cells down the center of the neck begin to degenerate, leaving an opening for the sperm to swim through. The **ventral canal cell** near the egg remains until just before the egg is ripe. If anything destroyed the egg, this cell would become a second egg. If not needed, it disintegrates.

Rainwater or heavy dew provides the water necessary for fertilization. The sperm swim along the surface film, attracted to the archegonia by chemicals secreted into the water (**chemotaxis**). Fertilization takes place in the archegonium and the zygote remains there, nourished by its parent. A **foot** is produced at the base of the archegonium for attachment and nourishment, a short **stalk** extends from this, and the rest of the sporophyte forms a **capsule**. Inside this capsule, meiosis takes place to form numerous 1N **spores**. Among the spores, long twisted **elaters** form. In humid conditions the elaters coil tightly, but when it is dry (ideal for spore dispersal via wind) the elaters expand, pushing the spores apart and rupturing the spore case to release the spores. Half of the spores will germinate to become male plants and half to become female plants.

1. Use a stereoscope to examine fresh and preserved specimens of *Marchantia* gametophyte thallus, gemmae cups, antheridial receptacles, and archegonial receptacles. Refer to Figure 16.1

2. Now, examine prepared slides of the *Marchantia* gemmae cups on a thallus, the antheridial receptacles, archegonial receptacles, and mature sporophyte. Refer to Figure 16.2.

3. Note various types of preserved and living liverworts displayed in the lab.

4. Learn the life cycle of the liverwort.

FIGURE 16.1 Liverwort, *Marchantia*, Colony

1. Archegonal head (n) 2. Gametophyte thallus (n) 3. Gemmae cup (n)

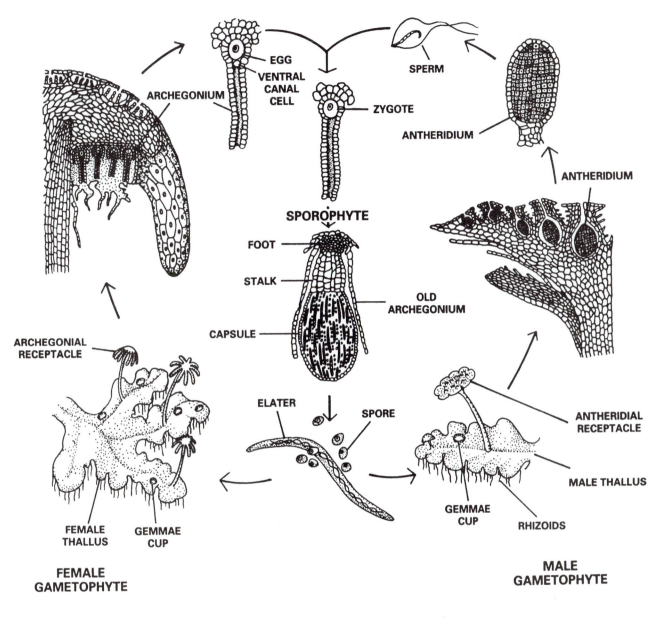

FIGURE 16.2 *Marchantia* Life Cycle

CLASS MUSCI — Mosses

Like the liverworts, the typical moss has a dominant gametophyte. Both germinate from spores to form threadlike bodies called **protonemata**. These bodies soon thicken to form the stemlike and leaflike structures and rhizoids of the mature gametophyte.

Some moss plants are **monoecious** (one house), having male and female structures in the same plant. The moss we have chosen to study is, however,

dioecious (two houses) with separate male and female plants. It is at the top of the leafy stalk that the reproductive structures are formed.

In the male, **antheridia** very similar to those of the liverwort are formed, but stand free on the surface, surrounded by protective filaments called **paraphyses**. **Sperms** are produced here by mitotic cell division, just as in the liverwort, and released into rain or dew, following the chemical attractants released by the female.

At the top of the female plant, **archegonia** (shaped very much like liverwort archegonia) stand upright, surrounded by paraphyses. Here, too, there is a **ventral canal cell** for insurance. The sperm swims down the neck canal, fertilizes the **egg** and the zygote begins to grow, fed by the female parent.

This sporophyte pushes a **foot** into the parental tissue for support and nourishment and begins to elongate. At first the archegonium grows to accommodate it, but later the top is torn off by the rapidly enlarging sporophyte and its remains ride along on top of the capsule as the **calyptra**. The **capsule** is elevated by a long thin **stalk**, almost as tall as the gametophyte parent. When the capsule is mature, the calyptra falls off. Inside the capsule, meiosis has produced a number of 1N spores. Just under the caplike lid (**operculum**) are a ring of pointed flaps known as the **peristome teeth**. When humidity is high, the teeth swell and bend over the opening to retain the spores, but when conditions are dry (ideal for wind dispersal of spores) they dry out, bend back, pushing off the operculum, and allow the spores to fall away to be caught by the wind.

1. Look at the prepared slides of moss archegonia and antheridia. Find these structures as well as the paraphyses and the egg. Refer to Figure 16.3

2. With a stereoscope, examine preserved moss gametophytes with sporophytes. Note the sporophyte's stalk, capsule, calyptra, operculum, and peristome teeth.

3. Observe various types of living and preserved mosses.

4. Learn the life cycle of the moss.

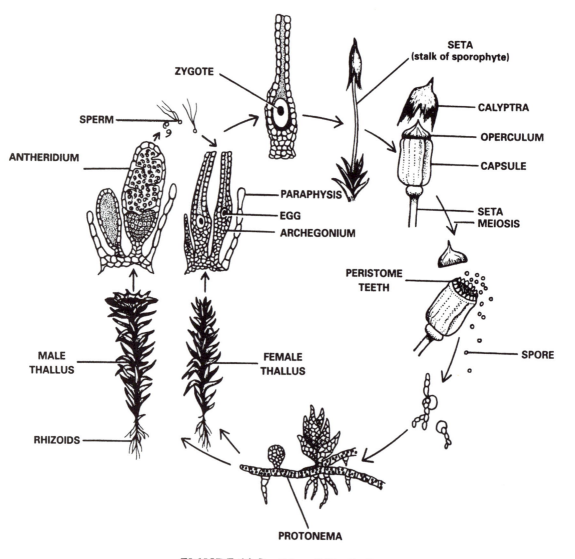

FIGURE 16.3 **Moss Life Cycle**

REVIEW QUESTIONS

1. How can you distinguish a moss from a liverwort?_____

_____

2. Why must bryophytes live in moist environments? _____

_____

3. What is the function of:

Gemmae _____

Paraphyses _____

Protonemata _____

Peristome teeth _____

Elaters _____

4. What is chemotaxis? _____

5. In which generation, gametophyte (1N) or sporophyte (2N), do the following belong?

Spore _____   Egg _____

Calyptra _____   Capsule _____

Archegonium _____   Zygote _____

Protonema _____   Paraphysis _____

6. Why are mosses and liverworts restricted in size and habitat? _____

_____

7. What is meant by the term "alternation of generations"? _____

_____

8. Which generation in the Division Bryophyta is the most conspicuous? _____

_____

9. Define:

Rhizoid _____

Thallus _____

CROSSWORD PUZZLE — LIVERWORTS AND MOSSES

ACROSS

3. Egg-producing structure
6. Haploid
7. The vegetative body of a bryophyte
9. Structure bearing sex organs
14. Female gamete
15. Moss spores produce _____
17. Rootlike structure
19. 1N and haploid have the _____ meaning.
20. Thallus liverworts possess neither____ nor roots.
21. Moss leaves are the _____ of photosynthesis.
22. 2N
25. Necessary for bryophyte reproduction.
27. If an archegonium is a "she," and an antheridium is a "he," then a capsule is an ____.
28. A member of the Hepaticae
29. The gametophyte is the _____ generation.
30. Male gamete

DOWN

1. The nucleus of each sporophyte cell is _____.
2. Part of a bryophyte sporophyte
3. _____ receptacle
4. The sporophyte "gets a little _____ from its friend" the gametophyte.
5. A genus of liverworts
8. A moss can make neither _____ nor fruits.
10. Gemmae are often found in a _____.
11. Paraphyses surround the moss archegonia and antheridia for _____.
12. A rolling stone gathers neither a liverwort nor _____.
13. Below the operculum _____ the peristome.
16. Sperm are the _____ gametes.
18. A fertilized egg.
21. Capsule produces_____.
23. Eat a *Marchantia* and cure your _____.
24. Moss sperm swim in a film of _____.
26. Economically important moss.

Kingdom Plantae
Vascular Plants Without Seeds: Ferns (Division Pterophyta) and Others

OBJECTIVES

After completion of this exercise, the student should be able to do each of the following:

- Explain what a vascular plant is and how it differs from all previous plants studied.
- Identify and discuss **rhizome** and **frond** of a fern.
- Identify fern **sporangia** with **annulus**, **lip cells**, and **spores**.
- Identify fern **prothallia** bearing **archegonia**, **antheridia**, or a young sporophyte.
- Discuss and draw alternation of generations in the ferns.
- Answer the review questions at the end of this exercise.

INTRODUCTION TO VASCULAR PLANTS

The vascular plants comprise the highly diversified plants which have exploited land habitats more successfully than any other plant group. They are characterized by a dominant sporophyte which is differentiated into **true roots**, **stems**, and **leaves**, each made up of various types of specialized tissues. All parts of the body are interconnected by way of special conductive or **vascular** tissue which carries water and solutes throughout the plant. One type of vascular element, the evolutionary basis for all the others is the **tracheid**. Many vascular plants attain great size, acquiring both a highly efficient conduction system and the development of supportive mechanical tissue that enables the plant to assume an upright habit of growth.

Vascular plants are divided into 2 categories: those that produce seeds and those that do not. A *seed* contains a dormant embryo sporophyte with a stored food supply allowing it to survive long periods of adverse growing conditions. Primitive vascular plants such as ferns (and their relatives) do not produce seeds.

The vascular plants without seeds include *Equisetum* (horsetails) in the Division Sphenophyta, *Lycopodium* (club mosses) in the Division Lycophyta, *Psilotum* (whisk ferns) in the Division Psilophyta and the ferns (Division Pterophyta). The first two examples have their sporangia clustered at the tips of the stems and usually have small leaves on the rest of the stem. Whisk ferns have no leaves or roots and the 3-lobed sporangia are attached at intervals along the branching stem.

Also in these groups, sexual reproduction is still dependent upon water for the sperm must swim from the male to the female sex organ. In the higher vascular plants (the seed plants) this swimming sperm stage is replaced by a new male gametophyte

137

form — **pollen** — which can be dispersed by wind or by other organisms. The emergence of this type of male gametophyte was a major evolutionary breakthrough in the conquest of land, for it set seed plants free from dependence on water for completion of their life cycles.

DIVISION PTEROPHYTA — Ferns

The fern structure with which we are most acquainted is the vascular sporophyte. Ferns, once the dominant vegetation during the "coal-making age" approximately 300 million years ago, are distributed throughout the world in various forms and shapes. Some ancient ferns attained great size, but today only a relatively few tropical forms remain treelike. The sporophyte can exist indefinitely, spreading by underground stem branches, **rhizomes**, without passing through the sexual stage.

1. Examine the various herbarium samples of fern types, noting the variation in the **frond** (leaf) of several species. There is a general trend toward greatly dissected compound leaves. Note the reproductive structures of species where present. In some ferns the sterile (photosynthetic) frond and the fertile fronds are completely separate, while in others, the sterile and fertile leaves are either on the same leaf stalk, or the reproductive structures are borne on the undersides of the photosynthetic tissue.

Distinguish between the **rhizome**, the **leaf stalk**, **blade** (broad green part) and a **pinna** (leaflet) of a fern frond, and examine the pattern of **sori** (sporangium clusters) distribution of several ferns.

2. Examine slides of fern sori and sporangia. Within these **sporangia** meiosis occurs to produce 1N spores. The row of cells down the back of the helmet-like sporangium is called the **annulus**. Since these cells are thin-walled on the outer edge, water loss during spore maturation causes this row to shrink, eventually causing the

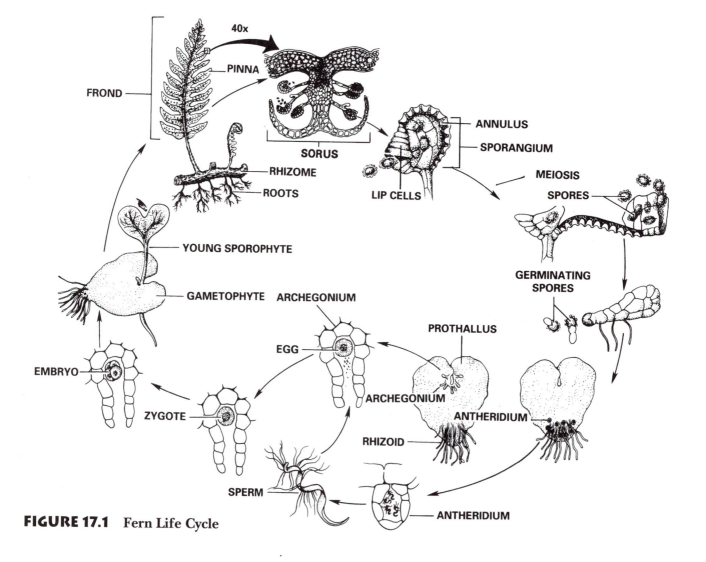

FIGURE 17.1 Fern Life Cycle

sporangium to crack along the large **lip cells** and catapult the spores away from the parent plant. Refer to Figure 17.1

3. Upon germination those spores do not grow directly into the leafy fern seen on the herbarium sheets. The first strand of cells to emerge is thread-like and is known as the **protonema**. Later, there are divisions in two planes, producing a flat heart-shaped or bi-lobed structure called the **prothallus** — the mature gametophyte plant. **Rhizoids** arise on the under surface, as do the reproductive organs.

The **antheridia** are produced first, release **sperm** and then drop off. Then the **archegonia** are produced. This staggered production of gametes promotes cross fertilization between the prothallia of different monoecious parent plants. Some fern species are dioecious.

Obtain a slide of fern **prothalli** which show gametophytes in both sperm and egg producing stages. Refer to Figures 17.2 and 17.3.

4. The young **zygote** develops in the archegonium and for a short time takes nourishment from the gametophyte. Soon the **sporophyte** takes on the recognizable leafy-fern look, crushing the prothallus as it grows.

5. Learn the life cycle of the fern.

6. Observe the other groups of seedless vascular plants (horsetails, club mosses, and whisk ferns). Refer to Figure 17.4. Their life cycles are similar to that of the ferns.

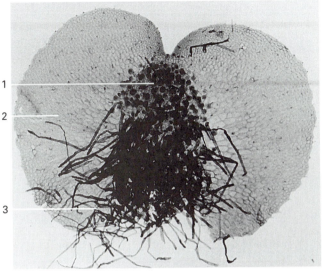

Photo courtesy of Samuel R. Rushforth

FIGURE 17.3
Fern Gametophyte Showing Antheridia

1. Antheridia　　2. Prothallus　　3. Rhizoids

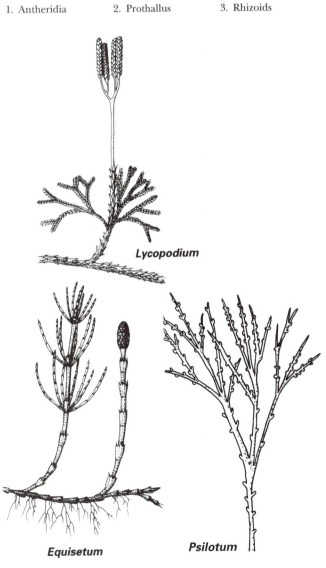

Lycopodium

Equisetum　　　*Psilotum*

FIGURE 17.4　**Various Fern Allies**

FIGURE 17.2
Fern Gametophyte Showing Archegonia

1. Archegonia　　2. Prothallus (n)　　3. Rhizoids

Photo courtesy of Samuel R. Rushforth

X40

REVIEW QUESTIONS

1. What are the advances of ferns over bryophytes? _____

   _____

2. What is the function of the annulus? _____

   _____

3. What is a rhizome?_____

   _____

4. In which generation, gametophyte (1N) or sporophyte (2N), do the following belong?

   prothallus _____   pinna_____   sporangium_____

   spores _____   frond_____   antheridium _____

5. With what cell does the gametophyte generation begin?_____

   Is it haploid (IN) or diploid (2N)? _____

   Within which fern structures is it formed? _____

6. With what cell does the sporophyte generation begin? _____

   Is it haploid (IN) or diploid (2N)? _____

   Within which fern structure does it begin its development? _____

7. Why don't the archegonia and antheridia mature at the same time on a prothallus?_____

   _____

8. What does an archegonium produce? _____

   An antheridium? _____

9. What happens to the gametophyte as the sporophyte matures?_____

10. Small brownish bumps may be evident on the undersurface of the pinnas. What are these called and what

    is their function? _____

11. Some fern species are dioecious, other monoecious. What is meant by these terms?

    Dioecious _____

    Monoecious _____

12. Is the fern plant you see in pots the gametophyte or sporophyte? _____

13. Which generation contains xylem and phloem? _____

14. What is the prothallus? What does it resemble? _____

    _____

Kingdom Plantae
The Gymnosperms: Division Coniferophyta and Others

GYMNOSPERMS

The gymnosperms are a group of seed plants including the ginkgos (Division Ginkgophyta), cycads (Division Cycadophyta), gnetophytes (Division Gnetophyta) and conifers (Division Coniferophyta). The ginkgo is a tree with fan-shaped leaves and round smooth seeds. It is often planted along city streets, around public buildings, and on college campuses. Ask your instructor if there are any ginkgo trees on your campus. Cycads resemble small palm trees (except that they produce large cones) and grow in tropical and subtropical areas.

The most common of the gymnosperms are the conifers. Although much of the earth's surface is covered by conifers, the number of species is relatively few. Many groups of the division are complete or partially extinct. We will study *Pinus*, a pine, which is one of the more modern forms of gymnosperms.

The conifers include our well-known evergreens: pine, spruce, fir, hemlock, redwood, as well as the southern conifers. They include the largest and oldest of all living things known. A bristlecone pine growing on the rim of the Grand Canyon has been dated as being over 2000 years old and is the oldest known living thing on earth.

In conifers, the gametophytes live as parasites on sporophyte plants. Sporophytes which bear male gametophytes on one plant and female gametophytes on another are termed **dioecious** (= two houses). Many conifers are dioecious, but more are **monoecious** (= one house), bearing male and female gametophytes on the same sporophyte plant.

The reproductive structures (sporangia) are located on specialized leaves called **sporophylls**. In both gymnosperms and angiosperms two types of spores are produced. Four large **megaspores** (1N) are meiotically produced in a **megasporangium** or **ovule.** Three of the four megaspores die, leaving one megaspore to grow and become the female gametophyte. This gametophyte contains two or three archegonia, each containing one egg.

Small, numerous **microspores** (1N), the second type of spore, are meiotically produced in a

microsporangium. Then, these microspores or **pollen** undergo mitosis to form two nuclei; one, the **generative nucleus**, will form two sperm; the other, the **tube nucleus**, will synthesize the **pollen tube**.

The **megasporophylls** are grouped together in clusters called **cones**, as are the **microsporophylls**. The **ovulate cones** (megasporangiate) are borne on the sides of the upper branches and last one to two years before opening and dropping. This is the kind of cone usually meant when one says "pine cone." The **pollinate cones** are produced at the tips of the lower branches and last only a few months.

The wind blows the pollen into the cracks of the female cone and the pollen tube grows toward the interior. The tube nucleus works itself to death making cytoplasm, while the generative nucleus follows it down the tube, eventually dividing into two **sperm** nuclei. Just outside the archegonia the pollen tube ruptures and the sperm swim through the cytoplasm to the eggs. Although all eggs may be fertilized, usually one embryo will grow faster than the others and the weaker embryos will abort. The surviving embryo produces 6, 8, or more **cotyledons** to store food, then goes dormant. The gametophyte tissue is dissolved and absorbed into these cotyledons. The surrounding megasporophyll tissue dries and flattens to produce a winged **seed coat**. After the seed is blown from the opened female cone and germinates, it will use the food stored in the cotyledons to supply nourishment until it can produce roots and leaves of its own.

1. Learn the life cycle of the pine. (See Figure 18.1)

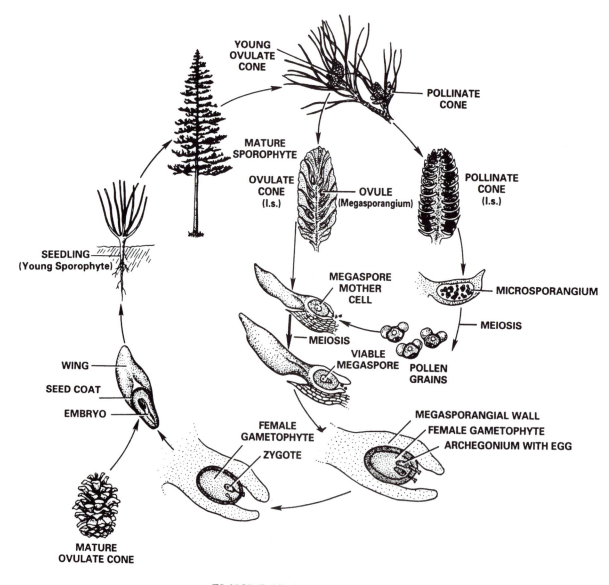

FIGURE 18.1   Pine Life Cycle

2. Examine prepared slides of ovulate and pollinate cones of *Pinus*.

3. Prepare a slide of pine pollen. Notice the air bladders on both sides of the pollen grain.

4. Observe various preserved materials which include immature and mature ovulate cones, pollinate cones, and pine seeds.

5. On demonstration is a slide showing pine ovules. Be able to identify the female gametophyte, the archegonia, and the megasporangial wall.

6. On demonstration is a slide showing a pine seed. Identify the embryo, the gametophyte, and the seed coat.

7. The pine needle or leaf is a result of adaptation to cold and dry climates. During winters, frozen water is useless to plants. Therefore, antifreezing resin is made and stored in **resin canals** of the leaf, stem, and root of the pine. In addition, resin has insecticidal and fungicidal properties.

a. Obtain a prepared slide of a cross-section of a pine needle (see Figure 18.2). Observe the outer layer of thick cells, the **epidermis**. This layer has numerous sunken pairs of **guard cells** (surrounding an opening, the **stomate**) which regulates gas exchange and prevents excessive water loss. Just below the epidermis is another thick cell layer, the **hypodermis** which supports the leaf and helps retain water. The middle of the leaf is the photosynthetic area called the **mesophyll**. The central portion of the leaf is surrounded by a ring of cells, the waterproofing **endodermis**. Inside this ring of cells is the **xylem** which conducts water and **phloem** which carries food and hormones. Xylem cells have thicker walls and usually stains a dark red color. Phloem sits next to the xylem, has thinner walls, and usually stains a bluish-green.

8. Observe preserved specimens of various genera of gymnosperms.

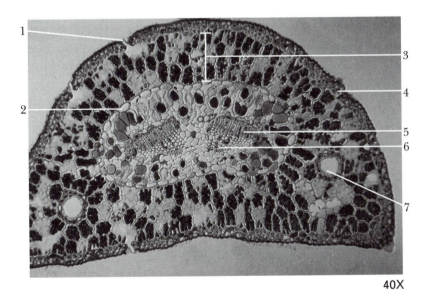

1. Stomate with guard cells
2. Endodermis
3. Photosynthetic mesophyll
4. Epidermis with hypodermis below
5. Phloem
6. Xylem
7. Resin duct

40X

FIGURE 18.2 Pine Needle (X.S.)

REVIEW QUESTIONS

1. What does the word "gymnosperm" mean? _____

2. What are the economic uses of gymnosperms? _____
   _____

3. Seeds are found in which type of cone: megasporangiate or microsporangiate? _____

4. What does a microspore develop into?_____

5. What is the function of resin canals in pine needles?_____

6. What is the difference between pollination and fertilization?_____
   _____
   _____

7. What is another name for a megasporangium?_____

8. In what ways are the male and female pine cones different?

   Size _____

   Location on tree _____

   Life span _____

9. With what cell does the gametophyte generation (1N) begin in each cone?

   Female_____

   Male _____

10 With which cell does the sporophyte generation begin? _____

11. Is a Christmas tree a sporophyte or gametophyte? _____

    Is it haploid or diploid? _____

12. Name a cone-bearing tree that loses its leaves in the winter? _____

13. By what method are pine cones pollinated? _____

14. How can you tell when the seeds on a pine cone are ripe? _____

15. When the pine seed germinates it depends on food stored in the _____ for nourishment
    until roots and leaves are formed.

16. A ripe ovule is called a _____.

17. Give an example of each group of gymnosperms. _____

18. MATCHING: Match the function with the correct term.

    _____ 1. Tube nucleus   a. synthesizes the pollen tube

    _____ 2. Generative nucleus   b. male and female gametophytes all on same sporophyte

    _____ 3. Microsporangium   c. produces the microspore and pollen grain

    _____ 4. Megasporangium   d. gives rise to two sperm nuclei

    _____ 5. Monoecious   e. produces the megaspores and female gametophyte

Kingdom Plantae

Division Anthophyta, The Angiosperms
Class Monocotyledonae
Class Dicotyledonae

OBJECTIVES

After completion of this exercise, the student should be able to do each of the following:

- Identify the parts and their functions of a typical **angiosperm flower**, **fruit**, and **seed**.
- Describe the life cycle of an angiosperm.
- Identify an angiosperm as a **dicot** or a **monocot** on the basis of number of flower parts, number of cotyledons, and pattern of leaf venation.
- Examine a flower or fruit and determine the **carpel number**.
- Explain the **carpel theory**.
- Answer the review questions at the end of this exercise.

FLORAL ANATOMY

Although the flowers of angiosperms or flowering plants show a tremendous diversity (80% of all species of plants) the structures which make up the flower are basically the same throughout. The following parts are recognized although not all may occur in any one flower.

The supporting stalk of the flower is called the **pedicel**.

A **receptacle** is at the center base of a flower and represents the end of the stem. In some cases it may form a dome for the other appendages or a cup with some appendages on the rim and others on the bottom. It is not usually a large or very noticeable part of most flowers.

The **calyx**, composed of **sepals**, makes up the lower or outermost whorl of floral appendages. Usually these are small and leaf-like, although different in shape from the other foliage leaves of the plant. Their major function is the protection of the immature inner parts, but in some cases they look and function as petals as well (lily).

The **corolla**, composed of **petals**, arises inside and usually above the sepals, usually being brightly pigmented and very broad. These serve as additional protection in early stages and later to attract pollinators. Some serve as "landing pads" for the pollinator, and may have glands (nectaries) at the base which produce sweet nectar and/or volatile oils to provide the pollinator with a scent trail to the flower. (Those angiosperms which are wind pollinated usually do not have petals.)

The male reproductive structures are the **stamens**, arising inside the petals. Each consist of a slender stalk or **filament** and variously shaped, enlarged sacs called **anthers**. Inside the anther, special groups of cells undergo meiosis to produce male gametophytes — **pollen grains**.

One or more **pistils** or **carpels** may occur in the center of the flower. A pistil is composed of: (1) a **stigma** which produces a sticky substance which catches pollen grains and induces their germination; (2) a **style** through which the pollen tubes grow toward the ovules; and, (3) an **ovary** — the basal region which contains one or more ovules (megasporangia). It is within the ovule that one of the four megaspores will produce a female gametophyte. In angiosperms the female gametophyte contains eight nuclei at maturity and is called an **embryo sac**.

1. Label the flower illustration (Figure 19.1) with these terms: anther, filament, ovary, ovule, pedicel, petal, pistil, receptacle, sepal, stamen, stigma, and style.

2. At fertilization the egg fuses with one of the two sperm produced by a pollen grain. A zygote (2N) results. However, the remaining sperm is not wasted but unites with the two polar nuclei of the embryo sac to form a triploid (3N) organism called the **endosperm**. The endosperm will devour the remaining embryo sac and provide food for the young (2N) **embryo**. Study Figure 19.2 — the typical angiosperm life cycle.

3. Dissect a fresh or preserved flower using dissecting utensils and a stereomicroscope. Fill in the following information concerning the dissected plant flower:

Plant name _____

Number of parts:

   sepals _____

   petals _____

   stamens _____

   pistils _____

4. Examine prepared slides of the lily ovary and flower buds.

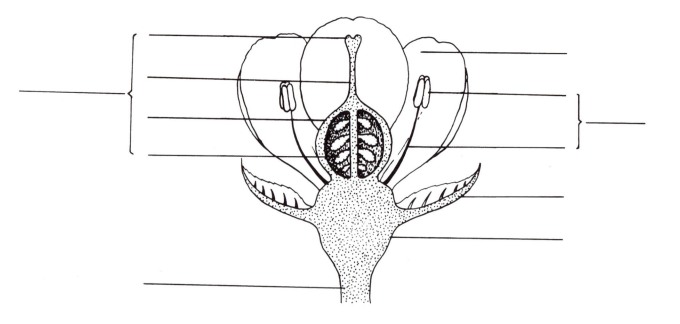

FIGURE 19.1 Generalized Floral Anatomy

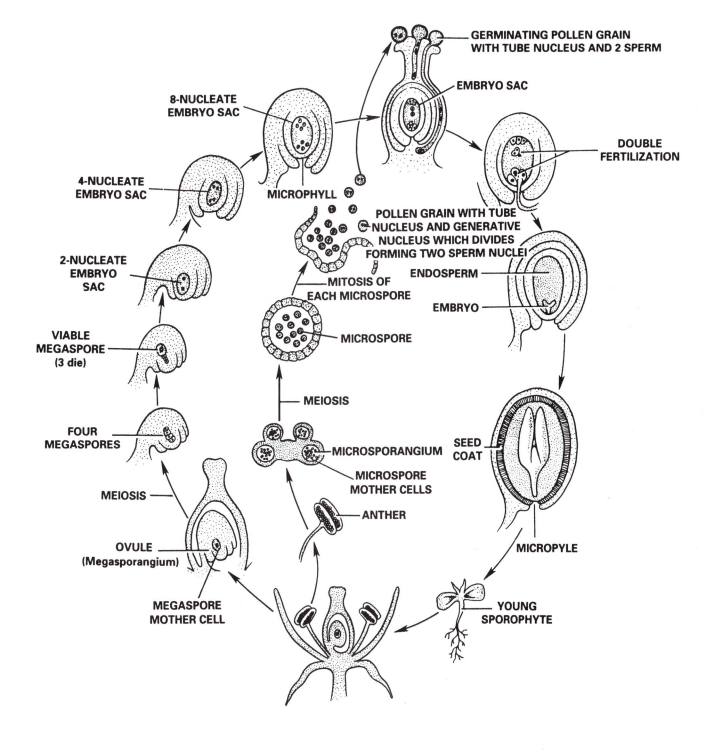

FIGURE 19.2 Angiosperm Life Cycle

FRUIT

The **fruit**, a unique characteristic of the angiosperms, develops from the ovary of the flower and sometimes includes the receptacle. Study the various fruits on demonstration and carefully note from what types of flowers they arose. Refer to Figures 19.3–19.10.

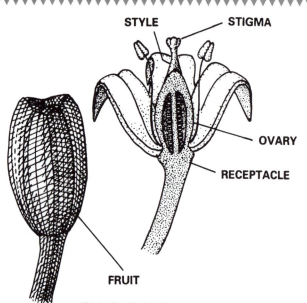

FIGURE 19.3 Lily Flower and Fruit

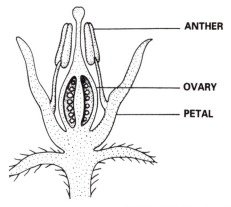
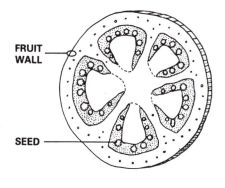

FIGURE 19.4 Tomato Flower and Fruit

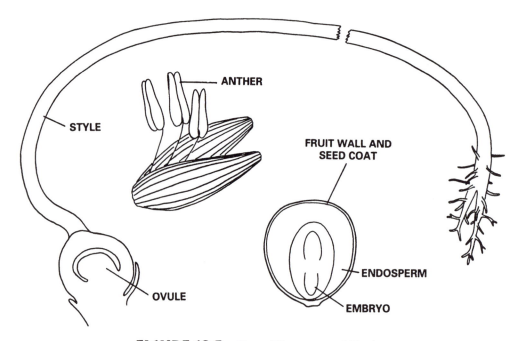

FIGURE 19.5 Corn Flowers and Fruit

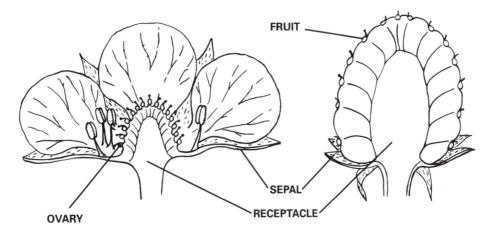

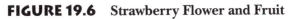

FIGURE 19.6 Strawberry Flower and Fruit

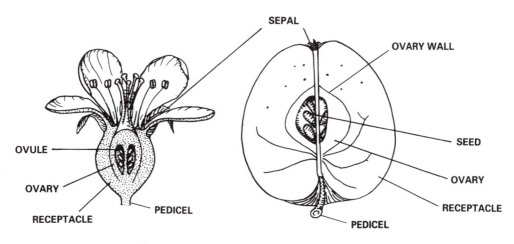

FIGURE 19.7 Apple Flower and Fruit

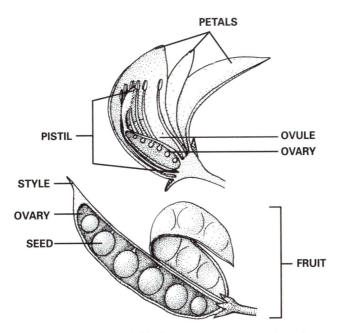

FIGURE 19.8 Pea Flower and Fruit

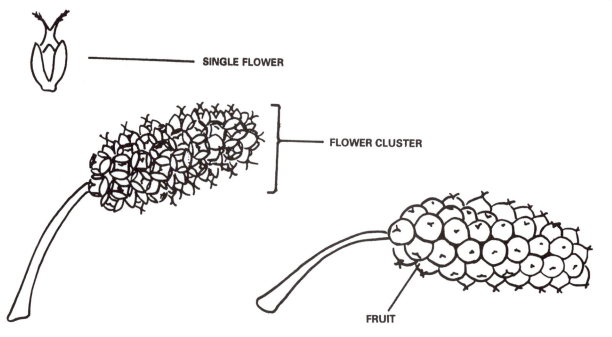

SINGLE FLOWER

FLOWER CLUSTER

FRUIT

FIGURE 19.9   Mulberry Flowers and Fruits

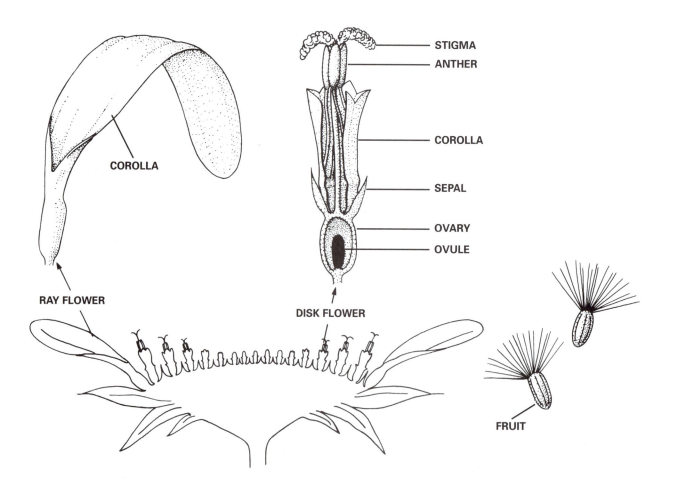

COROLLA

STIGMA

ANTHER

COROLLA

SEPAL

OVARY

OVULE

RAY FLOWER

DISK FLOWER

FRUIT

FIGURE 19.10   Composite Flower and Fruits

SEED

1. Examine a soaked bean seed. Identify the **micropyle**, the minute opening on the surface through which the pollen tube grew and the **hilum**, the adjacent elliptical area where the ovule was attached to the ovary. Next, carefully peel off the **seed coat**. The part of the bean remaining is the embryo. Separate the two large fleshy **cotyledons** which serve as an area for food storage in the bean. Now identify the **epicotyl**, a small portion of the embryo that lies above the attachment of the cotyledons. At the tip of the epicotyl is the **plumule** which will produce new stem and leaves. Beneath the cotyledons is the **hypocotyl** which terminates in the **radicle** or embryonic root. Label Figure 19.11 with the terms given above.

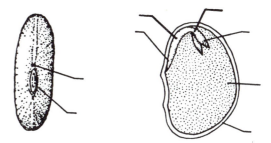

FIGURE 19.11 Bean Seed

2. Observe bean seedlings on demonstration. Identify which parts of the seedling came from the embryonic structures mentioned previously. Label Figure 19.12 with the following terms: hypocotyl, seed coat, primary root, foliage leaves, epicotyl, and cotyledon.

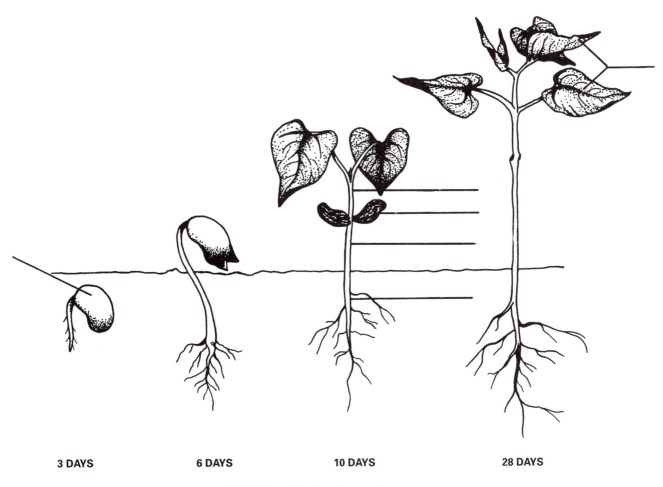

3 DAYS   6 DAYS   10 DAYS   28 DAYS

FIGURE 19.12 Bean Germination

CARPEL THEORY

The evolution of angiosperms is still a mystery to scientists. However, it is believed that they descended from plants whose sporangia were borne on the edges of leaves. Then, due to the increasing selective pressures by the rapidly-evolving insects, the leaf slowly wrapped around and enclosed the megasporangia. This leaf today is known as a **carpel**, and the ovules are no longer naked but are covered (angiosperm). (See Figure 19.13.)

The numerous carpels in the flower through millions of years of evolution often fused in various groupings to form the pistil. Thus, some flowers have pistils containing one carpel while other pistils of flowers have numerous carpels. Reexamine several fruits and flowers by viewing their cross sections or counting the stigmata. How many carpels do they have?

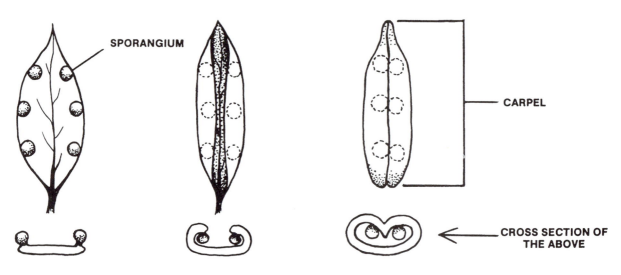

FIGURE 19.13   Carpel Theory

MONOCOTS AND DICOTS

Another evolutionary event was the creation of a group of plants so unique that they are given the name **monocotyledonae** to distinguish them from the other flowering plants, the **dicotyledonae**. These class names are frequently shortened to **monocot** and **dicot** for convenience.

Monocots and dicots differ from each other in the following ways:

MONOCOTS

number of cotyledons . . . . . one
venation of leaves . . . . parallel
root system . . . . usually fibrous
flower parts. . . . . 3's or multiples of 3
stems . . . . mostly herbaceous

DICOTS

number of cotyledons . . . . . two
venation of leaves . . . . netted
root system . . . . either fibrous or taproots
flower parts. . . . . 4 or 5, or multiples
                                         of 4 or 5
stems . . . . herbaceous or woody

1. Observe on display some examples of monocots and dicots. Examples of monocots are lilies, grasses, sugar cane, corn, wheat and rice. Examples of dicots include lima beans, peas, oaks, maples, and buttercups.

In the next exercise, you will also see differences in the internal anatomy of the roots and stems of monocots and dicots.

REVIEW QUESTIONS

1. Angiosperms compose what percentage of all the plant kingdom? _____

2. What structure of a flower is an immature fruit?_____

3. The embryo sac (mature female gametophyte) typically has _____ nuclei.

4. Double fertilization in the angiosperms results in the zygote and the _____.

5. A lily (monocot) has its flower parts in multiples of _____.

6. What structure does the radicle in a seed produce in the mature plant? _____

7. What does "angiosperm" mean? _____

8. One or more carpels make up: a. a pistil   b. a corolla   c. a calyx   d. a cone

9. What does the term "dicot" actually mean? _____

10. What is the major function of the calyx? _____

11. What are the functions of the corolla?

    a. _____

    b. _____

12. The microsporangium is located in which part of the stamen? _____

13. Why is the sticky substance on top of the stigma important?

    a. _____

    b. _____

14. The megasporangium are located in which part of the pistil? _____

15. The male gametophyte is called a _____.

16. The female gametophyte is called an _____.

17. How does the endosperm differ from the embryo? _____

    _____

18. Angiosperms belong to what division? _____

19. Distinguish between monocots and dicots. _____

    _____

20. Carpels are believed to have formed in response to? _____

21. Define the terms hypocotyl and epicotyl._____

    _____

Vegetative Structures of Angiosperms

EXERCISE
20
EXERCISE EXERCISE EXERCISE

OBJECTIVES

After completion of this exercise, the student should be able to do each of the following:

- Describe the external characteristics of the root, stem, and leaf.
- List the functions of **root**, **stem**, and **leaf**.
- Explain how the **root**, **apical**, and **cambium meristems** function.
- Identify the internal structures of roots, stems, and leaves.
- Explain what causes **growth rings** and determine the age of a tree from a section of its trunk.
- Name the various types of roots based on their origin or structure.
- Explain the differences between a monocot and dicot stem.
- Answer the review questions at the end of this exercise.

ROOTS

The major functions of roots are: absorption of water and minerals, anchorage, and storage of food in the form of starch and other carbohydrates. Roots have few distinctive external features except root hairs for absorption and a root cap. The **root cap** protects the meristem as the root pushes through the soil. The root cap cells are replenished internally as they are sloughed away on the outside.

The **root meristem**, like all meristematic regions, has the property of **totipotency**, the capacity for continuous division and growth. It is in this area where new root and root cap cells are produced by mitotic cell division. As these immature cells begin to grow they elongate. Eventually these cells differentiate into their predetermined form and functions. Below, the meristem remains intact and busily produces new cells.

1. Examine a prepared slide of an onion (*Allium*) root tip. Note the root cap, root meristem, and zone of elongation.

2. On demonstration are roots which possess **root hairs**, minute structures for absorbing water and soil minerals.

3. Label Figure 20.1 with these terms: zone of elongation, zone of maturation, root cap, root hair, and apical meristem or zone of cell division. The above-mentioned terms will be discussed by your instructor.

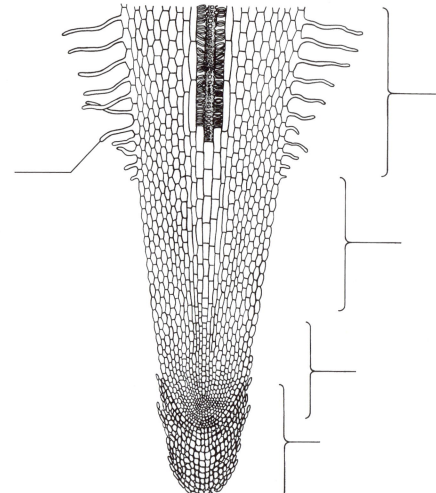

FIGURE 20.1
Root Tip (L.S.)

4. Examine a prepared slide of a mature dicot root such as the buttercup, *Ranunculus*. Refer to Figure 20.2. In the area of maturation, the root cells are in their final form. The root is covered by a protective cell layer, the **epidermis**. It is this epidermal layer which also produces outgrowths of its cell walls and cytoplasm called root hairs.

Inside the epidermis, the **cortex** of a dicot root shows its storage function in the many pinkish starch grains present in many of the cells. This tissue also helps transport water and minerals from the root hairs to the xylem. The innermost layer of the cortex is the **endodermis**, a single layer of thick-walled cells which act as waterproofing and support tissue for the vascular cylinder. Just inside the endodermis is a **pericycle** of one or more cells in thickness. This layer is important because it produces the secondary or branch roots and the vascular cambium of the root at a later time.

Surrounded by the pericycle is the vascular tissue or **stele**, composed of xylem and phloem.

The xylem cells, when mature, die and become hollowed cell walls (similar to drinking straws) in order to conduct water with dissolved minerals upward to the leaves. In the typical dicot root, the xylem is in the center and around the xylem are bundles of **phloem**. Phloem cells conduct food and hormones. Phloem consists of large **sieve tube cells** through which solutes are conducted and the **companion cells** which maintain and control the sieve tube cells. Since the food and hormones are conducted by active transport, phloem cells must remain alive in order to function. Between the xylem and phloem a few inconspicuous meristematic cells occur. This is the **cambium**, which may eventually manufacture new secondary xylem and phloem.

5. Look at the demonstration slide showing **lateral** or **branch root** origin. Remember, the pericycle becomes temporarily meristematic and produces these secondary branch roots.

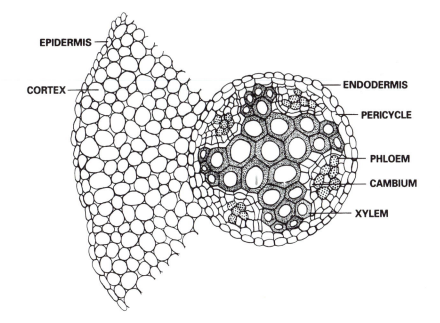

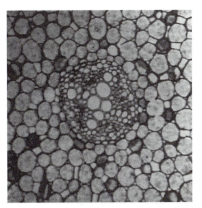

Photomicrograph (100x)
Courtesy of Turtox, Inc.

FIGURE 20.2 Mature Dicot Root (X.S.)

6. Roots can be classified structurally either as **fibrous roots** or **taproots** (Figure 20.3). They can also be grouped according to their origin; **primary roots** arise from the radicle of the embryo, **secondary roots** arise from the pericycle, and **adventitious roots** (Figure 20.4) arise from the stem or leaves. Study the various types of roots on demonstration.

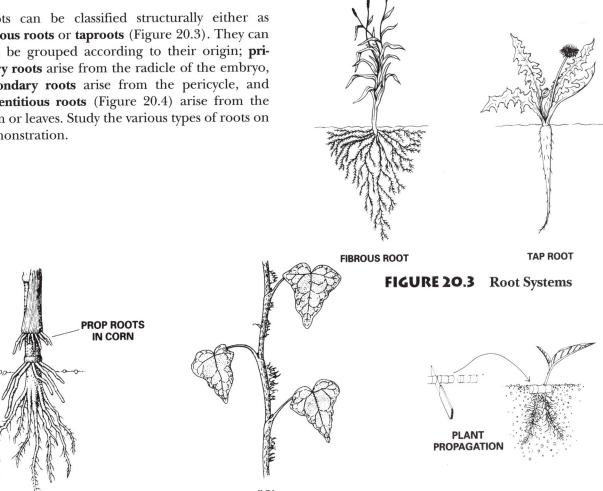

FIBROUS ROOT

TAP ROOT

FIGURE 20.3 Root Systems

PROP ROOTS
IN CORN

IVY

PLANT
PROPAGATION

FIGURE 20.4 Adventitious Roots

STEMS

The major stem functions are: conduction of food, water, and minerals between root and leaves; production and display of photosynthetic leaves; and storage of food.

1. Stems have several distinct external features which you should examine and be able to identify using Figure 20.5 as a guide. Leaves and buds arise at the **nodes**. The areas in between nodes are called **internodes**. Buds are immature stems, those which arise along the sides of the stem producing branches of the stem, and those located at the tips of stems producing stem elongation. These are known as **lateral** and **terminal buds**, respectively. **Terminal bud scars** are formed each spring when the scales covering the **terminal bud** fall off. When leaves drop off in the fall, they leave **leaf scars** as well. Stems are also covered by a number of light or dark spots. These disruptions in the bark are called **lenticels**, and function in gas exchange.

2. The shoot apex is covered by young leaves which arch over the meristem. This is more difficult to see in fresh material, so we will use a longitudinal section of an *Elodea* bud. Obtain such a slide and try to identify the parts labeled in Figure 20.6.

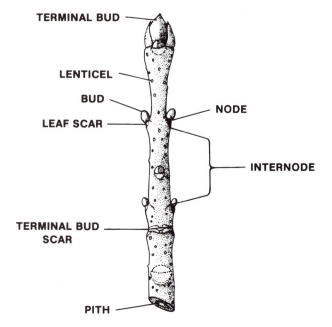

FIGURE 20.5 **Features of a Dormant Twig**

Photomicrograph (100x)
Courtesy of Turtox, Inc.

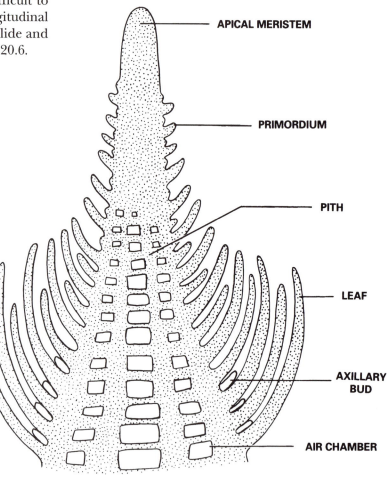

FIGURE 20.6 *Elodea* Bud

There is no cap over the stem and the zones of elongation and differentiation are not as conspicuous as in the root. At about ground level the core of vascular tissue of the root branches out to form numerous vascular bundles. This leaves a large parenchyma core called the **pith** in the center. *Elodea* is a rooted plant even though it is most often seen as stem cuttings. Being an aquatic plant it has numerous air chambers in the pith and the cortex to aid in flotation.

At the extreme tip is the primary meristem of the stem. On the flanks of this tall meristem are the leaf **primordia** appearing as bumps in young stages. As the leaves mature another "bump" appears where the leaf joins the stem. These are the **axillary buds** and may become new vegetative or flowering stems.

3. Study a prepared slide of a cross-section of both a monocot (Figure 20.7) and a dicot stem (Figure 20.8).

The monocot stem covered by an outer **epidermis**, has a scattered arrangement of the vascular bundles. The individual **vascular bundles** resemble faces, with **xylem** for "eyes, nose, and mouth," **phloem** for the "forehead," and **supporting cells** or **sclerenchyma cells** for "hair and beard." Surrounding the vascular bundles is pith or **ground parenchyma**, an area of large, white **parenchyma cells**. **Air spaces** may also be present in the parenchyma. Although parenchyma cells are found in many areas of a plant (root cortex, epidermis, and leaf mesophyll) and have various functions, the function of the pith parenchyma cells is storage of food. Sugar cane contains sucrose in its pithy stem.

4. In the young dicot stem (usually the smaller one on the slide) the outermost layer of cells is the **epidermis**. These cells have a waxy, non-living covering called the **cuticle**. Just inside is a narrow, parenchymous **cortex** for storage. The

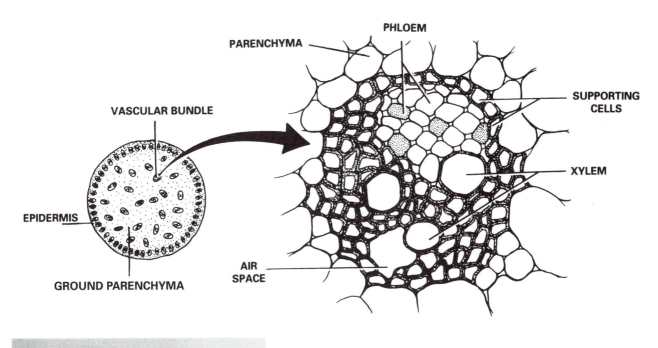

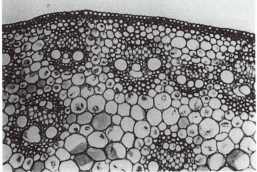

Photomicrograph (100x)
Courtesy of Turtox, Inc.

FIGURE 20.7
Monocot Stem (X.S.)

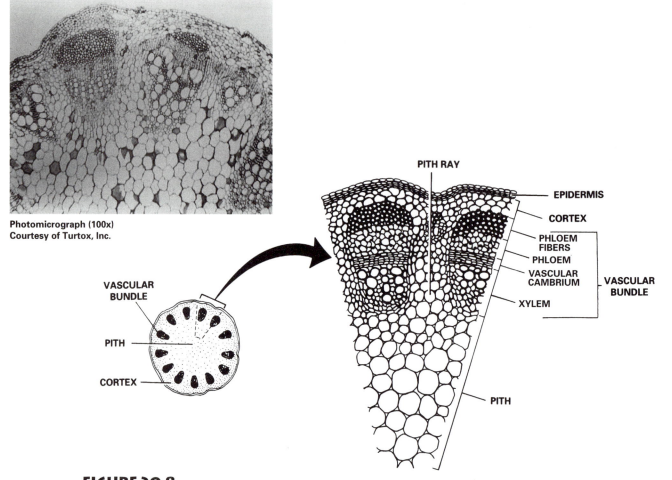

Photomicrograph (100x)
Courtesy of Turtox, Inc.

VASCULAR BUNDLE

PITH

CORTEX

PITH RAY

EPIDERMIS

CORTEX

PHLOEM FIBERS

PHLOEM

VASCULAR CAMBRIUM

XYLEM

PITH

VASCULAR BUNDLE

FIGURE 20.8
Young Dicot Stem (X.S.)

stele of the stem is broken up into vascular bundles with **pith rays** running between them and the large, central **pith**.

The inner portion of the vascular bundles is the **xylem**, separated from the phloem by a small area of the meristematic **vascular cambium**. Just outside of the conducting phloem area is a cap of **phloem fibers** for support of the young stem.

5. In many of the dicots the plant lives more than one or two years and the stem usually becomes woody. The cambium stretches to connect the vascular bundles and begins to manufacture new xylem and phloem. In the areas of the world with seasonal climates, the growth of the vascular cambium changes likewise. In the spring, warm temperatures and plentiful rainfall and sunshine produce a need for numerous large water-conducting **vessels** (an advanced type of xylem cell shaped similarly to a plumbing pipe). As summer begins, the weather, although hot, brings fewer inches of rain. Thus, the cambium

manufactures fewer and fewer vessels that are smaller in diameter. Then in the autumn, the leaves stop photosynthesis and fall from the trees. No wood is produced in the fall or winter.

The contrast between the larger and lighter-shaded "**spring wood**" and the smaller, darker "**summer wood**" result in **annual rings** or **growth rings**. In wet years, the growth rings will be wide; in drier years, these rings will be narrow. Thus, examination of tree rings can trace the condition of past years in the location where the tree grew. Radioactive carbon dating is often checked and calibrated with the results of tree ring analysis.

Pick out a *Tilia* 2-year stem slide and find all of the parts labeled in Figure 20.9. *Tilia* is commonly known as Linden or basswood.

NOTE

The term bark *refers to all tissues external to the vascular cambium.*

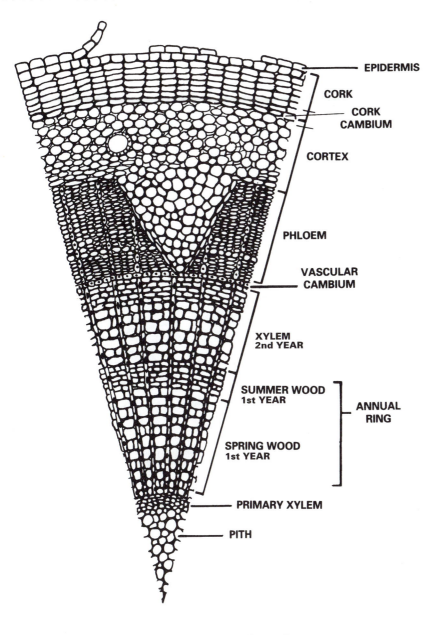

Photomicrograph (100x)

Courtesy of Carolina Biological Supply Company

FIGURE 20.9 *Tilia* Two-Year Stem (X.S.)

LEAVES

1. The major function of leaves is usually photo-synthesis. However, other minor functions may be protection, secretion, storage of food, or even the capture of small animals for nitrogen. Note the various leaves on display and refer to Figure 20.10.

2. Obtain a prepared slide of a cross-section of a typical dicot leaf. Locate the single-layered **epi-dermis**, noting those cells of the upper epidermis have a waterproofing **cutin** covering. Now locate the **guard cells** and the **stomata** on the lower epidermis.

Between the upper and lower epidermis, the **mesophyll parenchyma** is found. There are one or more layers of closely packed **palisade paren-chyma** just below the upper epidermis, where most of the photosynthesis takes place. Below this is an area of loose **spongy parenchyma**, pro-viding numerous air chambers for gas exchange via the stomata.

In the center of the leaf is the major **vein**, or **vascular bundle**. Such veins branch out through the parenchyma to distribute needed water and minerals and pick up sugars.

Farther out from the midvein you may see streak-like clusters of cells. This is where the slice

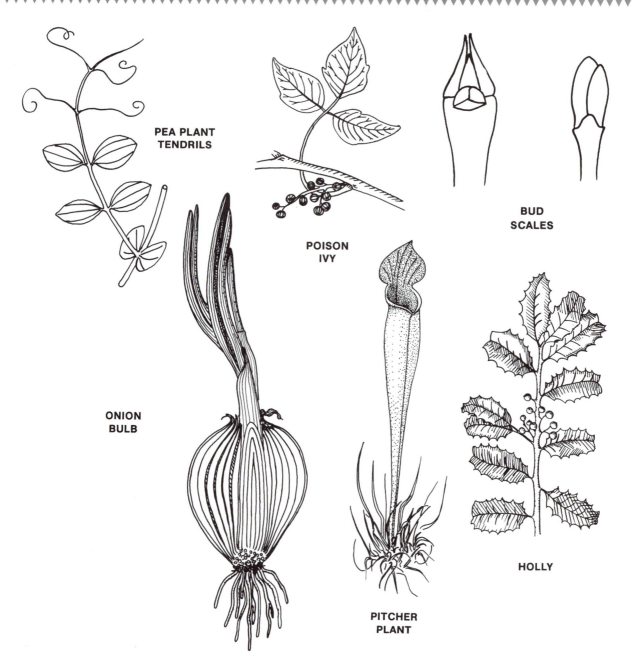

PEA PLANT
TENDRILS

POISON
IVY

BUD
SCALES

ONION
BULB

PITCHER
PLANT

HOLLY

FIGURE 20.10 Leaves Specialized For Functions Other Than Photosynthesis

caught a vein curving out to the edge of the leaf. You may be able to see the rings of the xylem vessels very well in one of these side veins. Each such vein contains the **xylem** for water and mineral transport and the **phloem** for food and hormone distribution and transport. Xylem will be the large redwalled cells on the top of the vein,

while the phloem appears as densely cytoplasmic, greenish, and on the bottom of the vein.

3. Label the illustration of the dicot leaf cross-section (Figure 20.11) with these terms: cuticle, epidermis, palisade parenchyma, spongy parenchyma, stoma, vascular bundle, guard cells, and air space.

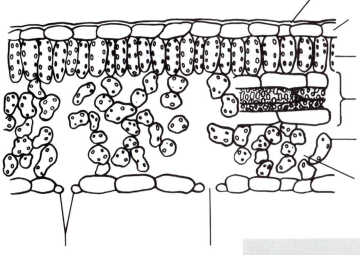

The Common Hedge Privet,
Ligustrum, **Leaf.**   X100

1. Upper epidermis
2. Palisade mesophyll
3. Xylem trachery elements
4. Bundle sheath
5. Xylem
6. Phloem
7. Spongy mesophyll
8. Lower epidermis
9. Trichome

FIGURE 20.11   Dicot Leaf (X.S.)

REVIEW QUESTIONS

1. What is totipotency? _____

   Name at least four areas in plants which possess **totipotency.** _____

   _____

2. Epidermal outgrowths for absorption in roots are _____.

3. Roots arising from an African violet leaf can be termed _____.

4. What is the function of lenticels? _____

   _____

5. If monocots possess no cambium, can they produce wood? _____

6. In southern California where the winters are warm and rainy, while the summers are hot and dry, would trees produce growth rings?_____

   Why? _____

7. In tropical rainforests with constant rainfall throughout the year, do the trees produce prominent growth rings? _____Why? _____

   _____

8. The major areas of photosynthesis in an angiosperm leaf are:

   a. epidermis

   b. palisade parenchyma

   c. spongy parenchyma

   d. only b & c

   e. vascular bundles

9. List the major function of roots. _____

   _____

10. What purpose is served by the root cap? _____

    _____

11. The cortex of the dicot root is responsible for _____

    _____ and _____.

12. MATCHING: Match the characteristic or function with the correct term.

    _____ 1. Endodermis   a. xylem and phloem

    _____ 2. Pericycle   b. meristematic tissue, producing new xylem and phloem

    _____ 3. Epidermis of root   c. secondary or branch roots originate here

    _____ 4. Stele   d. conduct food and hormones

    _____ 5. Phloem   e. support tissue for vascular cylinder and acts in waterproofing

    _____ 6. Xylem   f. produces root hairs

    _____ 7. Cambium   g. carry water and dissolved minerals

13. Compare the origin of primary, secondary, and adventitious roots. _____

    _____

14. How does the arrangement of vascular bundles differ between monocot and dicot stems?_____

    _____

Water Movement

EXERCISE
21
EXERCISE · EXERCISE · EXERCISE

OBJECTIVES

After completion of this exercise, the student should be able to do each of the following:

- Describe stomatal structure and how the stomatal opening is influenced by water.
- Diagram a stoma and label its parts.
- Determine how environmental conditions affect the rate of transpiration.
- Answer the questions throughout the exercise including the review questions at the end.

INTRODUCTION

Multicellular organisms possess many adaptive advantages, but these are not bestowed without cost. Many cells in such organisms have no free access to the external environment. Therefore, some kind of transport system must provide the cells with nutrients and remove wastes if necessary. Transport systems are truly the pipelines of life — in man, circulatory failure is a major cause of death.

The movement of materials in plants is not circular as in animals. Since much of the ascending fluid is lost by evaporation from leaves (**transpiration**), the upward movement of water and dissolved salts exceed the downward movement of organic molecules (**translocation**). Transpiration is accomplished mainly by the tissue known as xylem, translocation by the phloem. Much of the movement in both cases relies on the diffusion of molecules between cells.

OSMOSIS IN PLANT CELLS

In addition to moving the materials from cell to cell within the plant, osmosis is responsible for the opening and closing of the stomata which are surrounded by guard cells, thus controlling the amount of water loss in a plant (Figure 21.1). We will demonstrate changing the shape of guard cells and active transport across membranes with living materials.

If a cell is placed in a hypotonic solution (high water, low solute), water will move into the guard cells until the walls stop further expansion. This full cell is called **turgid.** When the cell gains water and becomes turgid, the stomata open. If this same cell is then placed in a hypertonic solution (low water, high solute), the water will leave the guard cells causing the **protoplast** to shrink away from the walls, causing a condition called **plasmolysis** which was discussed in Exercise 5. When the cells lose water and become **flaccid**, the stoma closes.

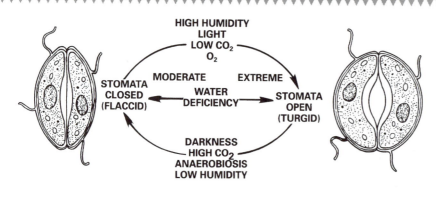

FIGURE 21.1 Guard Cell Action

ACTIVE TRANSPORT OF MEMBRANES

1. Make two wet mounts from two leaves from the aquatic plant *Elodea*. Mount one in water and the other in a drop of formalin, to kill the cells.

2. Place a drop of neutral red dye on each and observe the results with the compound microscope.

3. Now plasmolyze the cells by adding a drop of 5% NaCl and observe.

4. Diagram the cells as you see them in each case.

OSMOSIS AND STOMATA REGULATION

Just as the movement of water plasmolyzed the *Elodea* cells, the same reaction occurs in the guard cells of leaf epidermis. Refer back to the diagrams in Figure 21.1. Since the walls of the guard cell are unequally thickened, the thin side stretches out or collapses a bit as water movement makes the cells turgid or plasmolyzed.

1. Detach a *Tradescantia* ("Wandering Jew") leaf from its stem.

2. Bend the leaf until it snaps. Using forceps, remove a small piece of both the lower epidermis and the upper epidermis.

3. Make a wet mount of epidermis from both sides of the leaf.

4. Examine under a microscope to notice the numerous stomata. Draw what you see. Refer to the photograph, Figure 21.2.

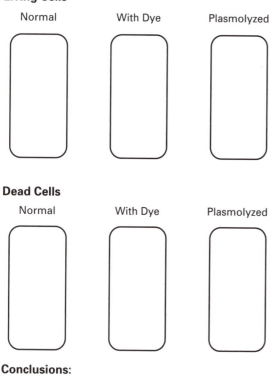

Living Cells

Normal   With Dye   Plasmolyzed

Dead Cells

Normal   With Dye   Plasmolyzed

Conclusions:

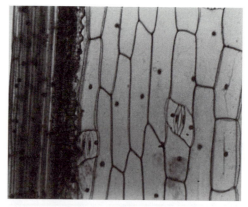

FIGURE 21.2 Photomicrography of *Tradescantia* Epidermis (430x)

Are stomata more numerous on one surface than the other? _____

If so, which one? _____

Are the stomata arranged in a pattern or do they occur randomly? _____

5. Place a few drops of 5% NaCl on one edge of the coverslip.

6. Draw the solution under the coverslip by using a paper towel to absorb the fluid from the opposite side of the coverslip.

7. Diagram a stoma treated with NaCl.

Are the stomata open or closed? _____

What is the shape of the guard cells? _____

TRANSPORT UPWARD

Transpiration is the loss of water from plants and occurs mostly from the stomata of the leaves. What mechanisms operate to bring water from the soil to the tops of trees? The best model available today as an explanation combines the effects of two phenomena: root pressure and the cohesion-adhesion-tension theory.

A. Root pressure may be measured directly using a well-watered potted plant whose top has been cut off a few inches above the soil level. A pipette is then attached to the stem by means of a short piece of tubing. Osmotic pressure will force the water out of the cut end of the stem and up the tube. Measurements of the height of the water column should be noted twice during the lab using a demonstration plant for data. (Next time you pull up a dandelion and get only the top part, notice the milky sap oozing from the root; of course a milkweed works even better.)

Initial reading: _____

Second reading: _____

Final reading: _____

B. Cohesiveness of water may be very simply demonstrated by placing a drop of water on the desk top and pulling it with your fingertip. The drop follows the finger rather than flattening out all over the desk top. Another demonstration of cohesion and adhesion would be the rise of water in a micropipette or capillary tube when the tip is submerged in water (or chlorophyll extract).

C. Transpiration rates may be determined quantitatively by using a device called a **potometer**, several of which have been assembled in the lab for your use. Essentially a potometer is a system of water-filled tubes to which a fresh plant has been attached. As water is transpired from the leaves of the plant, the fluid level in the graduated capillary tube drops correspondingly. Thus, the volume change in the capillary tube is a direct measure of the volume of water transpired during a given time interval. The system is refilled from the large burette tube by opening the stopcock at its lower end until the level in the capillary tube again rises to the top calibration. **The stopcock must be closed during measurements.**

Rate of water loss is directly related to both temperature and air movement over a leaf surface. A thermometer suspended alongside the leaves gives some measure of the temperature at the leaf surface. This temperature should be recorded at the beginning and end of each measurement period. A lamp and a fan will be used to simulate the effects of sun and wind, respectively.

1. Select a branch and shave the bottom while underwater, to expose fresh xylem.

2. Fit the branch into the potometer tube, taking care that there is no bubble at its base.

3. Allow the setup to equilibrate for 5–10 minutes.

4. Take initial readings of the temperature and water level in the capillary tube of the potometer (NOT of the burette).

5. With the system under normal laboratory conditions, wait 10 minutes, then take readings of temperature and amount of water transpired.

6. Repeat steps 3, 4, and 5, using first the lamp at leaf level, then the fan pointed at the leaves, and finally both lamp and fan instead of normal laboratory conditions as done the first time.

Environmental Conditions	Initial		Final		Minutes Measured	Ave. Temp.	Transpiration Rate (ml/hr)
	Temp	ml	Temp	ml			
Normal							
Heat (lamp)							
Wind (fan)							
Heat + Wind							

Under which condition was the rate of transpiration greatest? _____

Do you think the number or size of leaves might affect the rate of transpiration? _____

Why? _____

_____

REVIEW QUESTIONS

1. What is the advantage of closed stomata when water supply is low? _____

   _____

2. How might the distribution of guard cells help prevent excess water loss? _____

   _____

3. Consider all four conditions and explain how each condition caused an increase or decrease in transpiration. _____

   _____

   _____

4. Which would be harder on agricultural crops, hot-windy days or hot-windless days? _____
   Why? _____

5. Why was it important to sever the stem (Transpiration rate) while under water? _____

   _____

6. What was the name of the instrument used to measure transpiration? _____

7. Compare transpiration to translocation. _____

   _____

8. As the guard cells become more turgid what happens to the stomatal opening? _____
   Flaccid? _____

9. How did the shape of the guard cells change when a NaCl solution was used? _____
   _____ Did the stomata open or close? _____

Kingdom Animalia
Porifera, Cnidaria, and Ctenophora Phyla

OBJECTIVES

After completion of this exercise, the student should be able to do each of the following:

- Compare the Phyla Porifera, Cnidaria, and Ctenophora as to their level of organization (cellular, tissue, organ, system).
- Describe the feeding and nutrient processing of a sponge and a hydra.
- Identify the anatomical structures of the sponge and the hydra.
- Define the terms given in boldface other than the anatomical structures.
- Identify the phyla and classes of each of the animals in the jars on display.
- List the basic cell types in the sponge and the function of each.
- Compare and contrast the two body forms of the Cnidaria.
- Describe the alternation of asexual and sexual reproduction shown by *Obelia.*
- Answer the review questions at the end of this exercise.

THE ANIMAL KINGDOM

The animal kingdom is composed of multicellular, obviously motile, heterotrophic organisms which reproduce sexually by oogamy and produce multicellular embryos. You may wish to refer to Exercise 12 and review the outline of animal taxa to be studied in the next several exercises.

PHYLUM PORIFERA

The phylum Porifera contains animals known as sponges. Porifera means "bearing pores." Sponges are multicellular, but the cell aggregates of this phylum do not form true tissues. However, differentiation of cells has occurred so that there is a division of labor such as reproduction, feeding, and water circulation.

Many of these animals have body shapes resembling vases (See Figure 22.1). The walls of sponges have numerous tiny openings called **ostia,** through which currents of water enter carrying food and oxygen to the central cavity of **spongocoel.** The movement of water is produced by the beating of the flagella of **collar cells,** or **choanocytes,** in the **flagellated chambers,** or **radial canals.** The choanocytes capture and digest food particles brought in by the water currents with the aid of collar-like structures surrounding their flagella. The water entering the

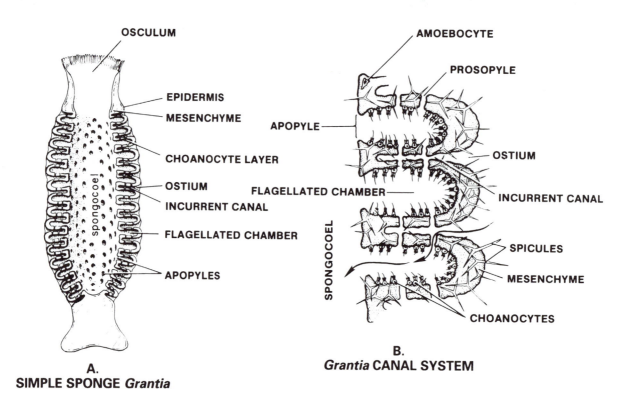

OSCULUM

EPIDERMIS

MESENCHYME

CHOANOCYTE LAYER

OSTIUM

INCURRENT CANAL

FLAGELLATED CHAMBER

APOPYLES

spongocoel

A.
SIMPLE SPONGE *Grantia*

AMOEBOCYTE

PROSOPYLE

APOPYLE

OSTIUM

INCURRENT CANAL

FLAGELLATED CHAMBER

SPICULES

MESENCHYME

CHOANOCYTES

SPONGOCOEL

B.
Grantia **CANAL SYSTEM**

FIGURE 22.1   *Grantia* (L.S.)

ostia circulates through the **incurrent canal, proso-pyles** (openings between the incurrent canals and radial canals), **radial canal, apopyles** (openings between the radial canals and the spongocoel), and **spongocoel** and exits through the opening at the top of the animal. This opening is called the **osculum.**

The body wall consists of three layers. The outer layer consists of flat epithelial cells, among which are contractile cells called **pinacocytes** which regulate the sizes of the ostia. The middle layer consists of gelatinous non-living matrix containing living mesenchyme cells called **amoebocytes,** which are capable of amoeboid movement. Amoebocytes have many functions. They collect food from the flagellated collar cells, secrete the gelatinous matrix, collect wastes, produce **spicules,** and can differentiate into any of the other cell types. The spicules, which are minute crystal-like structures composed of calcium salts or silicious material, form the supportive skeleton of the sponge.

Sponges are unique in that it is believed that they have evolved from a completely different group of flagellates than did other animals. For this reason, they are considered to be an evolutionary dead end.

Reproduction in sponges occurs asexually by **budding, fragmentation,** and in freshwater forms, by **gemmule** formation. A gemmule consists of a ball of amoebocytes surrounded by a capsule consisting of spicules and dead cells. Sexual reproduction in sponges involves the fusion of eggs and sperm formed from amoebocytes. The zygote develops into a free-swimming ciliated larva.

Obtain a prepared slide of the cross section of *Grantia.* Make a sketch of what you see, labeling the central cavity, or spongocoel, incurrent canal, radial canal, and choanocytes. Refer to Figure 22.2. Also note the small openings, or apopyles, leading from the radial canals into the spongocoel.

Obtain another slide showing the **spicules** which form the skeleton of the sponge. Make a sketch of *Grantia* spicules below.

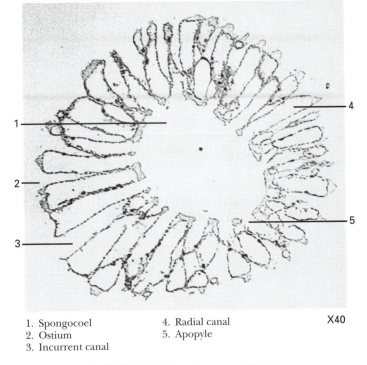

1. Spongocoel
2. Ostium
3. Incurrent canal

4. Radial canal
5. Apopyle

X40

FIGURE 22.2   *Grantia* (X.S.)

The classification of sponges is determined largely by shape and chemical composition of the skeleton. The phylum Porifera is subdivided into three classes:

1. **Class Calcarea** — Calcareous or chalky sponges. Example: *Grantia*

2. **Class Hexactinellida** — Glass sponges, composed of siliceous spicules. Example: Venus Flower Basket

3. **Class Demospongiae** — Commercial or bath sponges. Skeleton includes proteinaceous **spongin** fibers. Example: *Spongia*

Examine the specimens of sponges in the display jars, noting the class to which each belongs.

PHYLUM CNIDARIA (Coelenterates)

The animals of this phylum demonstrate the tissue level of organization. Although the cell aggregates function as tissues, no true organs are present. The body of the animal is composed of two epithelial layers: an outer **epidermis** and an inner **gastrodermis,** which lines the **gastrovascular cavity**, or **coelenteron.** Between these two layers is a gelatinous layer of **mesoglea.** The epidermis is a well developed layer of closely-packed cells which function to protect the organism and obtain food. The gastrodermis is also well developed and serves in digestion and internal transport. The mesoglea is poorly developed, varying from a jellylike substance containing a few cells which serve in coordinating the actions of the organism and the production of gametes, to a true cellular layer in the most advanced members of the group.

The most distinguishing characteristic of this phylum is the possession of specialized cells called **cnidoblasts.** These cells contain stinging structures called **nematocysts** which are used for defense and food capture. A nematocyst is a capsule containing a long coiled thread which shoots out and either traps and holds the prey or injects a toxic substance which paralyzes the prey or predator.

Another characteristic of these animals is that they are polymorphic, displaying different body forms at different points in the life cycle of the organism. There are two basic body forms; the **polyp** and the **medusa** (Figure 22.3). The medusa is generally a more active swimming form where as most polyps are **sessile,** remaining attached to some substrate. Both forms are found in the life cycle of many coelenterates. Some coelenterate colonies may be

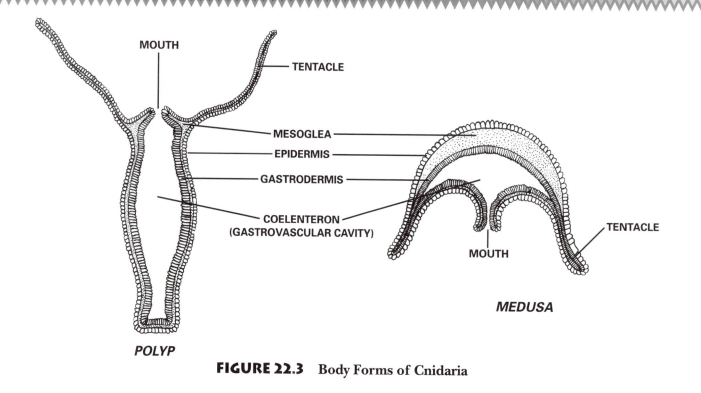

FIGURE 22.3   Body Forms of Cnidaria

composed of both forms of individuals. Both forms are basically **radially symmetrical**.

The phylum Cnidaria is subdivided into three classes: Class Hydrozoa, Class Scyphozoa, and Class Anthozoa.

CLASS HYDROZOA

Many members of this class develop into sessile polyp colonies such as *Obelia*. Their life cycles characteristically involve a regular alternation between asexual (polyp form) and sexual (medusa form) reproduction, but the polyp form is generally dominant.

Hydra is an exceptional member of this class. It is a mobile individual polyp rather than a sessile or colonial form. Reproduction occurs asexually by budding or sexually by the production of sperm and eggs, and only polyps are produced. There are no medusae produced at any time.

Obtain a prepared slide of a *Hydra*. Note the following parts (Refer to Figure 22.4) as you find them; the **basal disc**, for attachment, the cylindrical **body**, the circle of **tentacles**, the elevated **hypostome** at the base of the tentacles, a **mouth** in the center of the hypostome, **buds**, **spermaries** (swellings just beneath the tentacles), **ovaries** (swellings in the lower portion of the body), **nematocysts**, **gastrovascular**

cavity, **epidermis**, **mesoglea**, **gastrodermis**, and **flagellum**.

Prepare a wet mount of living *Hydra* using a concave depression slide. **Do not cover with a cover slip.** For the best results, reduce the amount of light entering the slide. Do not jar the table or microscope, as any disturbance will cause the *Hydra* to contract. Feeding the *Hydra* will be optional. After viewing the living *Hydra*, return it to the container marked **Fed Hydra.**

Obtain a prepared slide of the colonial *Obelia* and find the parts listed on Figure 22.5. This colony forms from the repeated budding of a single individual. The various buds take up either reproductive (**gonangium** or reproductive polyp) or feeding (**hydranth** or feeding polyp) duties that are performed for the entire colony. The reproductive polyp, or gonagium, will produce medusae which will in turn produce eggs or sperm. When fertilized, or when egg and sperm unite, the zygote will go through various stages of embryonic development and will give rise to a ciliated larvae, the **planula**. The planula attaches itself to the substratum and gives rise to a new asexual hydroid colony.

Another example of a Hydrozoan is *Gonionemus*, which has a dominant medusa stage. For years, *Gonionemus* was thought to lack the polyp stage because it was so small (1 mm.) that no one had been able to detect it. The polyp cannot only produce medusae

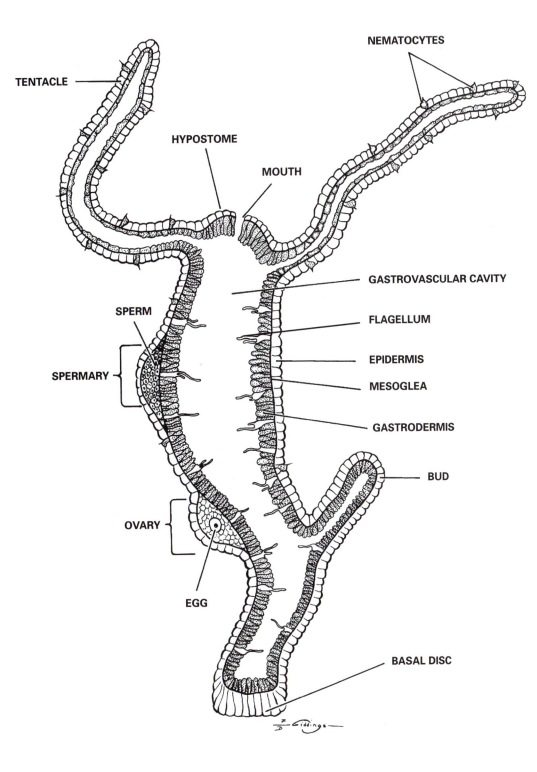

NEMATOCYTES

TENTACLE

HYPOSTOME

MOUTH

GASTROVASCULAR CAVITY

SPERM

FLAGELLUM

SPERMARY

EPIDERMIS

MESOGLEA

GASTRODERMIS

BUD

OVARY

EGG

BASAL DISC

FIGURE 22.4 *Hydra* (L.S.)

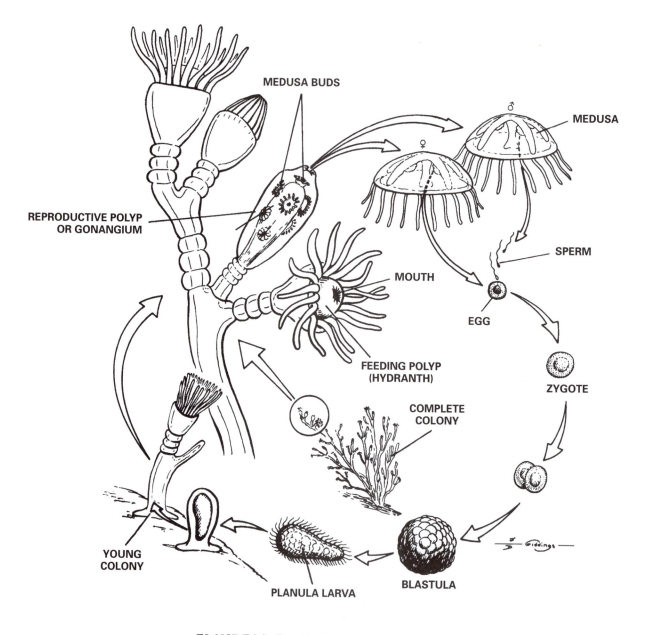

FIGURE 22.5 *Obelia* Hydroid Colony

by budding, but can also bud to produce other polyps. Upon the completion of its development, the medusa form of *Gonionemus* resembles a typical jellyfish, but unlike scyphozoan jellyfish, it has a muscular shelf-like **velum** around the margin of the "bell" which aids in swimming. Figure 22.6 illustrates a hydrozoan medusa.

Physalia, Portuguese man-of-war shows diversity in that it consists of floating colonies of specialized individuals. Each colony includes at least four types of polyps: the pneumatophore, or float, into which gas is secreted to render the colony buoyant, feeding polyps, defensive or stinging polyps, and reproductive polyps. Some species also have sensitive or feeling polyps.

Examine the hydrozoan specimens on display in jars.

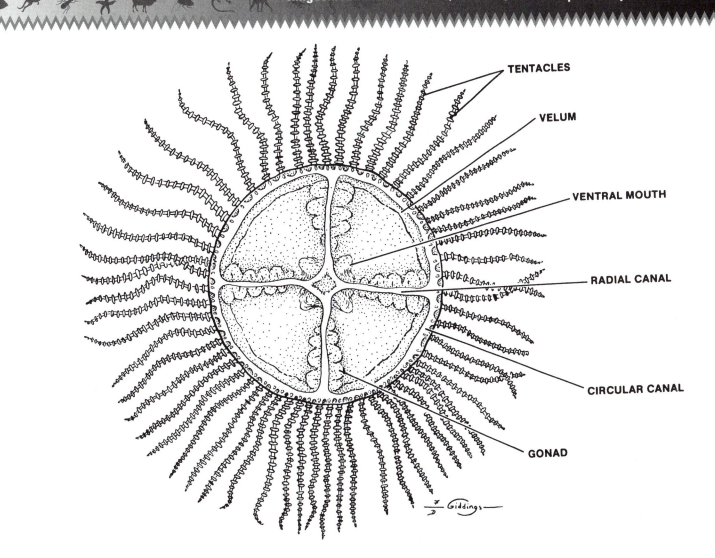

TENTACLES

VELUM

VENTRAL MOUTH

RADIAL CANAL

CIRCULAR CANAL

GONAD

FIGURE 22.6   *Gonionemus* Medusa

CLASS SCYPHOZOA

The members of this class are commonly called **jellyfish.** Most of them have both body forms in their life cycles, although the polyp form may be very small, as the medusa form is dominant. Scyphozoan medusae do not have a velum as do the hydrozoan medusae.

Examine the jellyfish on display in jars and refer to Figure 22.7.

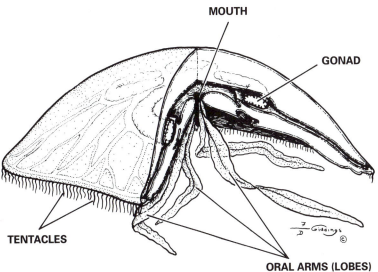

MOUTH

GONAD

TENTACLES

ORAL ARMS (LOBES)

FIGURE 22.7   Structure of Jellyfish, *Aurelia*

CLASS ANTHOZOA

All members of this class have the polyp body plan. There are no medusa forms. Many of them, such as the corals and sea pansies, live in colonies, whereas some such as the sea anemone, live independently. Examine the representatives of this class on display and compare them with the illustrations in Figure 22.8.

PHYLUM CTENOPHORA

Members of phylum Ctenophora are often known as the **comb jellies** (comb bearers) and **sea walnuts.** They comprise about 80 species of free-swimming marine animals with translucent gelatinous bodies. Ctenophores show resemblance to jellyfishes and at one time were classified with the coelenterates (see Figure 22.9). The unique characteristic is that they possess eight rows of swimming "combs", or **ctenes,** which are composed of fused cilia; unlike the cnidaria, they lack nematocysts, except for one species.

Observe the ctenophores on display.

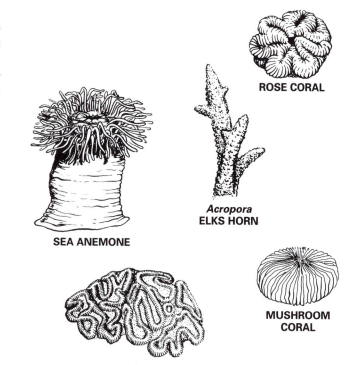

ROSE CORAL

Acropora
ELKS HORN

SEA ANEMONE

BRAIN CORAL

MUSHROOM CORAL

FIGURE 22.8   **Representative Anthozoans**

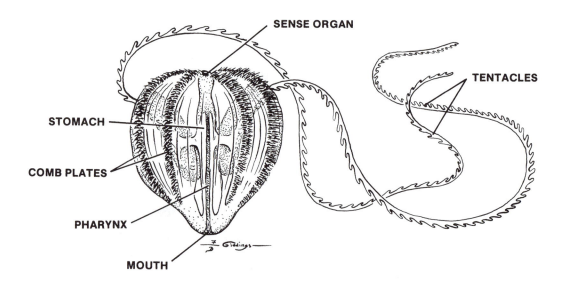

SENSE ORGAN

TENTACLES

STOMACH

COMB PLATES

PHARYNX

MOUTH

FIGURE 22.9   **A Ctenophoran**

REVIEW QUESTIONS

1. Two kinds of cells showing division of labor in a sponge are_____

   and _____.

2. Sponges reproduce by (describe each):

   a. _____

   b. _____

3. Why are sponges considered unique in the evolution of animals? _____

   _____

   _____

   _____

4. Identify the following:

   Gastrovascular cavity _____

   Nematocyst_____

   Planula _____

   Velum _____

   Polyp _____

   Medusa _____

   Amoebocyte _____

   Spicule_____

   Choanocyte_____

5. The unique characteristic of the Ctenophores is _____.

6. Members of the Phylum Porifera demonstrate a (multicellular, tissue, organ system) level of organization.

   _____

7. Members of the Phylum Cnidaria demonstrate a (multicellular, tissue, organ system) level of organization.

   _____

8. List the openings and structures a droplet of water would pass on its way through the sponge.

   Begin with the ostia and end with the osculum. _____

   _____

9. What cells are responsible for producing the water current? _____

10. List the functions of the amoebocytes. (Don't forget their reproductive functions.) _____

    _____

11. Name the two epithelial layers found in members of the Phylum Cnidaria and their functions. _____

_____

12. For what reason(s) are cnidoblasts important to the *Hydra*? _____

_____

13. How are polyps and medusas similar? How are they different?_____

_____

14. Why is the *Hydra* an exceptional member of the class Hydrozoa? _____

_____

15. MATCHING: Match the class with the correct organism or characteristic.

_____ 1. *Hydra*   a. Class Anthozoa

_____ 2. Jellyfish   b. Class Scyphozoa

_____ 3. *Gonionemus*   c. Class Hydrozoa

_____ 4. *Obelia*

_____ 5. Sea anemone

_____ 6. Coral

_____ 7. Both polyp and medusa forms with polyp form being dominant

_____ 8. Medusa form is dominant, polyp form greatly reduced

_____ 9. No medusa form, all polyps

KINGDOM ANIMALIA
PLATYHELMINTHES, NEMATODA, ROTIFERA, AND ANNELIDA PHYLA

OBJECTIVES

After completion of this exercise, the student should be able to do each of the following:

- List the advantages of the members of the phylum Platyhelminthes over the members of the phylum Porifera and phylum Cnidaria.

- Identify the anatomical structures of planarians, flukes, and tapeworms given in boldface in the text.

- List, cite an example of, and identify the major characteristics of each of the three classes of the phylum Platyhelminthes.

- Identify the phylum and class of each of the animals in the jars on display.

- Define the boldface terms referring to the earthworm.

- Describe two evolutionary advantages possessed by members of both the phylum Nematoda and the phylum Rotifera.

- Examine the *Ascaris* cross section slide and identify its component parts.

- Distinguish between a male and female nematode.

- Examine the rotifer slide and identify the "wheels" and the "forked foot."

- List, cite an example of, and identify the major characteristics of each of the three major classes of the phylum Annelida.

- Dissect an earthworm and identify the boldface anatomical parts referred to in the directions.

- Identify the structures referring to the earthworm cross section slide printed in boldface in the text.

- Distinguish between a **pseudocoelom** and a true **coelom** and explain the advantage of possessing a true coelom.

- Answer the review questions at the end of this exercise.

PHYLUM PLATYHELMINTHES

Members of this phylum are commonly referred to as the flatworms. As the name implies, they are flattened dorsoventrally. The flatworms show many advantages over the Porifera and Cnidaria, such as:

1. Bilateral symmetry.

2. A complex organ-system level of organization.

3. A mesodermal germ layer. They are therefore referred to as being **triploblastic** in general structure.

4. A central nervous system showing **cephalization**.

5. A distinct head with sense organs.

Other characteristics of this phylum include the absence of a body cavity, that is, they are **acoelomate**; the absence of an anus; and the combining of sexes within single animals (**hermaphrodism**). There are both parasitic and free-living forms.

The phylum Platyhelminthes is subdivided into three classes: Class Turbellaria, Class Trematoda, and Class Cestoda.

CLASS TURBELLARIA

This class includes the free-living flatworms, an example of which is *Dugesia*, the common planaria. Members of this class may be found under rocks or attached to submerged objects in the clear water of lakes, springs, and streams.

Obtain and examine a prepared slide of the freshwater planaria (see Figure 23.1). In your examination, note the general body shape, and give special attention to the head, **eyespots**, and **auricles** (lateral projections on the head which function as tactile and chemosensory organs in the anterior region). On this slide the digestive system is stained by feeding the worms India ink before preserving them; it clearly demonstrates the branching of the **gastrovascular cavity**. Identify the muscular **pharynx**, or **proboscis**, which is withdrawn into the **pharyngeal pouch** in the middle area of the body. This pharynx may be extended out of the body while feeding.

Prepare a wet mount of living planaria using a concave depression slide. **Cover with a cover slip**. Notice the general shape and the mode of locomotion of the flatworm. Feeding the planaria with bits of liver is optional. After viewing the planaria, return them to the container marked **Fed Planaria**. Also examine the preserved specimens on display in jars.

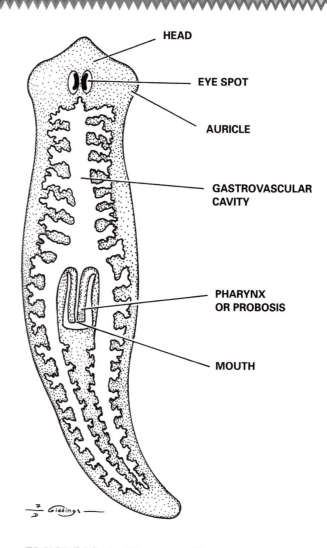

FIGURE 23.1   **Planaria: Digestive System**

CLASS TREMATODA

This class is composed of parasitic flatworms known as the flukes. They have evolved a thick, protective outer layer of nonliving substance known as the **cuticle** which is secreted by the epidermis and protects the organism from being digested by the host's digestive enzymes.

Examine the slide of *Clonorchis sinensis* (whole mount) under low power. *Clonorchis*, the Chinese liver fluke, possesses an anterior **oral sucker** and a large **ventral sucker**. Beginning anteriorly at the oral sucker, trace the digestive tract to the short muscular **pharynx**, and then to the forked **gastrovascular cavity**. The reproductive system includes a mass of **testes** located at the posterior end. Sperm that is produced here moves through a duct to the **genital pore** located just anterior to the ventral sucker. Eggs are produced in an **ovary**, anterior to the testes. In front of the ovary is a large convoluted duct called the **uterus** which serves as a storage area for fertilized eggs.

On Figure 23.2, label the anatomical structures of the "Chinese liver fluke" that are in bold face type in the last paragraph. Also examine the preserved specimens of trematodes in jars.

CLASS CESTODA

This class includes the tapeworms, all of which are parasitic. Examine the preserved specimens on display and obtain a prepared slide of *Taenia pisiformis*, the dog and cat tapeworm. This organism is transmitted by ingestion of infected fleas and may also infect humans in this way.

Examine the stained slide and, with the help of Figure 23.3 on the following page, identify the following areas and their associated structures.

1. **Scolex**, or head with **rostellum**, which attaches to the intestinal wall of the host, has rows of **hooks** and four **suckers**.

2. **Neck**, has no segmentation and represents a growing area.

3. The body, or **strobila**, consisting of units called **proglottids**.

The proglottids behind the neck are young, or **immature** and only contain the male sex organs while proglottids in the middle region are **mature** and contain both male and female sex organs. At the posterior end, the proglottids lack male reproductive organs as they have disintegrated and the proglottids in this region are **gravid**, or "ripe" sections filled with fertilized eggs which will become detached and pass out of the host via the feces, to be picked up by another successive host.

PHYLUM NEMATODA

This phylum includes the unsegmented roundworms. Some are parasitic, but most are free living. An advancement of this phylum over the ones previously studied is that animals in this phylum possess a **complete digestive tract** which has two openings, a **mouth** and an **anus**, allowing for oneway passage of ingested food. A second advancement is the presence of a body cavity. This cavity is not a true **coelom** since it does not lie between layers of mesoderm. The cavity, called a **pseudocoelom**, is found between the outer body wall and the digestive tube. A unique characteristic of this group is that each animal has a limited number of cells in its body. This specific number of cells is characteristic of its species. Animal growth in this phylum beyond the embryo stage is due to cell growth rather than cell multiplication as in other phyla. Their musculature consists mainly of longitudinal muscles which cause them to move in a whip-like fashion.

1. Examine the display jars containing specimens of *Ascaris*, a common roundworm found as a parasite in humans and pigs. These worms feed on the contents of the intestinal tract. Some damage may be done to their host by their thrashing motion or by clogging of the intestinal tract by large numbers. The male is smaller than the female

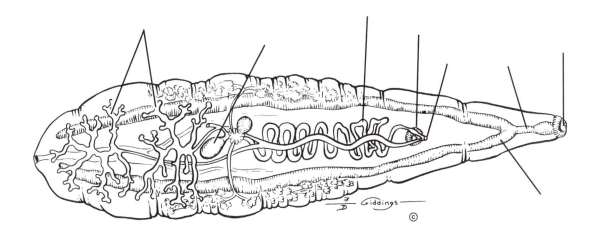

FIGURE 23.2   *Clonorchis sinensis*

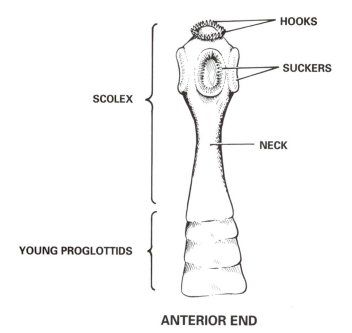

ANTERIOR END

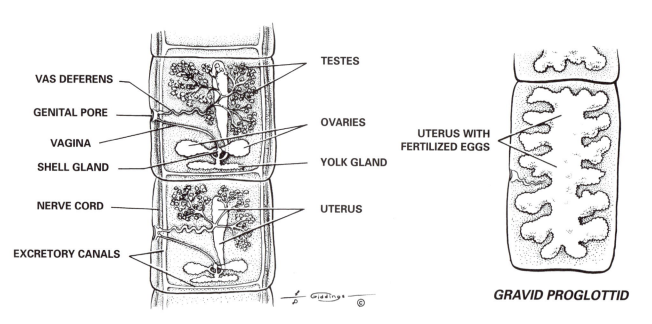

MATURE PROGLOTTIDS

GRAVID PROGLOTTID

FIGURE 23.3 *Taenia pisiformis*

and is curved at the posterior end. Also at the posterior end of the male are **spicules**, tiny bristles that aid in copulation. The **cuticle**, or outermost covering of the worm is important in protecting the animal from the digestive enzymes of its host and in acting as an exoskeleton to which muscles are attached. Sketch the male and female *Ascaris* below.

2. Examine a *Ascaris* male cross section slide and a female cross section slide. Refer to Figure 23.4 and identify the following parts: **pseudocoel, testis, ventral nerve cord, dorsal nerve cord, cuticle, intestine, ovary, eggs,** and **uterus.**

3. Examine slides of *Ancylostoma*, a human hookworm. The anterior end of the hookworm has hooks with which it attaches to the intestinal wall of the host. Unlike *Ascaris*, the hookworm feeds on its host's blood. You can distinguish the male from the female by looking at the posterior end of the worm, which in the male is modified for copulation by way of a fanlike structure called a **bursa.** Sketch the male and female hookworms indicating the anterior and posterior ends.

MALE *Ascaris*   FEMALE *Ascaris*

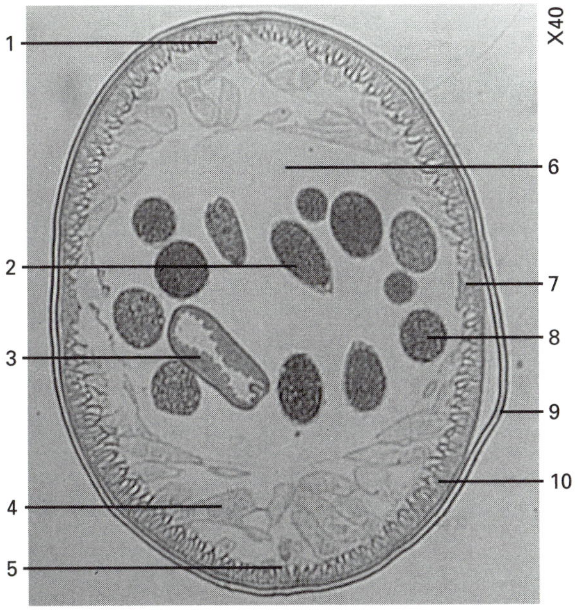

X40

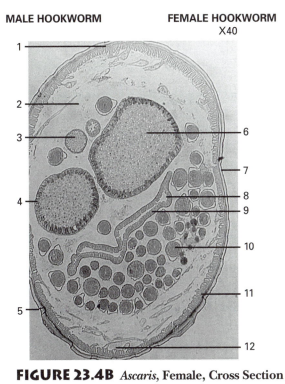

MALE HOOKWORM   FEMALE HOOKWORM
X40

FIGURE 23.4A *Ascaris,* Male, Cross Section

1. Dorsal nerve cord
2. Vas deferens
3. Intestine
4. Longitudinal muscle cell body
5. Ventral nerve cord
6. Pseudocoel
7. Lateral line
8. Testis
9. Cuticle
10. Contractile sheath of muscle cell

FIGURE 23.4B *Ascaris,* Female, Cross Section

1. Dorsal nerve cord
2. Pseudocoel
3. Oviduct
4. Uterus
5. Cuticle
6. Eggs
7. Lateral line
8. Lumen
9. Intestine
10. Ovary
11. Longitudinal muscles
12. Ventral nerve cord

4. Examine a slide of the voluntary muscle fibers of rat or man containing specimens of encysted *Trichinella spiralis.* This roundworm forms cysts in the muscle of hogs and may infect humans who eat insufficiently cooked pork. The cyst causing trichinosis you observe on the slide is only one stage in the life cycle of this animal. Try to locate the **muscle tissue of host**, **cyst wall**, and **worm.**

Other representatives of this phylum such as pin worms, filarial worms, the human whip worm, vinegar "eels," and heartworm found in dogs, to mention only a few, may be discussed in class.

5. Examine pond water for free-living nematodes. You should be able to identify them by their whip-like thrashing motion (refer to Figure 23.5). Attempt to isolate one on a slide and examine it more closely. Write a brief description of the nematode.

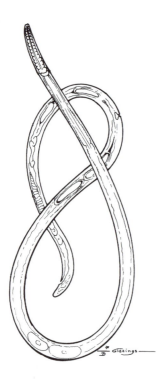

FIGURE 23.5   **Free-Living Nematode**

PHYLUM ROTIFERA

The rotifers or "wheel-bearing animalcules" are usually less than 1 mm. in length and most are found in freshwater. Most of the 1,500 species of Rotifers are free-living, a few are parasites, and a few inhabit salt water. Members of this phylum possess both a complete digestive tract and a **pseudocoelom** (pseudo — false; coelom — body cavity). The distinguishing characteristics of the animals included in this phylum are:

1. "Wheels" of beating cilia on the anterior end of the body.

2. Absence of external cilia elsewhere.

3. A chewing **pharynx** or **mastax** which is used for grinding the ingested food particles.

4. "Forked foot" on the posterior end of the body.

5. Growth beyond the embryo stage is by cell growth rather than by cell multiplication.

Examine a slide of a Rotifer whole mount. Identify the "**wheels**" and the "**forked foot**." Use Figure 23.6 to help you.

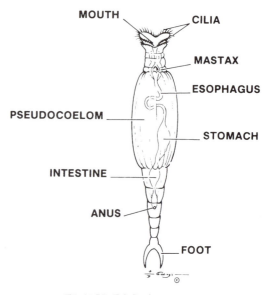

FIGURE 23.6   **Rotifer**

Examine pond water for aquatic rotifers. If they are moving quickly, add a drop of Protoslo to the slide under the cover slip. Write a brief description of a rotifer.

PHYLUM ANNELIDA

This phylum consists of approximately 7,000 species of **segmented** worms which are divided among 4 classes: **Polychaeta**, the marine bristleworms and sandworms; **Hirudinea**, the leeches; **Oligochaeta**, the earthworms; and **Archiannelida**, a small, primitive group of tiny marine worms. Study the specimen jars on display for the first three classes.

A unique feature of the phylum Annelida not previously seen in the organisms studied, is the presence of a true body cavity or **coelom**. The coelom is completely lined by mesodermal tissue. With a true body cavity there is an independence of the digestive tract from the muscles of the body wall. Therefore, rhythmic contraction of these muscles, called **peristalsis**, allows food to be moved through the digestive tract without the movement of the entire animal.

CLASS POLYCHAETA (Many Bristles)

The polychaetes represent the largest class of annelids. They possess fleshy tentacles on the head, and two fleshy appendages, **parapodia**, on each segment except the first and last. Many **setae**, or chitinous bristles, are found on the parapodia which are used for swimming, burrowing, crawling, and gas exchange. See Figure 23.7. Examine the polychaetes on display.

CLASS HIRUDINEA

The leeches lack tentacles, parapodia, and setae. Many of the members of this class are parasitic and inhabit freshwater. An interesting feature of the leeches is that they possess two muscular suckers: small anterior suckers which surround their mouths and large posterior suckers used in locomotion and attachment.

Examine the slide of a leech and locate the structures indicated in Figure 23.8. Examine the leeches on display.

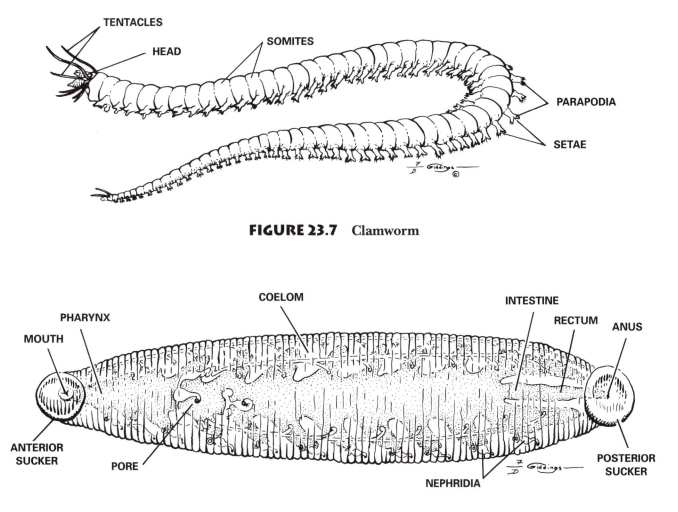

FIGURE 23.7 Clamworm

FIGURE 23.8 Leech

CLASS OLIGOCHAETA (Few Bristles)

Members of this class inhabit damp soil and freshwater. They are somewhat degenerate annelid types with the head and locomotor structures greatly reduced. They lack parapodia, having only setae.

The earthworm will be studied extensively in lab. Before you start dissecting, **read the directions carefully**. Remember that the diagrams are intended to help you find the anatomical parts, but you may ask your instructor for assistance. It is essential that you be able to identify the parts in the dissected animal and not just memorize the figures.

External Anatomy of the Earthworm

Obtain a specimen of *Lumbricus* (the "nightcrawler" earthworm) and run your fingers over the surface of its body. Do you note any differences between the resistance to the motions of your fingers in different directions? This is caused by the presence of the **setae** on the ventral and ventrolateral surfaces. Place the worm on moist paper towelling in a dissecting pan. Refer to Figures 23.9, 23.10, and 23.11 as you perform the dissection.

The most obvious feature of the earthworm is its segmentation. The very small first segment overhangs the mouth and is known as the **prostomium** (meaning in front of the mouth). Note that it does not have a corresponding ventral portion. Beginning with the next segment (the first complete one) we will assign numbers to the segments for convenience. Several segments, beginning with number 32 or 33 are swollen because of large hypodermal glands responsible for the formation of the **cocoon**. These swollen segments comprise the **clitellum** which is located anteriorly. This structure is not as obvious from the ventral surface as it is from the dorsal surface. The anus is located at the end of the last segment.

With the aid of a **stereomicroscope**, **determine the number and location of the setae** and their orientation on a segment. Locate the openings of the sperm ducts or vasa deferentia which lie ventrolaterally on segment 15; they are a pair of transverse slits lying between two swollen lips. The oviduct openings are similarly located on segment 14, but are less conspicuous.

Place the worm, **ventral side down**, in a dissecting pan. Carefully pin your specimen with pins through the prostomium and the posterior segment. Make a short, longitudinal incision in the dorsal midline, forward through the body wall from a short distance back of the clitellum to the anterior end. Be very careful not to cut through more than just the body wall, noting the **septa** (thin membranes). Pin the body flat by placing the pins in every fifth segment, and lean each pin toward the outer edges of the pan so that your view will be unobstructed.

Internal Anatomy

Circulatory System

The earthworm has a circulatory system consisting of five pairs of **hearts** surrounding the **esophagus**, one pair each in segments 7 through 11. The hearts are usually black. The **dorsal blood vessel** located along the middorsal line above the digestive tract, carries blood anteriorly. If you cannot see it at this stage, do not cut away to find it; search later. Smaller blood vessels will be seen on the outer surface of the gut and the inner surface of the body wall.

Digestive System

Identify the **mouth** and the **buccal cavity** which extend through the first three segments. More conspicuous is the **pharynx**, a thick muscular organ with accessory lubricating glands inside, occupying segments 3 to 5. From this segment to segment 14 is the relatively slim **esophagus**. In segments 15 through

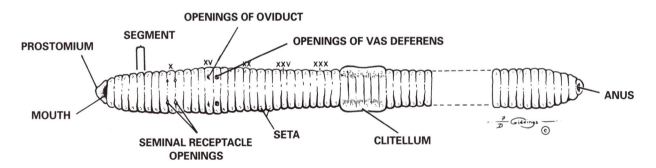

FIGURE 23.9   **Earthworm Ventral View**

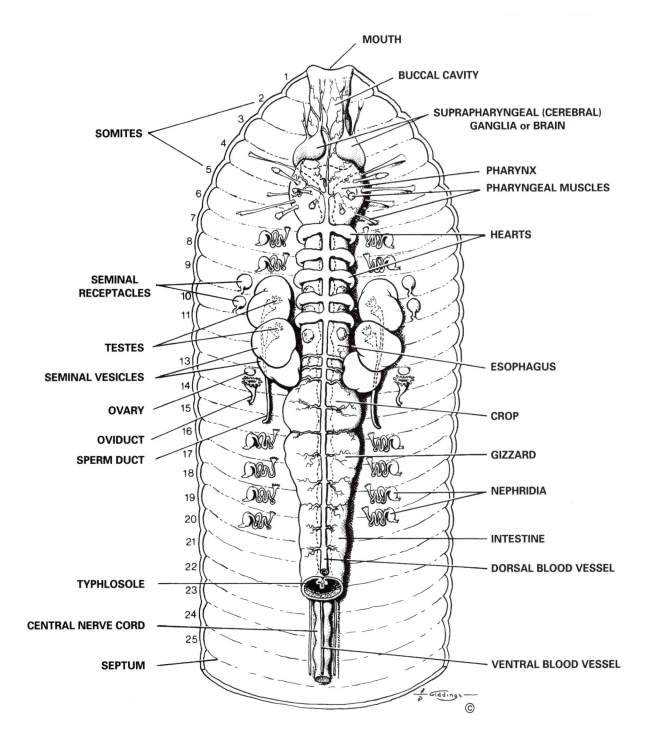

FIGURE 23.10 Earthworm: Dorsal Dissection

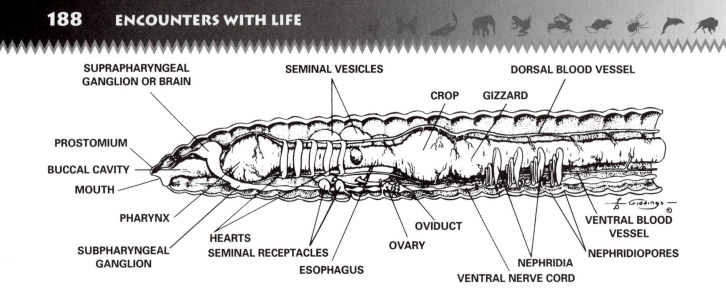

FIGURE 23.11 Earthworm: Lateral Dissection

17, the digestive tract expands into the **crop** where food is stored. Posterior to the crop is a thick-walled **gizzard**, a mastication organ Using your probe, feel the difference in the walls of the crop and gizzard. How are they different? _____

_____

The **intestine** is located beyond the gizzard and leads to the **anus**. Most digestion and absorption takes place in the intestine. It is well supplied with secretory cells. The inner surface area of the intestine through which absorption can take place is greatly increased by two devices: segmental constrictions and the **typhlosole**. The typhlosole is an internal longitudinal ridge of the intestinal wall. Use Figure 23.10 in order to help you find the typhlosole.

Excretory System

Each segment, except for the first three and the last, has a pair of white tubular excretory organs, the **nephridia**, which lie lateral to the gut. Each organ opens to the outside through its own duct and pore. These pores are difficult to see. The nephrida act like tubules in a human kidney. By filtration, reabsorption, and tubular secretion they yield a protein-free urine and maintain the steady state of the body.

Nervous System

Extend the middorsal incision from the clitellum towards the anus. Remove the intestine carefully to expose the **ventral nerve cord** located beneath the **ventral blood vessel**. **Ganglia** (singular, ganglion) are present along the cord in each segment and handle much of the coordination of these animals without the intervention of the main brain. The nerve cord and the ganglia are difficult to see in worms. You should see (unless you accidentally cut it away) the **brain**, composed of a pair of white ganglia above the pharynx in segment 3. These communicate with the ventral nerve cord through a pair of **circumpharyngeal connectives**.

Reproductive System

The most obvious portions of the male reproductive tract are the two three-lobed **seminal vesicles**, or sperm reservoirs, usually cream-white in color. Fastened ventrally and extending dorsally around each side of the esophagus, they include the two small **testes** within them. Sperm are freed from the testes and complete their development in the seminal vesicles. They are passed out through funnel-shaped mouths of the **vas deferens** or **sperm ducts**, which is also well hidden by the vesicles.

The female reproductive tract is composed of **ovaries**, **egg sac**, and **oviduct**, all difficult to distinguish. Despite the fact that these worms do not have separate sexes, they cannot fertilize themselves. Copulation must occur. But with both sexes in a single animal, any two worms which meet can copulate (obviously a convenient situation). A sperm transfer then occurs in both directions. After transfer, sperm is stored in **seminal receptacles** which are located in segments 9 and 10 until needed to fertilize eggs in cocoons.

Each earthworm produces a cocoon containing eggs. For each cocoon, a slime tube is secreted around the clitellum and anterior somites and within the cocoon forms as a separate secretion over the clitellum. The tube and cocoon then slip forward, and sperm to fertilize the eggs enter when the cocoon passes over the seminal receptacles. As the worm withdraws from the tube, the cocoon closes into a lemon-shaped case that is deposited in the damp soil. Each cocoon has several fertilized eggs of which one or two develop.

Cross Section of an Earthworm

In a prepared slide, note the following structures (Refer to Figure 23.12)

1. **Body Wall** (beginning with the outermost layer)
   a. **Cuticle**: thin external chitinous layer
   b. **Epidermis**: outer cellular layer, which contains epithelial cells

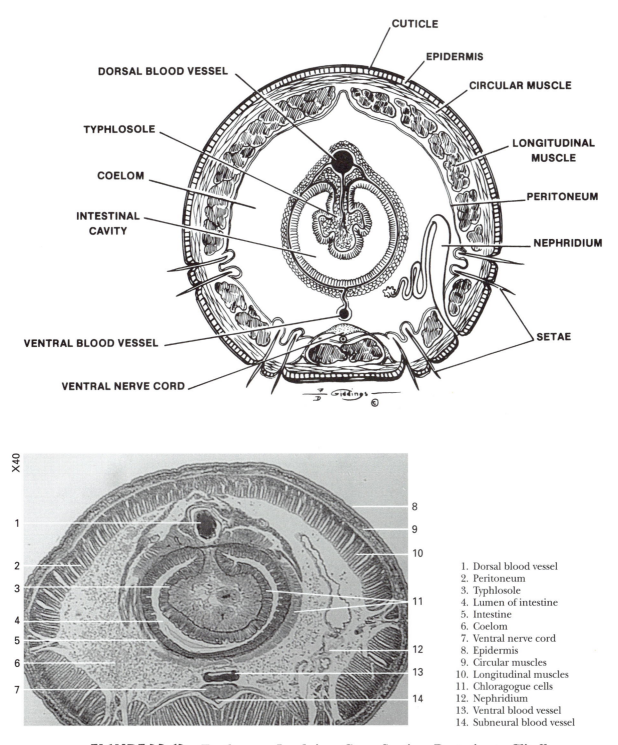

1. Dorsal blood vessel
2. Peritoneum
3. Typhlosole
4. Lumen of intestine
5. Intestine
6. Coelom
7. Ventral nerve cord
8. Epidermis
9. Circular muscles
10. Longitudinal muscles
11. Chloragogue cells
12. Nephridium
13. Ventral blood vessel
14. Subneural blood vessel

FIGURE 23.12 Earthworm *Lumbricus*, Cross Section, Posterior to Clitellum

c. **Circular Muscle Layer**: located just beneath the epidermis, with the fibers cut longitudinally in the section

d. **Longitudinal Muscle Layer**: these fibers are arranged in blocks of feather-like bundles extending toward the center; they are cut transversely

e. **Peritoneum**: a thin epithelial lining, separating the body cavity from the body wall.

f. **Coelom**: the body cavity

2. **Intestine**:

a. **Typhlosole**: dorsal invagination of the intestine

3. **Blood Vessels**:

a. **Dorsal Vessel**: just above the intestine

b. **Ventral Vessel**: just below the intestine

4. **Other Structures**:

a. **Ventral Nerve Cord**: positioned ventrally between the body wall and the ventral blood vessel

b. **Nephridia**: segmental excretory organs, within the coelom, between the intestine and the body wall; in these sections, only incomplete portions of nephridia can be seen, usually appearing as wavy lines.

c. **Setae**: 2 pair ventrally and 2 pair ventrolaterally projecting from the body wall.

REVIEW QUESTIONS

1. List the three major classes of flatworms, and give an example of each.

   a. _____

   b. _____

   c. _____

2. Why do tapeworms not need a digestive system? _____

   _____

3. Identify each of the following terms:

   Immature proglottid _____

   Mature proglottid _____

   Gravid proglottid _____

   Scolex _____

   Strobila _____

4. Distinguish between an incomplete and a complete digestive system, and give an example of an animal having each. _____

   _____

   _____

5. Distinguish between a pseudocoelom and a true coelom and give an example of an animal which possesses each. _____

_____

_____

6. List three examples of parasitic nematodes. _____

_____

7. List four distinguishing characteristics of the Phylum Rotifera. _____

_____

8. What is the function of the typhlosole in an earthworm? _____

9. What is the function of the seminal vesicles in an earthworm?_____

10. How many setae per segment are found in an earthworm?_____

11. What is the function of the earthworm's clitellum? _____

12. Diagram the digestive system of an earthworm, label all of its specialized structures, and give the function of each.

13. Describe four external features by which the ventral surface of an earthworm can be distinguished from the dorsal surface. _____

_____

_____

14. Describe three external features by which one can distinguish the anterior end of an earthworm from the posterior._____

_____

_____

15. Complete the following table:

	Porifera	Cnidaria	Platyhelminthes	Nematoda	Annelida
Level of organization	cellular	tissue			
Symmetry	radial	radial			
Tissue layer	none	epidermis gastrodermis			
Nervous system	none	none			
Body cavity	no	no			
Complete digestive tract	no	no			
Unique characteristic					

KINGDOM ANIMALIA

Arthropoda, Mollusca, and Echinodermata Phyla

EXERCISE 24 EXERCISE

OBJECTIVES

After completion of this exercise, the student should be able to do each of the following:

- Identify the phylum and class of each of the animals in the jars on display.
- Describe the characteristics contributing to the success of the arthropoda, particularly on land.
- List the three subphyla of the Phylum Arthropoda, list the classes belonging to each, and cite an example of each class.
- Dissect a crayfish and identify the parts listed in boldface print in this exercise.
- Distinguish between a male and a female crayfish.
- Identify each type of crayfish appendage.
- List the six classes of Mollusca and give an example of each.
- Identify the anatomical parts of a starfish listed in boldface in this exercise.
- Describe the operation of the **water vascular system** in a starfish.
- List the five classes of Echinodermata and give an example of each.
- List the features unique to each phylum and class in this exercise.
- Answer the review questions at the end of this exercise.

PHYLUM ARTHROPODA

The arthropods are probably the most successful group of animals ever to exist. About four-fifths of all living animal species are arthropods. Their success is attributed to a basic body plan characterized by **segmentation**, a hardened **exoskeleton** with **jointed appendages**, and a high degree of specialization in the brain and central nervous system allowing for a high degree of instinctive behavior. There is great diversity among the animals found in this phylum, though they are usually small in size. They are found in almost every conceivable environment.

The members of several of the classes of this phylum, such as the insects, centipedes, millipedes, and arachnids, are primarily terrestrial. They are better adapted to a land environment than any other invertebrates are, largely because of the following characteristics:

1. a **cuticle** which prevents water loss
2. efficient internal respiratory organs

193

3. **jointed appendages** with a hard, chitinous **exoskeleton**

Furthermore, some (insects) have developed wings, and their ability to fly has made possible their distribution over the earth. In addition, insects display great variation in specialized mouthparts allowing for chewing, biting, piercing, lapping, or sucking. Such variety allows many insects to share the same habitat without having to compete intensely for food.

Few arthropods are very large due to the restrictions of their exoskeletons. Because the exoskeleton surrounds the body, it must be **molted** periodically to allow for growth. Until the new exoskeleton hardens, the animal is helpless.

Below you will find a simplified scheme of classification for the Phylum Arthropoda. Refer to your textbook for a more detailed description of each taxon. Table 24.1 lists the general characteristics of the principal classes.

Phylum Arthropoda
    Subphylum Trilobita: no living representatives
    Subphylum Chelicerata
        Class Merostomata: horseshoe crabs
        Class Arachnida: spiders, mites, ticks, scorpions, harvestmen, and daddy longlegs
    Subphylum Mandibulata
        Class Crustacea: lobsters, crabs, crayfish, and shrimp
        Class Insecta: butterflies, bees, beetles, mosquitos, etc.
        Class Chilopoda: centipedes
        Class Diplopoda: millipedes

Examine the display jars containing arthropod specimens and attempt to determine the class to which each belongs using Table 24.1 as a guide to their characteristics.

CRAYFISH DISSECTION

Obtain a specimen of a crayfish, *Cambarus*. (See Figures 24.1, 24.2, and 24.3.

External Anatomy

Use Figure 24.1 for the external view. Note the **exoskeleton**. Anteriorly it forms the **carapace**, which

TABLE 24.1

Phylum Arthropoda: General Characteristics of the Principal Classes

Characteristic	Crustacea	Insecta	Arachnida	Chilopoda	Diplopoda	Merostomata
body divisions	usually cephalothorax and abdomen	head, thorax abdomen	cephalothorax and abdomen	head with body of similar segments	head, short thorax, long abdomen	cephalothorax and abdomen
paired appendages: antennae	2 pairs	1 pair	none	1 pair	1 pair	none
mouthparts (pairs)	mandibles–1 maxillae–2 maxillipeds–3	mandibles–1 maxillae–1 labia–1	chelicerae–1 pedipalps–1	mandibles–1 maxillae–2 maxillipeds	mandibles–1 maxillae–1 maxillipeds	chelicerae–1 pedipalps–1
legs	1 pair per somite or less	3 pairs on thorax	4 pairs on cephalothorax	1 pair per segment	2 pairs per segment	4 pairs on cephalothorax
gas exchange	gills or body surface	tracheae	book lungs and/or tracheae	tracheae	tracheae	book gills
principal habitat	salt or freshwater, few on land	mainly terrestrial	mainly terrestrial	all terrestrial	all terrestrial	marine

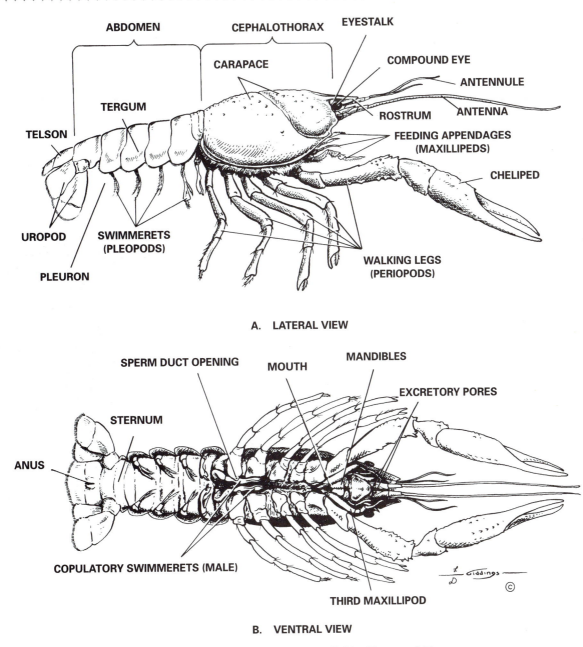

A. LATERAL VIEW

B. VENTRAL VIEW

FIGURE 24.1 *Cambarus* (Crayfish): **External Features**

covers the dorsal and lateral surfaces of the fused head and thorax, the **cephalothorax**. The posterior part of the body, or **abdomen**, is covered by segmentally arranged chitinous plates. These plates are named according to their positions: **tergum** = dorsal plate; **sternum** = ventral plate; and **pleuron** = lateral plate.

Examine the stalked **compound eyes**, a pair of **antennules**, a pair of **antennae**, the six pairs of mouth appendages, the large **claws** on the **chelipeds**, the **walking legs**, the **swimmerets** on the abdomen, and the broad **uropods** on the last abdominal segment. These, together with the **telson**, form the fan-shaped tail. All appendages are **serially homologous**. In the early development and in basic adult structure they are all alike, even though they often differ in detailed form and function. This basic structure may be examined in one of the swimmerets.

Remove the appendages on the left side of the crayfish one by one, starting at the posterior end and making sure that the whole appendage is taken off right at the base. Keep track of both the numbers

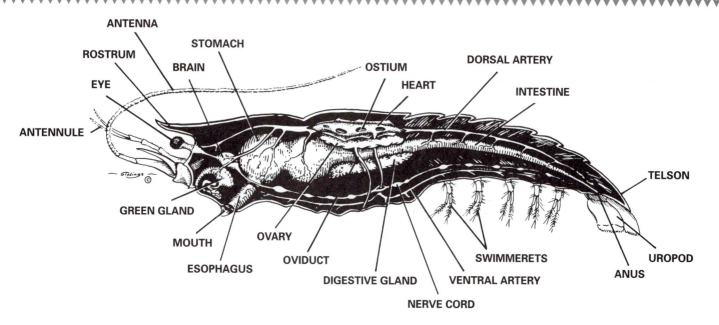

FIGURE 24.2   Longitudinal Section of Female Crayfish

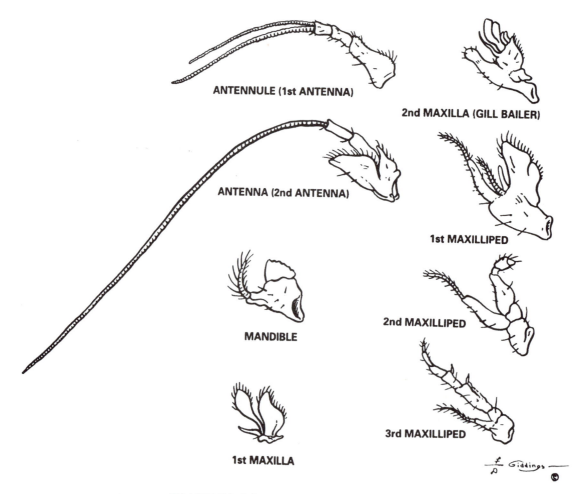

FIGURE 24.3   Crayfish: Head Appendages

and types of appendages removed, and lay them out in sequence on a sheet of paper. From the abdominal segments you should obtain, posterior to anterior, one telson, one uropod, and five swimmerets. In the female, the anteriormost swimmeret is small or absent; in the male, it functions in sperm transfer and is larger and anteriorly directed. To obtain the thoracic appendages, remove the left side of the carapace with scissors. This will expose a gill chamber and the **gills**, which are attached to all of the thoracic appendages except the first. Remove these thoracic appendages with their gills attached. There will be four similar walking legs and one large cheliped. Next remove the mouthparts. From posterior to anterior there are three **maxilipeds**, two **maxillae**, and one **mandible**. The maxillipeds are more obviously leg-like than the other mouthparts. They are used in sensory functions and in handling and tearing pieces of food. The maxillae are very thin and lie closely pressed to the hard, clublike mandible (jaw). Continue forward and remove in order the second antenna and the first antenna. The first antenna is referred to as an antennule also. Refer to Figure 24.3.

Note the opening at the base of the second antenna. This is the external opening of the **green gland**, the excretory organ of the crustaceans.

INTERNAL ANATOMY

Refer to Figure 24.2 in order to locate the internal organs. Carefully loosen the remainder of the carapace and the dorsal skeleton of the abdomen from the underlying membranous epidermis. Remove the exoskeleton and cut the epidermis to expose the internal organs. If muscles are in the way, do not tear them, but cut them with scissors. Cover the animal with water, and study the following:

The small **heart**, showing several openings, or **ostia**, is embedded in the pericardial cavity in the middorsal region. Anterior to the heart is the **stomach**, in the head region, a large sac containing a grinding structure, the **gastric mill**, in its wall. Follow the stomach posteriorly and trace the **intestine** to the **anus**, located ventrally in the last abdominal segment. Anteriorly, the stomach leads to a short **esophagus** which passes ventrally to the mouth. Locate the mouth and probe through it to the stomach. To each side of the stomach are the large yellowish **digestive glands**. Behind them, to each side of the heart, are the **gonads**. The **testes** are

difficult to distinguish from the digestive glands, but the **ovaries** are coarser in texture and darker in color (almost orange). Ducts from the reproductive organs lead to the exterior openings on the basal segments of the third pair of walking legs in the female, and of the fifth pair of walking legs in the male.

Starting in the abdomen and working forward, carefully remove muscles and other structures to expose the **ventral nerve cords** for their entire length. Note the segmental thickenings of ganglionic tissue. Try to identify the ring of nerve tissue encircling the esophagus and leading to the **brain ganglia**, which are the anterior portions of the ring. Also, locate the pad-like **green glands** (not green in preserved specimens) which lie beside the esophagus.

PHYLUM MOLLUSCA

The six classes of molluscs, containing an estimated 110,000 species, are represented by a tremendous variety of body forms, all derived from the same essential body organization. Molluscs are known as the "soft bodied animals."

Features unique to this phylum include:

1. A fleshy epithelial **mantle** that may secrete a calcareous shell.
2. A muscular ventral **foot**.
3. A dorsal **visceral body mass**.

Some additional characteristics of the phylum are:

1. An unsegmented body.
2. An open circulatory system.
3. Respiration by a single or many **ctenidia** (gills), by the mantle, or by the epidermis.
4. An extremely-reduced coelom.
5. Variety in feeding behavior and locomotion. The chiton, for example, is designed for algal grazing and adherence to wave-beaten rocks, the clam for filtering fine food material, the snail for gliding and protection, and the octopus for speed and predation.

See Figure 24.4 for examples of each of the molluscan classes, and examine the specimens on display in jars. Try to determine to which each belongs.

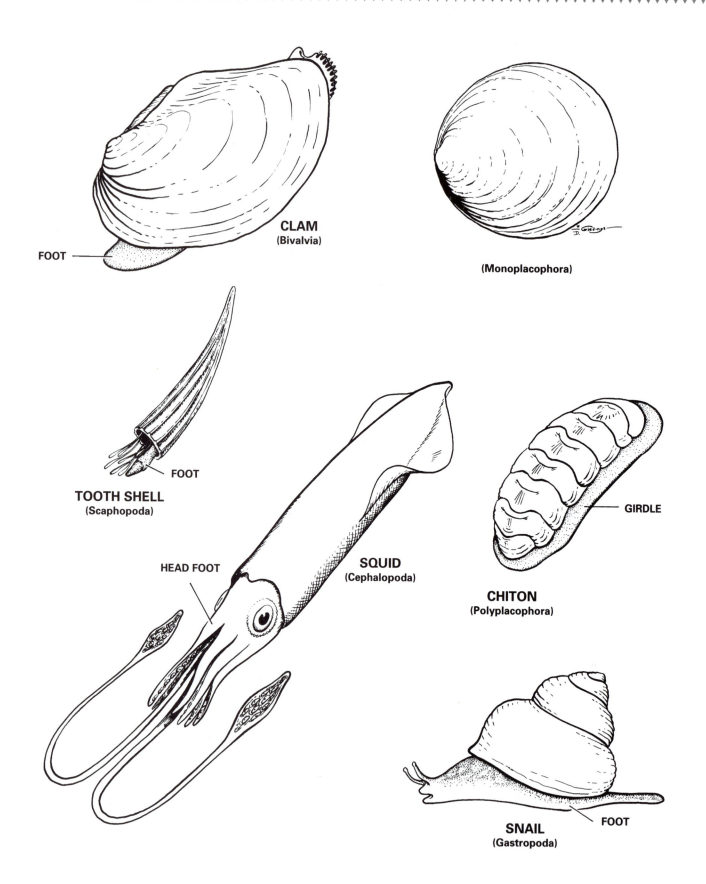

CLAM
(Bivalvia)

FOOT

(Monoplacophora)

TOOTH SHELL
(Scaphopoda)

FOOT

HEAD FOOT

SQUID
(Cephalopoda)

GIRDLE

CHITON
(Polyplacophora)

SNAIL
(Gastropoda)

FOOT

FIGURE 24.4 **Representative Molluscs**

CLASS POLYPLACOPHORA

These are the algal-grazing chitons occurring mainly in marine intertidal areas. Some preserved specimens are on display. Note the **foot**, **gills**, and the 8 linked calcareous plates that form a protective shell.

CLASS MONOPLACOPHORA

These are very primitive, somewhat segmented molluscs, known from 300 million year old fossils and from living specimens discovered in 1952 off the coast of Central America.

CLASS BIVALVIA

Bivalves, such as clams, oysters, mussels, and scallops comprise this group, all of which have shells comprised of two halves called **valves**. These dorsally hinged valves are tightly closed by well-developed adductor muscles. In general, bivalves are water-filtering organisms enclosed in shells.

CLASS GASTROPODA

This class includes snails, whelks, limpets, slugs, and nudibranchs. Except for the slugs and nudibranchs, most gastropods possess a single spiral shell. These molluscs have well developed heads with tentacles and a rasping **radula** — a structure which enables them to chew up vegetation.

CLASS CEPHALOPODA

The cephalopods are highly modified for motility and active predation. These, such as squid and octopus, possess a large head and eyes, a highly developed brain and central nervous system, 8, 10, or more arms equipped with rows of sucking discs, a mouth with horny beak and radula, and a large **siphon** for controlled, rapid movement. The arms, or tentacles, correspond to the foot in other molluscs.

CLASS SCAPHOPODA

The elephant-tusk shells are a small group adapted to life in mud or sand in marine waters. They are very seldom seen alive. Tusk, or tooth shells are identifiable by their tubular shells which are open at both ends.

DISSECTION OF A CLAM —
A Representative Bivalve Mollusc

EXTERNAL ANATOMY

Use Figure 24.5 as a guide. Examine the external shell. The more pointed end of the valves is the posterior end. Find the **posterior** and **anterior** ends. The valves or shells are secreted by the mantle and are hinged together dorsally by the **hinge ligament**. Locate the **dorsal** as well as the **ventral** regions of the clam.

INTERNAL ANATOMY

Refer to Figure 24.5 in order to locate the internal organs. Insert a scalpel between the lower edges of the valves and pry the valves apart enough to insert something to act as a wedge. Cut the **adductor muscles** which hold the valves together without damaging the internal organs. Separate the valves to expose the **mantle**.

Once opened, using the valve without the internal organs, locate the concentric **growth lines** on the outside and the remains of the adductor muscles. Also note the creamy texture (mother of pearl) on the inside of the valve.

On the other valve notice the adductor muscles and mantle which lies over the **visceral mass** and **foot** used for locomotion and digging. The visceral mass contains the digestive and reproductive organs. Locate the **incurrent** (more ventral) and **excurrent siphons** at the posterior end. The siphons function to draw water across the **gills** for respiration and feeding. Food in the incurrent water is trapped in mucus on the gills and is carried to the mouth by cilia on the gills for ingestion. Remove the mantle which covers the gills. Also carefully remove the left gills and slit the bottom of the foot to expose the **labial palps** which aid in moving the food trapped in mucus into the open mouth.

Parts of the digestive system are usually seen. Locate the posterior portion of the **intestine** and note that the **anus** discharges into the excurrent siphon. Trace the intestine forward. In the mid-dorsal region of the body, it will pass into the **pericardial cavity** or **sac** which contains the **heart**, also locate the **nephridium** or kidney which is beneath the pericardial cavity or **pericardium**.

ANODONTA (A FRESHWATER CLAM)

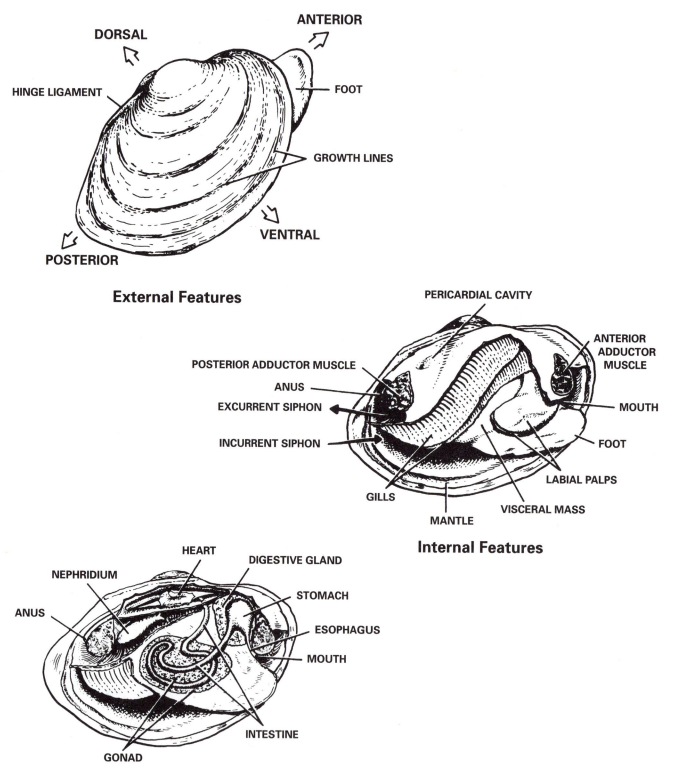

DORSAL

ANTERIOR

HINGE LIGAMENT

FOOT

GROWTH LINES

VENTRAL

POSTERIOR

External Features

PERICARDIAL CAVITY

ANTERIOR ADDUCTOR MUSCLE

POSTERIOR ADDUCTOR MUSCLE

ANUS

EXCURRENT SIPHON

MOUTH

INCURRENT SIPHON

FOOT

LABIAL PALPS

GILLS

VISCERAL MASS

MANTLE

Internal Features

HEART

DIGESTIVE GLAND

NEPHRIDIUM

STOMACH

ANUS

ESOPHAGUS

MOUTH

INTESTINE

GONAD

Dissection of the Visceral Mass

FIGURE 24.5   Clam

PHYLUM ECHINODERMATA

Included in this phylum are the spiny skinned animals, such as the sea lilies, starfish, brittle stars, sea urchins, and sea cucumbers. All of the species are marine. Adult forms are sessile or slowly creeping forms which are radially symmetrical around an oral-aboral axis. The larvae are free-swimming, bilaterally symmetrical forms. Echinoderms appear to have evolved from an ancestral line having bilateral symmetry such as the flatworm or some similar organism.

Features unique to this phylum include:

1. A **water vascular system**. This is a system of internal tubes communicating with the exterior by way of a sieve plate or madreporite, which regulates the amount of water in the system, and ending in a paired series of tube feet running the length of each ray, or along each section of the fused **endoskeleton**, or **test**, in sea urchins. The tube feet are extended by the contraction of muscular bulbs, **ampullae**, at their inner ends which force water into the tube feet, making them turgid. When the tube feet are brought in contact with a surface, the ampullae relax, permitting the echinoderm to adhere strongly to a surface without further use of energy. This enables them to withstand the crashing surf in the intertidal zone as well as to open the shells of bivalve mollusks for food.

2. Minute respiratory structures, skin gills, **dermal papillae**, or **dermal branchia**.

3. A calcareous endoskeleton of movable or fixed plates or ossicles.

4. Numerous hard spines that arise from the internal skeleton.

5. Many minute pincers, or **pedicellaria** which act to keep the body surface free of debris, aid in capturing food, and protecting the skin gills.

Some additional characteristics of the phylum are:

1. Ciliated organs.

2. Nervous system consisting of a circumoral ring and radial nerves to the arms.

3. Lack of cephalization.

4. Complete digestive tract.

5. Lack of segmentation.

6. Open circulatory system.

Living representatives are divided into 5 classes. Refer to Figure 24.6. Also examine the many specimens on display.

CLASS ECHINOIDEA

These spiny sea urchins and sand dollars are herbivorous echinoderms constructed as though their arms were folded back into a ball and fixed into a calcareous skeleton or test, then covered with long, sharp, movable spines and 3-jawed pedicellaria. The test is globular in sea urchins, and disc or heart-shaped in sand dollars. Tube feet are long, slender and equipped with suckers; mouth and anus are central or lateral. The large gut fills much of the test cavity, except during spawning periods.

CLASS HOLOTHUROIDEA

Sausage-shaped garbage collectors, the sea cucumbers are among the chief clean-up organisms of the ocean floor. With their oblong shape and warty skin, sea cucumbers are well-named. They vary in length from an inch to several feet, with body wall consistency from leathery to papery. Arms, spines, pedicellaria, and endoskeleton, except for scattered tiny plates in the body wall, are all absent. Tube feet are present. The mouth with tentacles is at one end of the body and the anus is at the other, the latter often bearing a complex called the **respiratory tree**, which functions in gas exchange.

CLASS CRINOIDEA

These stalked, flowerlike sea lilies and feather stars have 5 arms which display up to 10 or more branches, each bearing 5 branchlets or pinnules to form a cuplike central disc. No spines, pedicellaria, or suckers arise from tube feet lining the open ambulacral grooves. In a sea lily, the long jointed stalk with rootlike projections may attach the animal to the substrate. In feather stars, the adult may lack a stalk and be free-swimming with motile, gripping cirri and a mouth and anus on the upper, **oral**, surface.

CLASS OPHIUROIDEA

The brittle stars have a central disc to which highly flexible, jointed limbs are attached. Tube feet, confined to 2 rows and lacking ampullae, have a sensory function. Pedicellaria and anus are lacking; the madreporite is aboral.

CLASS ASTEROIDEA

This class includes the predaceous star-shaped or pentagonal sea stars. Some species, however, have as many as 50 arms. The ossicles are separate, permitting movement; short spines and pedicellaria are present. The oral surface is ventral, with 2 or 4 rows of tube feet lining the open ambulacral grooves in each arm; the madreporite is **aboral**. Sea stars are often called starfish, though this is an obvious misnomer. Fish are vertebrates.

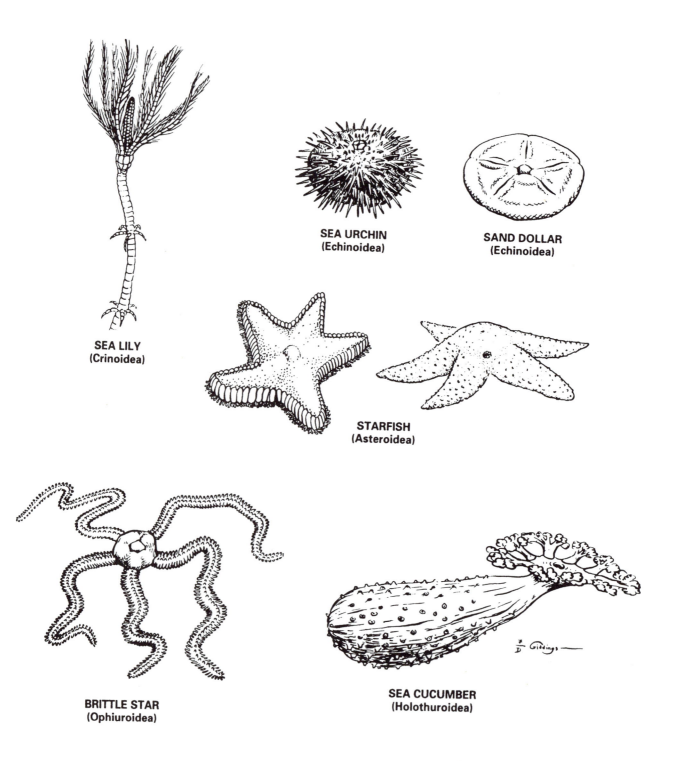

SEA URCHIN
(Echinoidea)

SAND DOLLAR
(Echinoidea)

SEA LILY
(Crinoidea)

STARFISH
(Asteroidea)

BRITTLE STAR
(Ophiuroidea)

SEA CUCUMBER
(Holothuroidea)

FIGURE 24.6 Representative Echinoderms

DISSECTION OF STARFISH —
Asterias forbesii

EXTERNAL ANATOMY

Use Figure 24.7 as a guide. Keep the specimen wet by adding some water to the dissecting pan. **Use the stereomicroscope** to aid you in these observations. Note the central disk and five arms, or **rays**. On the upper, **aboral**, surface locate the **madreporite**, a bright colored area near the edge of the disc at the junction of two rays. Can you find the red **eyespot** at the tip of each arm? This structure is difficult to see in preserved specimens. Observe the hard **spines** scattered over the surface. Located among the spines are the **dermal branchiae** and pincerlike **pedicellaria**.

On the oral or ventral surface, locate the **mouth**, surrounded by large spines. Is there any material protruding from the mouth? If so, what might it be? Running along the middle of each ray is an **ambulacral groove**, with rows of **tube feet**. Identify the suction cups at the ends of the tube feet.

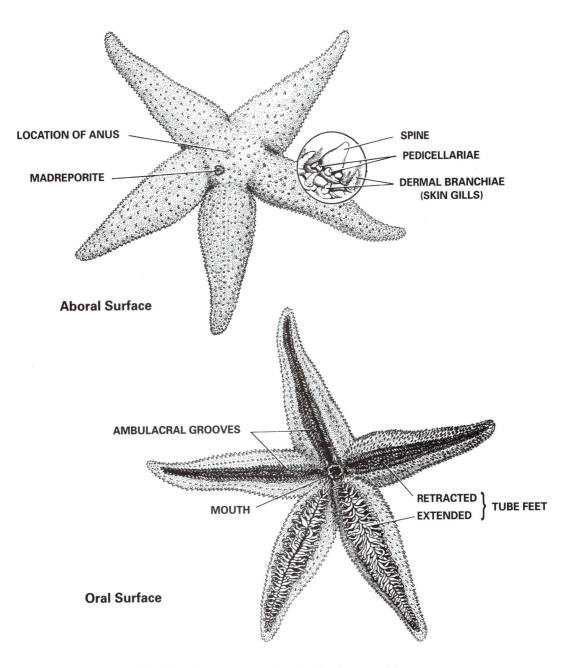

FIGURE 24.7 *Asterias* (Starfish): **External Features**

INTERNAL ANATOMY

Cut off about one-half inch of the top of the ray farthest from the madreporite and cut through the aboral wall of the ray along each side toward the central disc. Repeat this on an adjacent arm. Continue cutting around the central disc such that a circular area is removed from the disc, leaving only the madreporite and anus. Work carefully so that the delicate organs beneath are not macerated. Then, beginning near the tip of the ray, **carefully** remove the aboral skeleton in small sections, lifting and freeing it from the underlying tissue before actually cutting off the skeleton. Carefully examine Figure 24.8 showing a partial dissection.

The **mouth** leads into a short esophagus which is connected to a much-folded, sac-like **cardiac stomach**

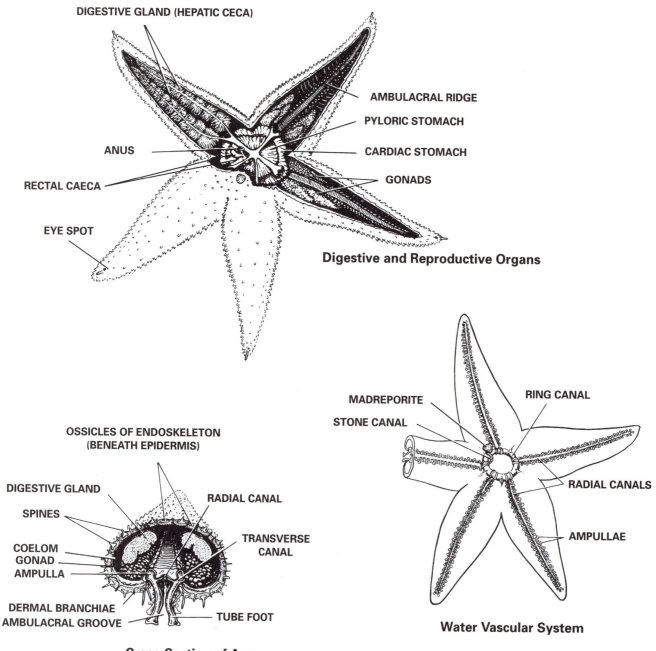

DIGESTIVE GLAND (HEPATIC CECA)

AMBULACRAL RIDGE
PYLORIC STOMACH
ANUS
CARDIAC STOMACH
RECTAL CAECA
GONADS
EYE SPOT

Digestive and Reproductive Organs

OSSICLES OF ENDOSKELETON (BENEATH EPIDERMIS)

DIGESTIVE GLAND
RADIAL CANAL
SPINES
TRANSVERSE CANAL
COELOM
GONAD
AMPULLA
DERMAL BRANCHIAE
AMBULACRAL GROOVE
TUBE FOOT

Cross-Section of Arm

MADREPORITE
RING CANAL
STONE CANAL
RADIAL CANALS
AMPULLAE

Water Vascular System

FIGURE 24.8 *Asterias* (Starfish): Internal Features

which is the portion that can stick out through the mouth of the starfish and start digesting the contents of an oyster. Above the cardiac stomach is another portion of the stomach known as the **pyloric stomach**. Connected to this are five **digestive glands**, each of which is located in a ray. Dorsal to the pyloric stomach are several **rectal caeca** plus a short intestine attaching the stomach to the **anus**. The rectal caeca and anus are difficult to find.

A pair of **gonads** is located in each ray. They are located underneath the digestive glands and are usually a deep brownish-red color and smaller in size than the digestive glands.

After finding the gonads and digestive glands, remove them from one of the rays. Notice the skeletonous **ambulacral ridge** running down the center of the ray. Along each side of the ridge are tiny reddish bulblike structures — these are the **ampulla**. Also, near where the arms are attached to the central disc, find the **two bands of muscle** per arm, responsible for movement of the arms.

To see the **water vascular system**, use Figure 24.8. Remove the stomach and try to find the **stone canal** leading orally from the madreporite to the **ring canal**, a hard ring around the mouth, from which branch 5 **radial canals** (one into each arm). Split the ambulacral ridge lengthwise to find the radial canal. It is connected to the **ampulla** and **tube feet** via short lateral canals.

REVIEW QUESTIONS

1. List the functions of the following appendages of arthropods:

   Cheliped _____

   Antenna _____

   Wing _____

   Maxilliped _____

   Maxilla _____

   Mandible _____

   Swimmeret _____

   Uropod _____

2. Use your textbook to help you find the function of the green glands in the crayfish. _____

   _____

3. What is the literal meaning of the word "arthropoda"? _____

4. In what respects do the Chelicerata and the Mandibulata differ? _____

   _____

5. What is the function of the water vascular system found in starfish? _____

   _____

6. Identify each of the following structures with respect to function and the phylum with which it is associated:

Radula _____

Ampulla _____

Madreporite _____

Siphon _____

Compound eye _____

Respiratory tree_____

7. What four characteristics have contributed to the success of Arthropods?

a. _____

b. _____

c. _____

d. _____

8. What anatomical differences help the arthropods to survive away from water?

a. _____

b. _____

c. _____

d. _____

9. All members of the Phylum Mollusca have 3 unique features. List these below.

a. _____

b. _____

c. _____

10. MATCHING: Match the class with the correct organism.

Common Name   **Class**

_____ Chitons   a. Gastropoda

_____ Squid, octopus   b. Bivalvia

_____ Oyster, clam   c. Cephalopoda

_____ Segmented and primitive   d. Polyplacophora

_____ Snails   e. Monoplacophora

_____ Elephant tusk shells   f. Scaphopoda

KINGDOM ANIMALIA
HEMICHORDATA AND CHORDATA PHYLA

EXERCISE 25

OBJECTIVES

After completion of this exercise, the student should be able to do each of the following:

- Explain the evolutionary relationships thought to exist between the echinoderms, hemichordates, and chordates.
- Identify the **proboscis, collar, trunk,** and **pharyngeal gill slits** in a prepared slide of *Balanoglossus.*
- Identify the phylum and class to which each of the animals in the jars on display belong.
- List the features unique to each phylum, subphylum, and class discussed in this exercise.
- List at least four adaptations exhibited by reptiles for life on land.
- List four specialized adaptations for flight exhibited by birds.
- Identify the chordate features which are lost by tunicates in the transition from larva to adult.
- Identify the **buccal cirri, dorsal nerve cord, notochord, gill bars, myotomes, atrium,** and **atriopore** in a prepared slide of an amphioxus.
- Identify any of the parts given in boldface in the dissection of the frog and give the function of each.
- Answer the review questions at the end of this exercise.

PHYLUM HEMICHORDATA

Present-day hemichordates are wormlike types known as **enteropneusts,** or **acorn worms.** They can be found burrowing in the sand and mud of tidal flats and shallow coastal waters. *Balanoglossus* and *Dolichoglossus* are common genera along the North American coasts.

Examine the plastomounts, slides, and preserved specimens in the jars on display. The body of the worm is divided into an anterior muscular **proboscis,** a short **collar,** and a posterior elongate **trunk.** **Pharyngeal gill slits** can be seen in the anterior region of the trunk. This is a characteristic hemichordates share with the chordates, along with the possession of a **dorsal nerve cord.** However, in hemichordates, the nerve cord is very short and is restricted to the collar region. A cartilaginous rod is also found in both hemichordates and chordates, but in the former it does not underlie the nerve

207

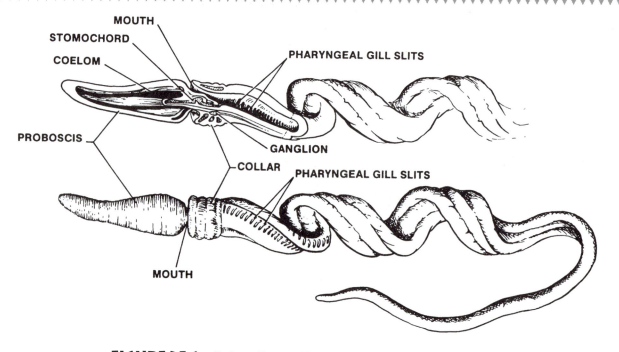

FIGURE 25.1 *Balanoglossus.* **(External View and Sagittal Section)**

cord as it does in the latter. For this reason, it is given the name **stomochord** in the hemichordates, while it is called a **notochord** in the chordates. Refer to Figure 25.1.

Acorn worms begin life as bilaterally symmetrical, ciliated larvae very much resembling those of the echinoderms. And their coeloms are divided, again a characteristic more resembling echinoderms than chordates. It is for all of the above reasons that the hemichordates, once classified as a subphylum of the Phylum Chordata, are now placed in a separate phylum by most taxonomists.

PHYLUM CHORDATA

This is the most advanced phylum, including a tremendous variety of organisms ranging from wormlike animals up to and including man. Despite the variation, all chordates are characterized by three unique features at some stage in their life histories. Sometimes these features may appear only in the embryonic life and disappear before the organism matures. These three features are as follows:

1. A **notochord**, a flexible supportive rod extending the length of the body dorsal to the digestive tract and furnishing skeletal support. The notochord is replaced by a backbone in the subphylum Vertebrata.

2. A **hollow nerve tube**, located **dorsal** to the notochord.

3. **Pharyngeal gill slits**, which function in respiration in the lower chordates but appear only in the developing embryo of higher chordates.

The three chordate subphyla are Subphylum Urochordata, Subphylum Cephalochordata, and Subphylum Vertebrata.

SUBPHYLUM UROCHORDATA

These are the tunicates or sea squirts. They are highly modified, sessile, filter-feeding animals with motile larvae. The larva shows all three chordate characteristics, but in the transition to adulthood, all of the notochord and most of the nerve cord are reabsorbed with the tail. An unusual feature of the adult is that the tunic surrounding the animal contains **cellulose**, a substance usually associated with plants. As in the acorn worms, the gills serve in feeding as well as in gas exchange. Food particles carried in by water entering the **incurrent siphon** are trapped in mucus on the **gill bars** as the water goes through the **gill slits** into the **atrium**. The water then leaves the atrium via the **excurrent siphon**. Refer to Figure 25.2 as you examine plastomounts and specimens of *Molgula* and its relatives in the jars on display.

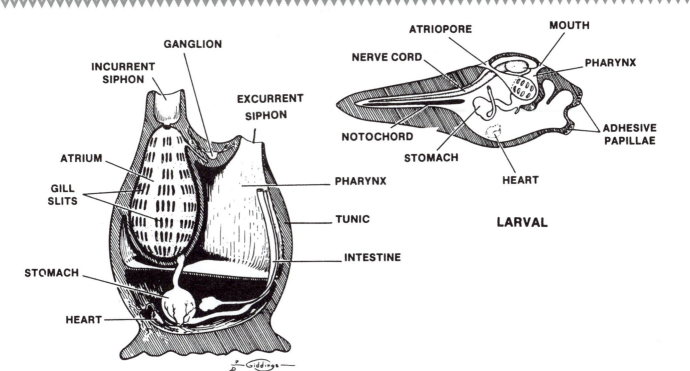

FIGURE 25.2 The Larval and Adult Tunicate

SUBPHYLUM CEPHALOCHORDATA

The lancets, or *Amphioxus*, are small, fishlike, mud or sand-dwelling filter feeders, important from a theoretical standpoint. They represent an evolutionary stage in which chordate characteristics are well developed, but vertebrate characteristics are absent. *Amphioxus* is a chordate of great interest, a prototype of how an ancestral vertebrate might have looked. Except for annelidlike ciliated nephridia, the lancet's organ systems are similar to those of vertebrates or other simple chordates.

Study the internal anatomy of *Amphioxus* by examining a stained slide specimen, while referring to Figure 25.3. Also examine the specimens in plastomounts and display jars.

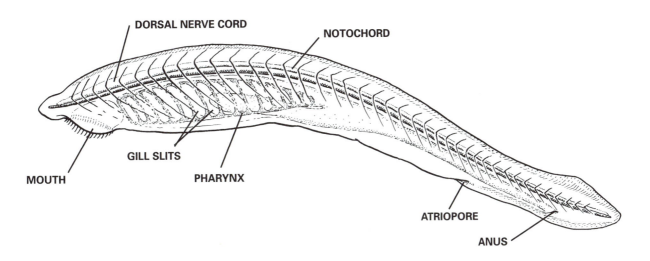

FIGURE 25.3 *Amphioxus*

SUBPHYLUM VERTEBRATA

This is the largest group of chordates and consists of animals in which the notochord is replaced by a vertebral column as the primary supporting structure. They have a brain consisting of five hollow vesicles, a jointed endoskeleton with a cranium to enclose the brain, a backbone of segmented vertebrae, and the concentration of sensory organs in the head region, a phenomenon known as **cephalization**.

The subphylum vertebrata includes Class Agnatha, Class Placodermi, Class Chondrichthyes, Class Osteichthyes, Class Amphibia, Class Reptilia, Class Aves, and Class Mammalia. (Refer to Figure 25.4.)

Agnatha

These jawless and limbless fishes are the most primitive vertebrates. The first records of vertebrates from early fossil deposits show they were abundant members of the sea-bottom fauna about 400 or more million years ago. Fossil agnatha (*Ostracodermi*) were characterized by heavy bony armor and probably a mud-sucking, filtering mode of feeding. Their highly specialized descendants, the lampreys and hagfishes (*Cyclostomata*), are still in existence. Survival of modern remnants of this ancient class is probably related to their specially adapted rasping and bloodsucking mouthparts, which enable them to feed on other fish. This parasitic way of life frees the lamprey from competition with later-evolved forms. Larval lampreys, or **ammocoetes larvae**, are good illustrations of the primitive vertebrate form, being strikingly similar to amphioxus.

Besides lacking jaws as adults, agnathans lack paired appendages, have a cartilaginous skeleton, and a single unpaired nostril. Examine the specimens on display.

Placodermi

These armored fish are now entirely extinct. The class is marked by development of a primitive type of jaw suspension that enabled the placoderms to replace the more primitive jawless fishes, although they themselves died out in the Permian period approximately 240 million years ago.

Chondrichthyes

This class of cartilaginous fish includes the sharks, rays, skates, and chimaeras. It is doubtful whether these fish preceded true bony fish in evolution, as is often assumed. Perhaps the two groups evolved in parallel fashion from placoderm ancestors.

The all-cartilaginous skeleton appears to be a comparatively recent adaptation (though still several hundred million years old). Another important adaptation in the group is the capacity to retain **urea** in the blood and body tissues, thus raising the internal osmotic pressure to a point nearly equal to that of the salt-water environment.

Chondrichthyes can be distinguished from agnathans by their smaller number of gill openings (5 pair), their paired nostrils, and their paired **pectoral** and **pelvic fins**. Another feature worthy of note is that their skins are covered by tiny **placoid scales**, structures which are tooth-like in their internal anatomy and which are thought to be the evolutionary precursors of the teeth, scales, hair and feathers characteristic of the following classes. Examine the specimens of chondrichthyes on display.

Osteichthyes

The enormously varied and numerous bony fish represent some 75 percent of described vertebrate species, including nearly all freshwater fish and the preponderance of marine species. The most distinctive external feature of this group is the possession of a gill cover, or **operculum**, operated by muscles and allowing the aeration of the gills in a bellows-like fashion while the fish is sitting still. The scales of bony fish are more elaborate and larger than those of the chondrichthyes.

One subclass of bony fish of special importance to students of evolution is the Choanichthyes (lobe-finned fishes), fish with nostrils connected to the mouth cavity and with paired fleshy or limb-like fins. These are presumed precursors of the amphibians and very ancient groups which have changed little since. Familiar lobe-finned fishes include the lungfish and the remarkable "living fossil," *Latimeria*, discovered in the fifties off the coasts of South Africa and Madagascar. Until the discovery, this fish was presumed to have been extinct for over 75 million years. Examine the specimens of bony fish on display.

Amphibia

Salamanders, frogs, toads, and caecilians (or apoda-tropical, limbless amphibians) were the first tetrapoda (possessing four limbs), or land-dwelling, limbed vertebrates. Amphi-, meaning "both," implies life both in water and on land.

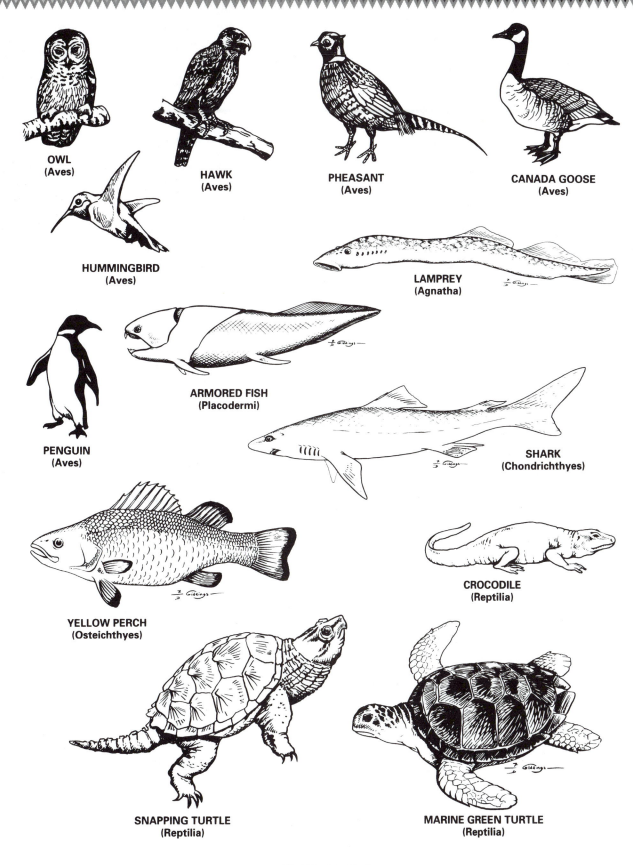

FIGURE 25.4
Representatives of the Classes of Subphylum Vertebrata
(Amphibians and Mammals Not Included)

Amphibians require abundant moisture, as their soft skin is not protected against water loss. Nearly all amphibians depend on water for egg laying and early development, though some frogs breed in damp moss instead of water. Certain toads are remarkably resistant to dessication; other amphibians such as the Mexican axolotl and a few frog species never leave the water. In general, we think of amphibians as transitional creatures, perfectly adapted neither to water nor to land, but somewhat adapted to both. Organisms like these could have been the ancestors of purely terrestrial animals.

Some characteristics of amphibians include:

1. Skeleton largely bony.

2. Ribs, if present, are not attached to the sternum.

3. Three chambered heart consisting of two **atria** and one **ventricle**.

4. Gas exchange by gills, lungs, skin, or lining of the oral cavity.

5. 10 pairs of cranial nerves.

6. **Poikilothermy** (dependence of body temperature on environment)

7. Fertilization external or internal. Mostly **oviparous** or egg layers. Eggs without shells.

8. **Holoblastic cleavage**, in other words, the whole zygote divides.

In a later section we will study the anatomy of the familiar leopard frog, *Rana pipiens*, a typical member of this class.

Reptilia

Snakes, lizards, turtles, crocodiles, and alligators, as well as numerous extinct forms (many different groups are linked by the common name "dinosaur") are included in this class. Reptiles are thought to have evolved from amphibian ancestors called labyrinthodonts. These organisms represent the completed transition from water to land. They are successful on land for a number of reasons:

1. The dry scaly skin helps to conserve moisture.

2. Internal fertilization to improve egg survival.

3. They form **amniotic eggs**, in which the embryo is enclosed in a fluid-filled sac. These are covered by leathery shells.

4. They have lungs for gas exchange in an air medium.

5. A free larval stage is lacking.

6. Nitrogenous wastes are excreted primarily as **uric acid**, an insoluble solid which can be stored within the egg shell.

7. More efficient limbs and superior muscular coordination.

8. Poikilothermy.

9. 12 pairs of cranial nerves.

10. Three chambered heart consisting of 2 atria, and a partially chambered ventricle, except in crocodiles and alligators which have a completely partitioned ventricle or two ventricles.

Aves

Birds have retained many of the features of the reptiles, from which they evolved, such as the formation of an amniotic egg, lungs, and internal fertilization. However, birds differ from reptiles in that they are **homeothermic**, that is, they can maintain a constant body temperature in the face of environmental fluctuation and thus can remain active throughout the year. A covering of modified scales, known as **feathers** aids in preventing heat loss by creating a lightweight layer of insulation. Birds also have four-chambered hearts.

Certain of the adaptations of birds are specializations for flight. Among them are:

1. The development of **pneumatic** (hollow) bones connected by way of air sacs to the lungs makes the bird much lighter.

2. The development of a one-way system for ventilation of the lungs which makes more rapid gas exchange possible.

3. The alteration of the forelimbs into **wings**.

4. The development of powerful **pectoral muscles** for operation of the wings.

5. The development of a **high-keeled breastbone** to increase the area available for muscle attachment.

6. The development of **uncinate processes** on the ribs which strengthen the ribcage and provide for more muscle attachment.

Examine the pigeon skeleton on display for the skeletal modifications listed above. Also examine the preserved specimens of birds in the jars on display.

Mammalia

Mammals evolved from a different subclass of reptiles than that which gave rise to the birds. Like birds, mammals are homeothermic. Gas exchange involves lungs assisted by a muscular **diaphragm**, and fertilization is internal. Some mammals, such as the duck-billed platypus and the spiny anteater lay eggs with shells, that is, they are **oviparous**. But most are **viviparous**, or bring forth live young. Mammals also have four-chambered hearts.

The distinguishing features of this class include:

1. The possession of modified scales known as **hair**, which can function as an insulating layer.

2. The production of milk by modified sweat glands known as **mammary glands**, providing a source of food for the young.

3. The possession of a highly developed nervous system.

4. The possession of a muscular diaphragm.

Examine the specimens of mammals in the jars on display. Then examine all jars containing vertebrate specimens and attempt to determine the class to which each belongs on the basis of its observable external features.

DISSECTION OF *Rana Pipiens*, LEOPARD FROG

The leopard frog is the species most frequently used for laboratory work. The bullfrog, *Rana catesbeiana*, will have been dissected for display in the lab. Use the bullfrog for comparison with the leopard frog.

EXTERNAL ANATOMY

Notice that the body of the frog can be divided into a **head** and a **trunk**. A distinct neck is absent (a characteristic retained from fish, since the presence of a neck would allow for independent movement of head and trunk — an unnecessary movement when swimming).

A large **mouth**, a pair of **external nares** (nostrils), and **eyes** are located on the head. Associated with the eye are an upper **eyelid** that is just a fold of skin and a lower eyelid, a transparent membrane, that can be extended over the surface of the eyeball. On the head, behind the eye, is a disc-shaped structure, the **tympanic membrane**; which functions as the eardrum.

Examine the front leg of the frog, noting the **upper arm**, the **elbow**, the **forearm**, the **wrist joint**, and the **hand** with 4 **fingers**. The finger closest to the body is thicker than the others and is therefore considered to be the thumb. It becomes swollen and deeply pigmented in the male during the breeding season. Looking at the hind leg, note the **thigh**, **knee**, **shank**, **ankle joint**, and the long **foot** with 5 toes. Just dorsal to where the hind legs join at the posterior end of the trunk is the opening common to both the digestive and the urogenital systems. This opening is the **cloacal opening**.

INTERNAL ANATOMY

Open the mouth, cutting through the angle of the jaw on each side, extending the cut posteriorly to the tympanic membrane. Pull the floor of the mouth ventrally, exposing the **buccopharyngeal cavity**. This cavity consists of the **buccal** (oral) **cavity** in the mouth and surrounded by jaws and the **pharynx**, which is a region between the buccal cavity and the esophagus, a tube leading to the stomach. With the help of Figure 25.5 locate and study the following structures: **maxillary teeth** on the margin of the upper jaw; the **vomerine teeth** located on the palate or roof of the buccal cavity (the teeth are used for holding food); the **internal nares**, openings from the nostrils; the **eustachian aperture** (opening) into the eustachian or auditory tube; the **vocal sac apertures** in males; the **glottis**, an opening into the trachea which leads to the lungs; the **esophagus** lying dorsal and posterior to the glottis; the large **tongue** which is attached at its anterior end.

Place the frog ventral side up in a dissecting pan. Fasten the frog to the pan with pins through the tip of the snout and all the limbs. Pick up the skin with forceps and make a longitudinal incision through it, about one-tenth inch to one side of the median line, from the cloacal opening to the lower jaw. Then cut the skin at right angles to the first incision, making these cuts just behind the forelimbs and just in front of the legs. Turn back the two skin flaps and pin them to the pan. Note how loosely the skin is attached to the underlying muscles and how evident the blood vessels are in the underside of the skin.

The dark line of the midventral region of the abdominal muscle layer is the **abdominal vein**. Keep this and other major blood vessels intact as long as possible. Make a longitudinal incision in the body wall on each side of this vein from the **pelvic girdle** to the **pectoral girdle** (the chest area). Make another set of right angle incisions and pin back the muscle flaps. Lift up the pectoral girdle, freeing the structures adhering to it, and cut through the bone.

The **coelom**, or body cavity, is now exposed. It is lined by a shiny layer of epithelial tissue,

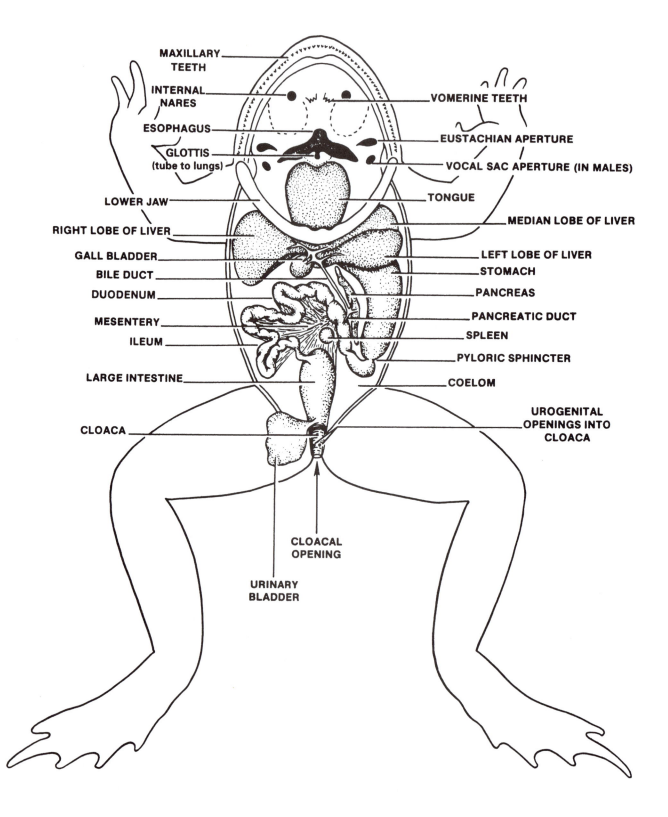

FIGURE 25.5   Buccal Cavity and Digestive System

peritoneum, which extends from the middorsal and midventral lines of the body to the organs. The membrane suspending the organs is called the **mesentery**. The coelom allows for the movement of the internal organs; the mesenteries restrict the movement of the organs and also contain the blood vessels and nerves involved with the organs. The frog's coelom is divided into two regions; the large **pleuroperitoneal cavity** in which respiratory, digestive, reproductive and excretory organs are located and the smaller **pericardial cavity** in which the heart is located.

Digestive System

Pass a probe down the esophagus and feel with your fingers when the probe enters the **stomach**, a large, sac-like structure in which food is stored and digestion is begun (refer to Figure 25.5). At the posterior end of the stomach is a thick muscular region, the **pyloric sphincter**, which regulates the passage of food from the stomach into the small intestine. Cut the stomach open and note the longitudinal folds. When the pyloric sphincter relaxes, food is allowed to move into the **duodenum**, the first section of the **small intestine**. Cut this tube open and note the folds here also. In addition to secretions from mucus glands in the lining of the small intestine, secretions from the **pancreas** (a whitish tissue lying in the loop between the duodenum and the stomach) enter the duodenum via the **pancreatic duct**. Secretions from the **liver** (a large, three lobed organ in the anterior part of the cavity) and the **gall bladder**, a storage sac for bile, located on the dorsal surface of the liver, enter the duodenum via the **bile duct**. Following the length of the long, thin, coiled small intestine, note that it joins the short fat **large intestine**, which joins with the **cloaca**. (To see this, carefully cut through the bony pelvic girdle and spread the hind legs apart.)

Respiratory System

Find the **glottis** and pass a probe into it until you can feel the probe when it enters the **lungs**. Use Figure 25.5 to help you find the lungs. In preserved frogs, the lungs are contracted to some degree. Enlarge the glottis by making a short longitudinal cut extending anteriorly and posteriorly from it. Spreading it open, you should be able to see a longitudinal fold on each side within the chamber; these are the **vocal folds** used in croaking. The openings to the lungs are beneath these folds.

Urogenital System

The excretory and reproductive organs are closely associated and together they comprise the urogenital system. Before you start to dissect your specimen, refer to Figure 25.6 and Figure 25.7 in order to determine the sex. The excretory structures are similar in the two sexes of the frog.

A pair of elongated organs, the **kidneys**, are directly attached to the dorsal body wall anterior to the cloaca. They are covered ventrally by the peritoneum. Each kidney is composed of several thousand tubules which filter waste material from the blood. Urine leaving the kidney is transported through the **ureter** to the dorsal surface of the cloaca. The **urinary bladder** is attached to the ventral surface of the cloaca, thereby necessitating the flow of urine across the cloaca before being stored in the bladder. On the ventral surface of the kidney is a light-colored band of tissue, the **adrenal gland**.

In the male (refer to Figure 25.6), find two small yellowish structures attached by short mesenteries to the ventral surface of the kidneys. These are the **testes**, the site of sperm production. Sperm are transported from the testes via tiny tubules, vasa efferentia, to the kidneys through which they pass before entering the **ureter**. The sperm are carried down to the end of the ureter where they are stored temporarily in an expanded portion called the seminal vesicle. At the time of **amplexus**, the mounting of the female by the male, sperm are released through the male's cloaca at the same time as the female releases the eggs. Fertilization is external. In the leopard frog note the rudimentary **oviducts** extending laterally alongside the kidneys. These are not found in the bullfrog.

In the female (refer to Figure 25.7), find two multi-lobed organs attached by short mesenteries to the ventral surface of the kidneys. These are the **ovaries**, the site of egg production. The ovaries vary in size according to the season. If ripe, the eggs will be black in color, with a little white on one surface. The **oviducts** are coiled tubes extending from the anterior part of the coelom to the cloaca. The anterior end has a small opening, the ostium, through which eggs released into the pleuroperitoneal cavity at the time of **ovulation** enter the oviduct. In traveling down the oviduct they become covered by layers of a gelatinous substance. They may be stored temporarily in an expanded area of the oviduct near its junction with the cloaca. The storage site is known as the **ovisac** or **uterus**. The sexual embrace of the male stimulates the release of the eggs from the ovisac into the cloaca and out into the water.

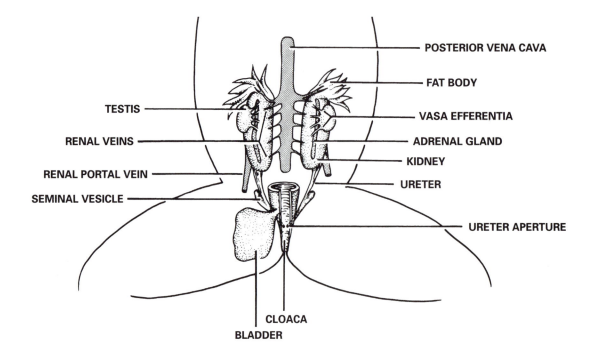

FIGURE 25.6   Male Urogenital System

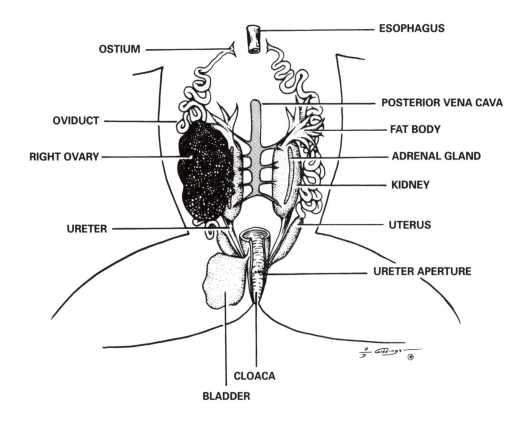

FIGURE 25.7   Female Urogenital System

Anterior to both the testes and the ovaries are **fat bodies**; these are yellowish structures with long, finger-like extensions. They function in storage of food material used during hibernation.

Circulatory System

This system consists of a **heart**, **blood vessels** (arteries, capillaries, veins), **spleen** (a small round structure on the left side of the mesentery supporting the intestine; functions in the manufacture and storage of blood cells), and **lymphatic network**. In this lab we will not make a detailed study of this system. Instead, familiarize yourself with the information given below, using Figures 25.8 and 25.9 as a guide to find the structures mentioned.

Heart

Carefully remove the **pericardium**, the membrane enveloping the heart, to expose the chambers of the heart and the blood vessels associated with it. Identify the cone-shaped muscular **ventricle** at the posterior end of the heart. Blood moves from the **ventricle** through the **conus arteriosus**, which divides

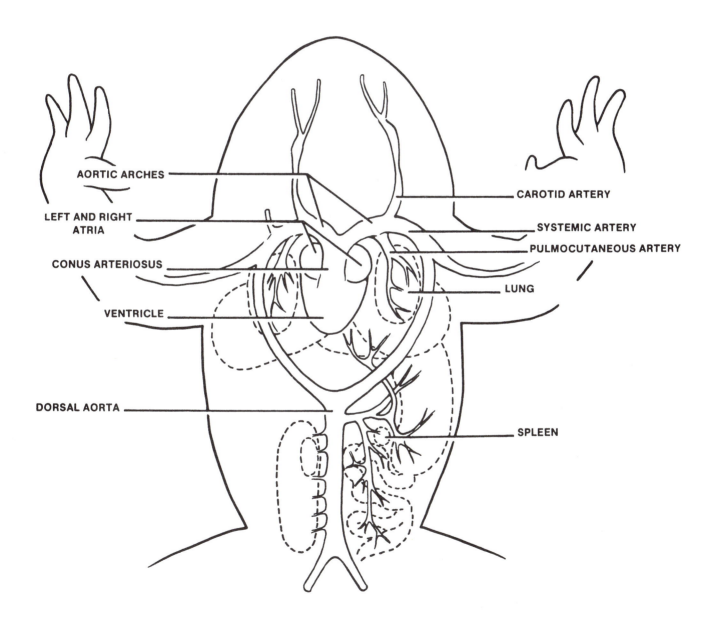

FIGURE 25.8 Arterial System: Ventral View

into a **right** and **left aortic arch**. On either side of the conus arteriosus are two thin walled chambers, the **atria**. Lift the ventricle toward the head, and note the dark thin-walled **sinus venosus** on the dorsal side. The sinus venosus is not in Figure 25.8 or Figure 25.9 but it drains into the right atrium. Entering the posterior end of the sinus venosus is the **posterior vena cava**, and entering on either side anteriorly is a pair of smaller **anterior vena cava**.

Vessels

Blood entering the atria is forced into the ventricle when the atria contract. Contraction of the ventricle and conus arteriosus then forces the blood to the body and lungs. Each aortic arch has three branches: (1) the **carotid arch**, carrying blood to the head region, (2) the **systemic arch**, carrying blood to most of the body. The two systemic arches eventually fuse middorsally to form the **dorsal aorta** with its many branches leading to the internal organs, legs, etc., (3) the **pulmocutaneous arch**, carrying blood to the lungs and skin for gas exchange. All of these are classified as **arteries**, vessels which carry blood away from the heart. After blood has been transported to the tissues and the cells adjacent to the capillaries have obtained their nutrients and released their wastes, the blood returns to the heart via the type of blood vessel known as a **vein**, a vessel which carries blood toward the heart. The major veins into which blood returning from body parts drains are the **anterior vena cava** and **posterior vena cava**. Note the smaller veins leading to these two from the head, kidneys, gonads, etc.

Lymph

This network of vessels picks up fluid between the cells and eventually empties it back into the blood. Contractions of body muscles adjacent to

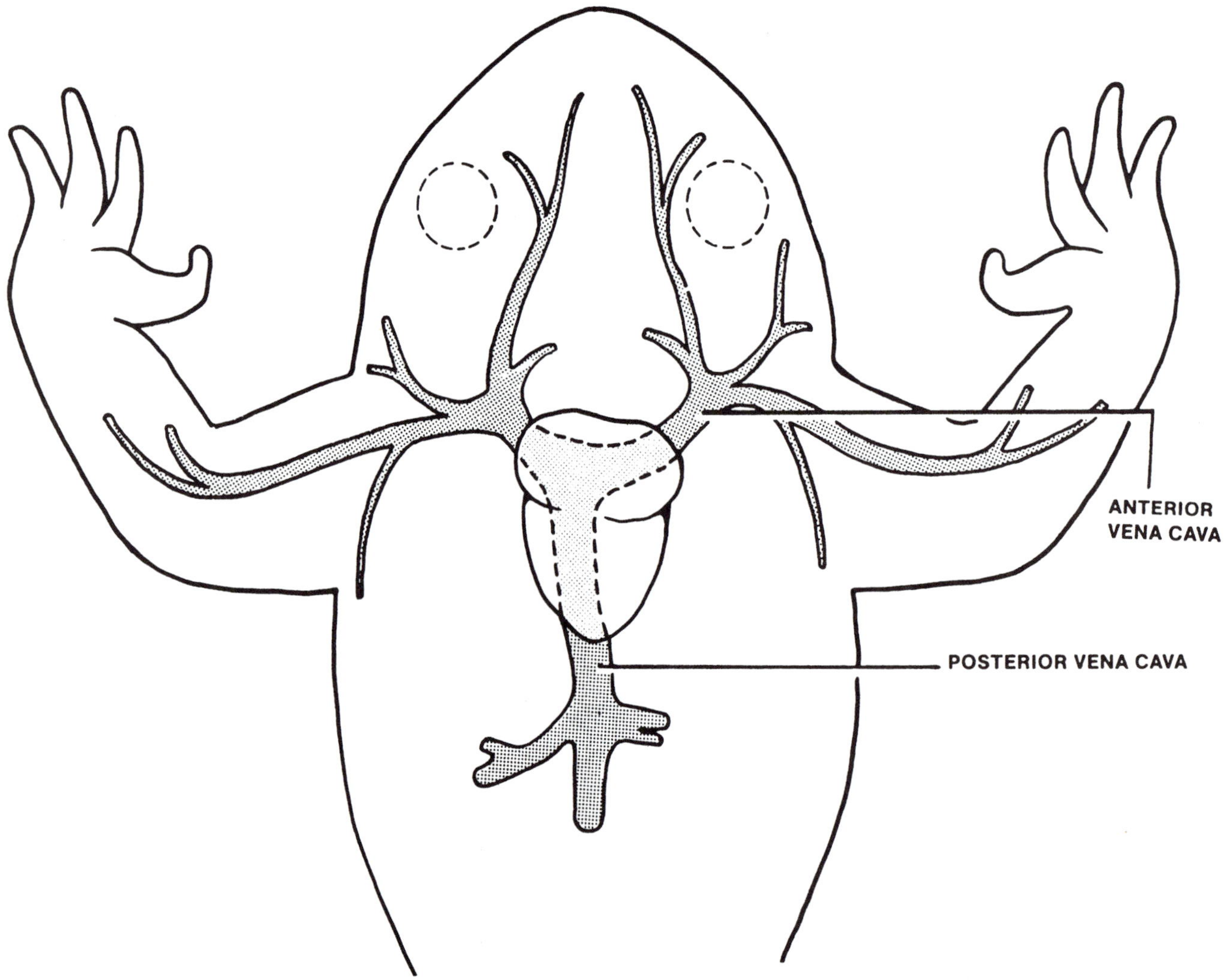

ANTERIOR VENA CAVA

POSTERIOR VENA CAVA

FIGURE 25.9 Venous System: Ventral View

large lymph vessels and actual pulsating "lymph hearts" in the lymph vessels move the lymph into the veins. Much of the lymphatic system is not very conspicuous; however, the frog does have several subcutaneous lymph sacs. These sacs account for the loose attachment of the skin to underlying muscle.

Nervous System

The nervous system of vertebrates can be divided into two basic parts: (a) The **central nervous system** which is composed of the five-lobed brain with the skull, and the spinal cord in the vertebral column; (b) the **peripheral nervous system** which consists of the cranial nerves and spinal nerves.

Place the frog on its back. Lift the digestive organs and kidneys and cut the mesenteries, leaving the dorsal aorta in place. Cut the intestine and urogenital ducts anterior to the cloaca and pull the entire mass of organs forward toward the head. Then cut the arches coming off the aorta and remove the organs by cutting through the posterior end of the pharynx. By removing all these internal organs, you will expose the ventral branches of the spinal nerves.

Spinal nerves leave the spinal cord and emerge between the vertebrae. Each of the 10 pairs of spinal nerves divides into two branches: an inconspicuous dorsal branch supplying muscles and skin on the back and the ventral branch supplying the ventral portion of the body wall.

Note that the second spinal nerve is the largest one of all; it forms a network with branches from the first and third spinal nerves. This network is the **brachial plexus** which supplies the muscles and skin of the shoulder and arm. The seventh, eighth, and ninth nerves also form a network called **sciatic plexus**, supplying the skin and muscles of the posterior abdomen, the pelvis, and the hind leg. Figure 25.10 will help you identify the two plexus networks.

To expose the **brain**, skin the head of the mid-dorsal region and remove the muscles from the posterior area of the skull. Carefully chip away the bone, exposing the brain located from about midway between the eyes to the back of the skull. Continue posteriorly and expose the spinal cord by cutting through the dorsal part of each vertebra and pulling the vertebra apart. Note that the central nervous system is covered by protective membranes (**meninges**). Remove these to see the brain parts more clearly. Refer to Figure 25.10 and identify the following parts starting at the anterior end.

1. **Cerebrum** or **telencephalon** — two hemispheres with the **olfactory lobes** extending anteriorly (the first cranial nerves extend from these lobes to the nasal cavities.

2. **Diencephalon** — a small medial structure covered by a thin vascular roof. The pineal gland or **epiphysis**, is located on the dorsal surface; the optic chiasma and **pituitary gland** are on the ventral surface.

3. **Optic lobes** or **mesencephalon** — two swollen lateral lobes on the dorsal surface.

4. **Cerebellum** or **metencephalon** — a transverse band of nervous tissue located behind the optic lobes.

5. **Medulla oblongata** or **myelencephalon** — the base of the brain, continuous with the spinal cord; it has a thin vascular roof.

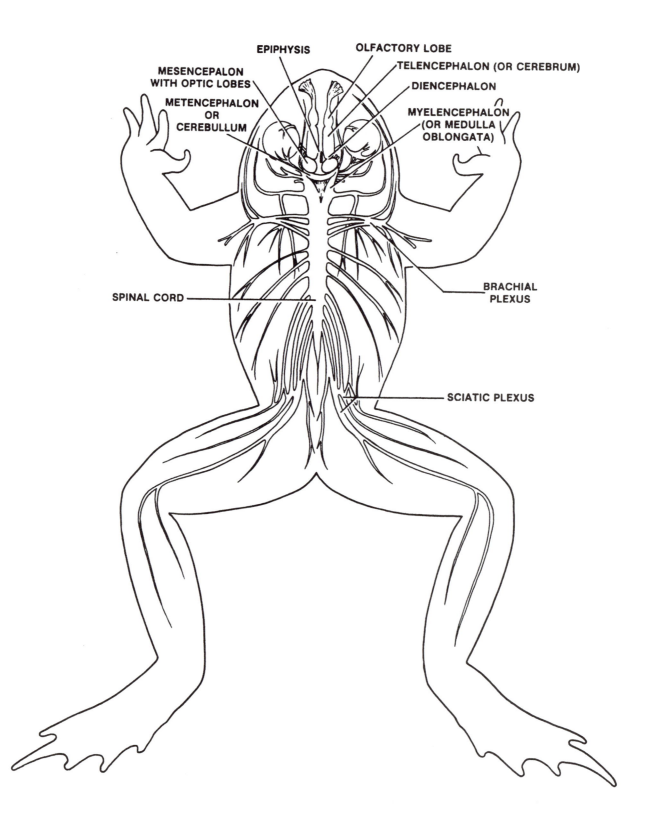

FIGURE 25.10   Nervous System: Dorsal View

REVIEW QUESTIONS

1. Give four reasons for placing the hemichordates in a separate phylum from the chordates.

   a. _____

   b. _____

   c. _____

   d. _____

2. List the identifying features of the Phylum Chordata. _____

   _____

   _____

3. What is the probable function of the folds in the frog's stomach? _____

   _____

4. Why would the ureter be slightly larger in the male frog than in the female? _____

   _____

5. A frog has to keep its mouth closed in order to breathe with its lungs. Explain why the same is not true of mammals. _____

   _____

6. What is an amniotic egg? _____

   _____

7. Compare poikilothermy with homeothermy. _____

   _____

8. Give the functions of each of the following:

   Cloaca _____

   Tympanic membrane _____

   Fat bodies _____

   Spleen _____

   Mesentery _____

   Maxillary teeth _____

9. Why would homeothermy be an advantage to an organism? _____

   _____

   How does poikilothermy limit an organism? _____

   _____

10. List a couple of characteristics that are unique to birds. _____

_____

_____

11. MATCHING: Match the subphylum/class with the correct organism.

	Subphylum/Class
_____ 1. *Amphioxus*	a. Urochordata
_____ 2. Snakes, lizards	b. Osteichthyes
_____ 3. Sea Squirts	c. Cephalochordata
_____ 4. Sharks, rays, skates	d. Agnatha
_____ 5. Bony fish, catfish	e. Chondrichthyes
_____ 6. Jawless fish, lampreys	f. Amphibia
_____ 7. Salamanders, toads, frogs	g. Reptilia
_____ 8. Cat	h. Aves
_____ 9. Birds	i. Mammalia

12. The Reptilia are very successful on land compared to the Amphibia. Complete the following table in comparing these specific characteristics.

	Amphibia	**Reptilia**
Skin		
Fertilization		
Eggs		
Gas exchange		

13. Most mammals give birth to live young. This is called _____.

14. Sweat glands, modified to produce milk, are the _____ glands.

15. Mammals breathe with the aide of a _____.

16. List some special skeletal adaptations in the birds.

a. _____

b. _____

c. _____

d. _____

Population Genetics and Evolution

OBJECTIVES

After completion of this exercise, the student should be able to do each of the following:

- Explain the Hardy-Weinberg Law in terms of gene frequencies in a population.
- State five conditions that must be met in order for the Hardy-Weinberg Law to operate.
- Calculate the proportion of AA, Aa, and aa genotypes in a population if the frequencies of "A" and "a" are known.
- Describe the effect of natural selection on stable gene frequencies.
- Explain the relationship between **natural selection**, **differential reproductive rates**, and the **adaptation of a population**.
- Discuss adaptations of plants to dry desert-like environments, aquatic environments, and competition for light.
- Explain the advantages of the carnivorous habit to plants such as the pitcher plant and the Venus flytrap.
- Define and give examples of: **mimicry**, **convergence** (convergent evolution), **divergence**, and **dispersal mechanisms**.
- Answer the review questions at the end of this exercise.

THE HARDY-WEINBERG LAW

Often a trait will be very rare even though it is caused by the presence of a dominant allele. For example, the trait of having 6 fingers on each hand is controlled by a dominant allele but this trait is rare in the human population. You may wonder why this trait does not become more common, and also perhaps why such recessive traits as albinism and colorblindness do not disappear entirely.

The Hardy-Weinberg Law states that under certain conditions of stability, both the allele frequencies and the genotype frequencies **remain constant** from generation to generation in a sexually reproducing population. This law applies regardless of whether the gene's expression is dominant or recessive. For example, the allele for having 6 fingers has a certain low frequency which does not vary noticeably and thus the trait remains rare.

This genetically stable condition is maintained under the following conditions: (1) a large population; (2) no mutations; (3) no organisms entering or leaving the population; (4) random mating; and (5) no natural selection.

To test the principles of the Hardy-Weinberg Law, the following experiment will be set up. An artificial population will be used to symbolize a group of sexually reproducing animals in which there are equal numbers of males and females. Some individuals of this population are homozygous for the dominant allele of a gene, some are homozygous recessive, and some are heterozygous. Let 'p' symbolize the frequency of the dominant allele in the population (p can vary from zero to one), and let 'q' symbolize the frequency of the recessive allele. Note that $p + q = 1$, so if p is already known, the $q = 1 - p$. It can be proved that if the population is very large and if there is no selection either for or against any particular genotype, the genotypes should be maintained in the population in the following proportions:

- Proportion of the population that is homozygous dominant (AA) $= p^2$ (pp)
- Proportion of the population that is heterozygous (Aa) $= 2 pq$
- Proportion of the population that is homozygous recessive (aa) $= q^2$ (qq)

To test if this relationship holds for a population of 64 breeding individuals in which there are equal numbers of recessive and dominant alleles ($p = q = \frac{1}{2}$), and where the parents are obtained by chance or at random, perform the following experiment.

Place 16 red toothpicks in a container. These represent the homozygous dominant individuals (AA). Sixteen red toothpicks are placed in the container because if $p = \frac{1}{2}$, then $p^2 \times 64 = 16$. Place 32 black toothpicks (Aa individuals) in the container. Why use 32 toothpicks?

Finally, place 16 yellow toothpicks (aa individuals) in the same container. Shake up the container and draw at random two toothpicks. These two toothpicks will be considered as parents which will produce 4 offspring, and the genotypes of the parents will determine the genotypes of the offspring according to genetic laws. The possible types of matings along with the kind of offspring which can be produced by such parents are as follows:

- red (AA) × red (AA) = 4 red (AA)
- yellow (aa) × yellow (aa) = 4 yellow (aa)
- black (Aa) × black (Aa) = 1 red (AA), 2 black (Aa), 1 yellow (aa)
- red (AA) × yellow (aa) = 4 black (Aa)
- red (AA) × black (Aa) = 2 red (AA), 2 black (Aa)
- yellow (aa) × black (Aa) = 2 yellow (aa), 2 black (Aa)

Therefore, if you should happen to draw two black toothpicks, place one red, 2 black, and one yellow toothpick in a pile which represents the individuals making up the next generation. Replace, in the container the toothpicks drawn as **parents** so that the probabilities of drawing will not be affected by removal of the toothpicks. Shake the toothpicks and draw a second pair which will also have 4 offspring according to the parents' colors. Continue the same procedure until the F_1 generation is composed of 64 offspring (by drawing 16 pairs of toothpicks). Record the composition of the F_1 generation in Table 26.1. Repeat the process, drawing from the F_1 generation to obtain the parents of the F_2 generation,

TABLE 26.1

Population Size is 64; $p = q = .5$
Each Mating Produces 4 Offspring

GENERATION	NUMBER OF REDS (AA)	NUMBER OF BLACKS (Aa)	NUMBER OF YELLOWS (aa)	FREQUENCY OF RECESSIVE ALLELE
original	16	32	16	64/128
F_1				
F_2				
F_3				
F_4				
Class Average of F_4				

these parents reproducing in the same way as their parents. Continue through the formation of the F_4 generation. Calculate the frequency of the recessive allele in each generation by doubling the number of yellow toothpicks and adding this to the number of black toothpicks (why is each black toothpick counted only once?). Record this in Table 26.1.

Has the original ratio of genotypes been substantially changed even though the breeding population is small?

The above experiments assumed random mating. Do you think that humans mate at random? List some of the factors which might disturb the randomness of human mating.

NATURAL SELECTION

Natural selection occurs when a certain trait is better or worse suited to a particular set of environmental conditions. This selection would tend to favor a particular genotype which produces a trait that is better suited to the environment, and would work for the elimination of a genotype which produces a trait that is not well adapted to its particular environment by allowing higher reproductive rates for the former than for the latter. As mentioned, natural selection will tend to upset the stable gene and genotype frequencies and work against the Hardy-Weinberg equilibrium. Since evolution occurs when natural selection works upon random variations in a population, it upsets the stability predicted by the Hardy-Weinberg Law.

To demonstrate natural selection, perform an experiment where no offspring are produced by homozygous recessive individuals. This is equivalent to selection against the recessive trait.

Make up the parent population as before. Draw at random two toothpicks from the container. Discard any yellow toothpick that is drawn, and draw again to replace it. Add 4 offspring to the next generation according to the genotypes of the parents as in the last experiment. There will be no yellow parents in this experiment but one of the offspring will be yellow when both parents are black (Aa). Continue drawing pairs until there are 64 individuals in the F_1 generation. Record the results in Table 26.2. Repeat the process through the formation of the F_4 generation.

Why is it so difficult to eliminate an inherited condition from a population of animals if the phenotype is determined by the homozygous condition of a recessive allele?

If all the individuals showing the phenotype determined by a dominant allele were eliminated (or prevented from reproducing) how many generations would it take to eliminate this allele from the population?

TABLE 26.2

Population Size is 64; p = q = .5
Each Mating Produces 4 Offspring
Homozygous Recessive Individuals (aa) Do Not Mate

GENERATION	NUMBER OF REDS (AA)	NUMBER OF BLACKS (Aa)	NUMBER OF YELLOWS (aa)	FREQUENCY OF RECESSIVE ALLELE
original	16	32	16	64/128
F_1				
F_2				
F_3				
F_4				

OPTIONAL: COMPUTER SIMULATION

The preceding "toothpick" demonstration of the Hardy-Weinberg Law picked 16 pairs of parents from a population of 64 organisms which would be considered a small population. Biologically, populations may have a far greater number of members. Demonstrating the law with this number of toothpicks would be very impractical! Fortunately, a computer program can be written to do the same thing that was done with the toothpicks but the computer can choose 160 or even 1600 sets of parents in only a fraction of a second longer than it can pick 16 sets. The secret of the program is a "random number generator" which picks pairs of numbers at random (like pairs of parents or toothpicks). The computer uses the position of the numbers rather than colors of toothpicks to determine the genotypes of the parents. For example, in the original population in Table 26.1, the computer would call numbers from 1 through 16 "AA", numbers 17 through 48 "Aa" and numbers 49 through 64 "aa". Once the computer has chosen 2 numbers (converted to genotypes of parents) the second part of the program simply determines the genotypes of their offspring, and repeats the process over and over until a certain number of parents are selected just like the toothpicks demonstration.

There are 2 programs available which perform the same function as the 2 toothpick experiments. HWLAW gives a simulation of a population according to the Hardy-Weinberg law, and NATSEL. demonstrates how natural selection works on gene frequencies. The student can call up either program and can plug in different values for gene frequencies (p and q), size of the population, number of parents to be selected at random (always ¼ of the population size), and number of generations to be calculated.

Your lab instructor will give you directions for using the computer terminals on this campus.

The technique of showing how biological events might occur using a computer is called computer simulation or modeling. In the past few years, modeling of entire ecosystems has become a very important area of research since this will allow ecologists to predict in advance what would happen if an ecosystem is modified.

ADAPTATION

Modifications which enable an organism to live in an environment are known as **adaptations**. Since these modifications arise as a result of natural selection working on favorable variations in a particular environment, these adaptations are the direct result of evolution.

In the following, some of the more noticeable adaptations and evolutionary concepts are discussed, and examples of each are given.

ADAPTATIONS TO DESERT-LIKE HABITATS

Desert plants of various families possess structural devices for collecting and retaining water. Examine the cacti and other succulents (Euphorbes, *Stapelia*, Aloe, etc.) on display.

In the cacti, the spines are modified leaves. What adaptive advantages do these leaves offer the cactus plant?

Also in cacti, the green, fleshy pods are modified stems. What functions do they serve?

Another dweller in desert-like environments is *Tillandsia usneoides* (Spanish moss). This epiphyte has no root system by which water can be obtained. Instead, it absorbs and retains rainwater by means of modified hairs or scales covering its stems and leaves.

Observe the plant, and examine these water-retaining scales under a microscope.

ADAPTATIONS TO GET THE SUNLIGHT

Sunlight is one of the necessary requirements for photosynthesis in plants. So any interference with obtaining sunlight places environmental, and thus evolutionary, pressures on the organism.

Below are listed several approaches which plants have evolved for securing sunlight. Give an example of a plant using each of these approaches:

1. grows tall _____

2. grows on other plants _____

3. climbs_____

4. grows on the forest floor in the early spring (before trees have leaves on them) _____

In the adaptation for climbing, plants have used every part — roots, stems, and leaves. Some woody plants have formed slender trunks and intertwining leaves and stems to ascend toward the sun. Woody vines are known as **lianas**. Give several examples of lianas.

_____   _____

_____   _____

_____   _____

Other modifications for climbing include **aerial roots** and **tendrils**. Examine the specimens on display and note their adaptations for obtaining sunlight.

ADAPTATIONS TO AQUATIC HABITATS

Plants must deal with some of the unique differences between terrestrial and aquatic environments in living in or on the water.

List some of the differences between terrestrial and aquatic environments.

_____

_____

_____

Observe some of the aquatic plants on display and list any characteristics which might aid their success in water.

_____

_____

_____

_____

_____

ADAPTATIONS FOR LIVING IN NITROGEN-DEFICIENT SOIL

Carnivorous plants grow primarily in habitats deficient in nitrogen. These plants secure nitrogen from the bodies of decaying insects and other small animals which have been captured by the plant's modified leaves. This organic nitrogen supply improves growth as compared with plants not furnished a supply of insects. In addition, all carnivorous plants have chlorophyll and hence carry on photosynthesis. Observe the different trapping devices used by the carnivorous plants on display.

CAMOUFLAGE AND MIMICRY

A major evolutionary advantage to organisms is the ability to escape notice from predators or prey. This is particularly true for animals. One means of doing this is to look similar to the surrounding environment or to some other organism. The former is known as **camouflage**, and the latter, **mimicry**.

Plants sometimes employ mimicry to increase their chances for successful pollination.

Examine the specimens on display which demonstrate camouflage and mimicry.

DIVERGENT EVOLUTION

Any population of organisms tends to diversify when it spreads (by its particular dispersal mechanisms) to various habitats. These new habitats offer new and unique environmental pressures to the expanding species. This is the force which creates the diversity which may be seen in practically every major group of organisms. The variations within a taxonomic group are examples of **divergence**, or **divergent evolution**.

Examine the members of the Bromeliaceae (pineapple family) and the Euphorbiaceae (spurge family) on display. These families show extreme diversity, even though each maintains its family resemblances (flower structure).

CONVERGENT EVOLUTION

Similar habitats may be found in various scattered regions of the world which offer similar environmental pressures to an organism. If two unrelated groups of organisms exist in similar habitats, it should not be surprising to find that the organisms have made similar adaptations in structure and behavior. This tendency of one group of organisms to develop resemblances to another of a different ancestry is known as **convergence**, or **convergent evolution**.

Cacti are found native only in the Western Hemisphere. However, other plant families have also made adaptations to fit arid environments elsewhere. Examine the cacti and euphorbes on display, noting their similarities.

DISPERSAL

The spread of plants from their parents to an area where they can germinate, grow, and mature, is called **dispersal**. Plants have evolved certain adaptations to ensure adequate dispersal which can be seen in the fruits and seeds on display. In each case, write down the specific modification of the seed or fruit and tell how it ensures seed dispersal.

REVIEW QUESTIONS

1. List the five conditions under which the Hardy-Weinberg equilibrium is maintained._____

_____

_____

_____

2. If 9% or a .09 frequency of the population cannot taste PTC (recessive trait), what is the frequency of the non-taster allele in the population?_____What is the frequency of the taster allele?_____

What percentage of the population will be homozygous dominant?_____

What percentage of the population will be heterozygous? _____

3. Name at least one adaptation found in plants to each of the following:

Desert conditions _____

Excessive shade at ground level or high population density_____

Deficiency of nitrogen in the soil _____

4. Give an example of convergent evolution among vertebrates._____

_____

5. Give an example of divergent evolution among mammals. _____

_____

Tissues, Organs, and Systems

EXERCISE **27** EXERCISE EXERCISE EXERCISE

OBJECTIVES

After completion of this exercise, the student should be able to do each of the following:

- Identify each of the following tissues and structures upon microscopic examination:

epithelium	**connective tissue**		**muscle tissue**
simple squamous	areolar	adipose	skeletal
stratified squamous	fibrous	lymph	smooth
cuboidal	blood	erythrocytes	cardiac
columnar	leukocytes	thrombocytes	striations
pseudostratified ciliated columnar	hyaline cartilage		intercalated discs
goblet cells	elastic cartilage		
fibroblasts	fibrocartilage		**nervous tissue**
collagen fibers	chondrocytes		neurons
basement membrane	lacunae		
elastic fibers	bone		

- List the major types of tissue found in animals and describe the unique structural characteristics of each.
- State the major function(s) of each type of tissue.
- Relate the structure of a tissue to its function.
- Identify the four layers of tissue found in a cross section of small intestine.
- Answer the review questions at the end of this exercise.

CELLULAR ORGANIZATION

The organization of multicellular organisms is very complex. However, the structure and function of each living thing is dependent on the structures and functions of its individual cells and the material which those cells produce. The cell, then, is the basic unit of structure and function in all living things.

The degree of complexity of cellular organization varies among animals, ranging from single-celled organisms to those composed of billions of different kinds of cells — cells which are organized into the following biological hierarchy:

1. A group of similar cells together performing a specific function, such as contraction, comprise a **tissue**.

229

2. An **organ** is a group of different tissues organized into a complex structural and functional unit.

3. A group of organs performing related functions are collectively an **organ system**.

4. Among higher animals, the **organism** (the entire living thing) is composed of all its organ systems working together, among higher animals.

In this laboratory period we will study cells, tissues, and one organ — the intestine. All of the material will be that of vertebrates, the most highly organized of all animals.

CELLS

The only cells which exist singly rather than as parts of a tissue in vertebrates are the sperm cells and egg cells.

SPERM CELLS

Obtain prepared slides of stained sperm cells of guinea pigs, frogs, and humans and examine under high power.

What is their shape and size? _____

_____

_____

_____

How do they seem to be structurally fitted to

perform their function? _____

_____

_____

EGG CELLS

Observe the demonstration of egg cells available in the laboratory. Compare the size of the egg cell to that of the sperm cells.

Given equal amounts of cytoplasm, estimate the relative number of eggs produced as compared to the number of sperm. _____

The sperm cell and egg cell both contribute material to the zygote, the first cell of a new organism.

What is contributed equally by both? _____

_____

What is contributed almost entirely by the egg cell?

_____

TISSUES

Cells become specialized, or differentiated, to perform specific functions. In multicellular organisms, groups of cells may become associated to perform a particular task. A group of similar cells involved in the same task is called a **tissue**.

Tissues are generally classified into one of four major types: **epithelial**, **connective**, **muscle**, and **nervous**. They resemble one another to the extent that each is composed of cells and intercellular material. There are many different subtypes, each with different functional modifications. The first two types of tissue will be studied in detail now; the last two will be studied in detail in later exercises. Use Figure 27.1 while examining the slides specified in this exercise.

EPITHELIAL TISSUE

The surface of the body, the coverings and linings of most organs, the secretory glands, and the sensory areas of sense organs are all composed of epithelial tissue. It may have one or more different functions including protection, absorption, secretion, and sensation. Structurally, it consists of cells closely packed together with very little, if any, intercellular (between cells) spaces or substance. Epithelial tissue which covers internal and external surfaces has a free surface and a basal surface in contact with other tissues, usually connective tissue. The **basement membrane**, made up of intercellular material and located at the basal surface, holds the epithelial tissue to underlying connective tissue. Epithelial tissue is subdivided into several categories based on the shape and arrangement of its cells.

Squamous Epithelium

This subtype consists of flat, pancake-like cells. If the cells are arranged in a single layer, the tissue is called **simple squamous epithelium**. These epithelia are unlikely to repair themselves readily. The peritoneum, pleural membranes, and linings of blood vessels consist of simple squamous epithelium. Several layers of flat cells are referred to as **stratified squamous epithelium**.

Stratified squamous epithelium is protective, guarding against dessication through loss of water.

If the epithelial surface is in contact with air, its outer cells usually produce keratin, a substance which helps reduce water loss. If the epithelial surface is constantly bathed with fluid, dessication is not a problem, and keratin is not produced. Stratified squamous epithelium is found at the outer surface of the skin, or epidermis. Examine prepared slides of simple and stratified squamous epithelium. Remember that you have already examined the squamous epithelium of your mouth in Exercise 4.

Cuboidal Epithelium

These cube-shaped cells with centrally located nuclei are found in various glands of the body, such as the thyroid, pancreas, and salivary glands. They function in secretion. Obtain a prepared slide containing a cross section of the thyroid gland or the kidney and locate the **cuboidal epithelium** lining the tubules.

Columnar Epithelium

This tissue consists of long, column-like cells with nuclei usually located at the bases of the cells. The linings of the stomach, intestine, and other parts of the digestive system are of this type. Their function is absorption. Some flask-like cells, called **goblet cells**, are seen interspersed among typical columnar cells. They contain mucus, which is poured out over the surface of this tissue. Obtain a prepared slide of a cross section through the frog intestine and identify the **columnar epithelium, goblet cells, and basement membrane**.

Some columnar epithelium has delicate hair-like structures called **cilia** projecting from the cells at their exposed surfaces. The trachea, bronchial tubes, and Fallopian tubes are lined with **ciliated columnar epithelium**. Wave-like movements of the cilia in the respiratory tract move a sheet of mucus upward, trapping bacteria and dust particles which enter in the air. Obtain a prepared slide of a cross section through the fallopian tube of the female reproductive system and identify the **simple ciliated columnar epithelial cells**, and **cilia**.

Pseudostratified Ciliated Columnar Epithelium

The cell nucleus may lie at any position within the cell in this type of epithelium, giving the tissue a layered appearance. However, all cells are actually in contact with the basement membrane.

Examine a prepared slide of a section of human trachea and identify the **pseudostratified ciliated columnar epithelium** lining the central cavity.

CONNECTIVE TISSUE

Connective tissue gives support to various organs, fills up spaces in the body, attaches organs to each other, and affords protection. The tissues of this type are characterized by cells which are interspersed in an abundance of nonliving intercellular substance termed the **matrix**. The nature of this intercellular material determines the type of connective tissue. The cells responsible for the production of the matrix are often called **fibroblasts**, since the matrix is usually composed of some fibers, primarily tough **collagenous fibers** which are thick, non-branching, and inelastic, and/or **elastic fibers**, which are thin and branching. Connective tissues are classified according to the predominant kinds of cells and/or fibers in addition to the nature of the ground substance. The ground substance may include liquid (such as in blood), a gel or semisolid substance (as in cartilage), or a solid composed of deposited minerals (as in bone).

Connective Tissue Proper

The intercellular matrix of this type of connective tissue always contains numerous fibers, which are composed of different types of protein. Connective tissue proper is very variable, generally containing several kinds of cells: fibroblasts secrete proteins; macrophages, common near blood vessels, become active when there is an inflammation; fat cells are highly specialized for fat storage. Tissue fluid is part of the connective tissue. It is derived from blood.

1. **Loose connective tissue** is characterized by the loose, irregular arrangements of its fibers, so that the tissue is fragile and easily torn.

Areolar Connective Tissue

This is characterized by the presence of both collagenous (thick) and elastic (thin) fibers between the fibroblasts. This type of tissue holds the skin to the body, and lends strength to walls of blood vessels and to many other organs. Examine a slide of **areolar connective tissue** and identify the fibroblasts, collagen fibers, and elastic fibers.

Adipose Connective Tissue

Fibroblasts which have large vacuoles of fat are found in this type of connective tissue. These cells are also called **fat cells**. In mature cells the fat vacuole fills up most of the cell. **Adipose tissue** lies under the skin, around the kidneys, in the orbit of the eye behind the eyeball, and in a

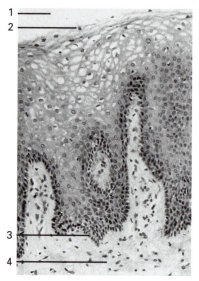

Stratified squamous epithelium, within the vagina of the female reproductive system.

1. Lumen of vagina
2. Free surface cells
3. Basal cells
4. Loose connective tissue

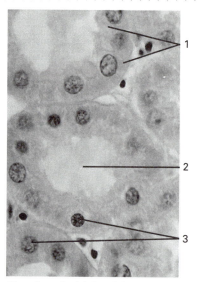

Simple cuboidal epithelium (kidney).

1. Simple cuboidal epithelial cells
2. Lumen of duct
3. Nuclei

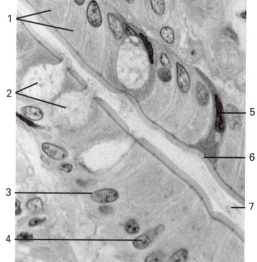

Simple columnar epithelium, within the ileum of the small intestine.

1. Simple columnar cells
2. Goblet cells
3. Nucleus
4. Nucleolus
5. Argentaffin cell
6. Striated brush border
7. Lumen

Simple ciliated columnar epithelium, within the uterine tube of the female reproductive system.

1. Simple ciliated columnar epithelial cell
2. Nucleus
3. Cell membrane
4. Lamina propria
5. Cilia
6. Lumen of uterine tube
7. Peg cell (secretory cell)

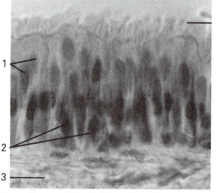

Pseudostratified ciliated columnar epithelium, within the trachea of the respiratory system.

1. Columnar cells
2. Filler cells
3. Connective tissue
4. Lumen of trachea
5. Cilia

Loose connective tissue.

1. Fixed macrophage
2. Capillary
3. Wandering macrophage
4. Fibroblast
5. Collagenous fibers
6. Matrix

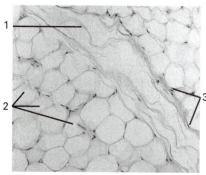

Adipose connective tissue.

1. Collagenous fibers
2. Adipocytes
3. Fibroblasts

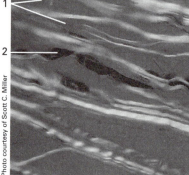

Photo courtesy of Scott C. Miller

Dense regular connective tissue.

1. Collagenous fibers
2. Fibroblast

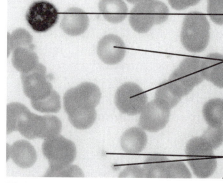

Photo courtesy of Scott C. Miller

Fluid connective tissue: blood.

1. Lymphocyte
2. Erythrocytes
3. Thrombocytes (platelets)

FIGURE 27.1 Vertebrate Tissues

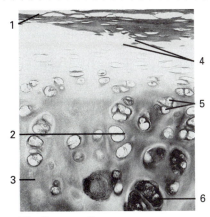

Hyaline cartilage.

1. Perichondrium
2. Chondrocyte in lacuna
3. Cartilagenous matrix
4. Chondroblasts
5. Lacunae with chondrocytes
6. Lacuna with necrotic chondrocytes

Photo courtesy of Scott C. Miller

Elastic cartilage.

1. Chondrocyte in lacuna   3. Matrix
2. Elastic fibers

Fibrocartilage.

1. Collagenous fibers irregularly arranged
2. Lacunae with chondrocytes

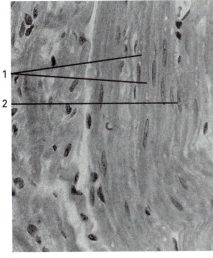

Bone tissue.

1. Interstitial lamellae   5. Osteons
2. Canaliculi   6. Central canal
3. Osteocyte   7. Lacuna
4. Lamellae of osteon

Skeletal muscle, longitudinal section.

1. Nuclei   3. Striations
2. Myofiber

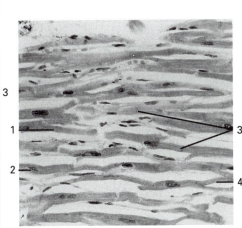

Cardiac muscle.

1. Cardiac myofiber
2. Nucleus of cardiac myofiber
3. Intercalated discs
4. Branching cardiac myofiber

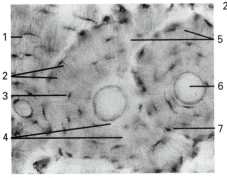

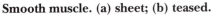

(a)

Smooth muscle. (a) sheet; (b) teased.

1. Smooth myofibers   2. Nucleus of smooth myofiber

(b)

Photo courtesy of Scott C. Miller

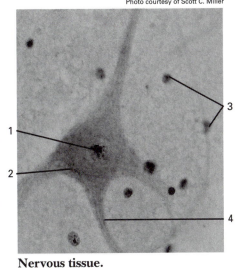

Nervous tissue.

1. Nucleus   3. Neuroglia
2. Cell body   4. Cytoplasmic extension

FIGURE 27.1 Vertebrate Tissues (*continued*)

number of other locations in the body where insulation or cushioning of delicate organs is needed. Examine a prepared slide of adipose tissue and identify the nucleus, vacuole, and cell membrane of a single fat cell.

2. **Dense**, or **fibrous**, **connective tissue** is characterized by fibers which fill up most of the intercellular spaces. This makes it stronger than loose connective tissue. In this type of tissue, which composes the tough **ligaments** and **tendons**, the fibroblasts are arranged in linear rows with masses of collagenous fibers packed between them. Tendons and ligaments will be studied further in Exercise 32.

Blood and Lymph

Blood and **lymph** have a liquid rather than a solid matrix. They are classified as connective tissues because they contain several types of formed elements which are derived from connective tissue structures. These include **erythrocytes**, or red blood cells; **leukocytes**, or white blood cells, and **thrombocytes**, or **platelets**. Blood will be examined more thoroughly in Exercise 29.

Cartilage

Cartilage is a specialized type of connective tissue characterized by a firmer type of intercellular material which gives it greater rigidity. **Chondrocytes** (cartilage producing cells) is the name applied to the fibroblasts of this connective tissue. They lie in cavities in the matrix called **lacunae**.

1. **Hyaline Cartilage**

   This glistening white material is found covering the ends of bones, reinforcing the nose, and attaching the ribs to the sternum. It is also seen in the larynx, trachea, and bronchial tubes. Depending upon where it is found, it may be referred to as articular, costal, skeletal, or embryonal cartilage.

2. **Elastic Cartilage**

   This is somewhat similar in appearance to hyaline cartilage, except that numerous elastic fibers are visible between the cells. This tissue is found in the external ear, Eustachian tube, and parts of the larynx.

3. **Fibrocartilage**

   Fibrocartilage resembles fibrous connective tissue except that the cells are not arranged in rows. Collagenous fibers are densely packed between the chondrocytes. The intervertebral discs which form cushions between the vertebrae are made of fibrocartilage, as is the pubic symphysis which connects the pelvic bones.

   Cartilage will be studied later in more detail in Exercise 32.

Bone

In this tissue, the intercellular matrix contains numerous fibers and water. It is also characterized by the impressive amount of inorganic salts, such as calcium carbonate ($CaCO_3$) and calcium phosphate ($Ca_3(PO_4)_2$) deposited in the matrix. An extensive study of bone will be done in Exercise 32.

Muscle Tissue

Muscle tissues contract to cause movement. Three types are **skeletal muscle**, **smooth muscle**, and **cardiac muscle**. Skeletal (usually voluntary) muscle is the type found attached to bones; smooth, or visceral (usually involuntary) muscle is found in the walls of tubular organs; cardiac muscle is restricted to the heart.

The different types of muscle cells (also called muscle fibers) vary in structure and arrangement. Both skeletal muscle and cardiac muscle cells have **striations**, or visible crossbands in the cell due to the arrangement of internal contractile elements, whereas smooth muscle cells lack striations. Skeletal muscle cells are larger, especially in length, than the other two types and contain many nuclei. In contrast, only one nucleus is found in each cell in the other two types of muscle. Each skeletal muscle cell extends the length of the whole muscle and attaches to tendons at either end. Smooth muscle cells overlap each other forming sheets of muscle. Cardiac muscle cells are connected end to end; thick bands called **intercalated discs** indicate the connections between cells. Muscle tissue will be discussed in detail in Exercise 33.

NERVOUS TISSUE

These tissues are specialized for the reception of stimuli and the transmission, interpretation, and coordination of nervous impulses within the nervous system. Refer to Figure 27.1 for a photograph of a nerve cell (**neuron**).

Bundles of nerve fibers are called **nerves**. The cell bodies of neurons are usually located in **ganglia**, the spinal cord, or the brain. Nervous tissue often includes accessory cells as well as the neurons. Nerves will be discussed in greater detail in Exercise 30.

AN ORGAN

Some organs in the body are composed largely of one type of cell. For example, the brain is made up largely of neurons, and muscles are composed mostly of muscle cells. However, many organs are composites of many different types of tissues. Two good examples are the intestine, discussed below, and the skin, which will be covered in Exercise 31.

Study a prepared slide of a cross section of small intestine and identify the following layers of tissue from the central cavity outward. Refer to Figure 27.2.

Mucosa. This tissue is made up of simple columnar epithelium that lines the central cavity and contains glandular epithelial cells which secrete mucus. This layer is arranged in deep folds called **villi** to increase the absorptive surface area of the intestine. Within the crevices of the villi are **digestive glands** which secrete digestive juices. In some places, the outermost part of the mucosa consists of a thin muscle layer.

Submucosa. The digestive glands extend down into this layer, which is a broad band of loose connective tissue containing blood vessels and nerves.

Muscularis externa. Two distinct layers of smooth muscle make up this layer. The first is an inner circular layer of cells in which long axes of the fibers encircle the intestine. The second is an outer longitudinal layer in which the muscle fibers run parallel with the long axis of the intestine.

Serosa. This is a very thin outermost layer of simple squamous epithelium. It is often difficult to distinguish between the serosa and the outer boundary of the muscularis externa.

Photo courtesy of Clifford E. Keeney

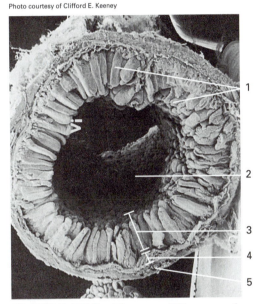

FIGURE 27.2 Electron Micrograph of the Ileum, Shown in Cross Section

1. Villi
2. Lumen
3. Mucosa
4. Submucosa
5. Muscularis

REVIEW QUESTIONS

1. How does a tissue differ from an organ? _____

2. In what main way does connective tissue differ in structure from the other three types?_____
   _____

3. How do collagenous fibers differ from elastic fibers? _____
   _____

4. List the specific tissues found in the small intestine. _____
   _____

5. List three types of matrix found in connective tissues and give an example of a connective tissue displaying
   each type. _____
   _____

6. Describe three types of epithelium and give the functions of each. _____
   _____

Fetal Pig Dissection

OBJECTIVES

After completion of this exercise, the student should be able to do each of the following:

- Locate and identify the anatomical parts of the fetal pig given in boldface in this exercise, and give the function of each.
- Define the terms of relative position and direction given in the introduction, and locate these positions and directions on any vertebrate.
- Distinguish between male and female fetal pigs on the basis of their external features.
- Locate the parts of the respiratory system seen in the fetal pig on the human torso model and color plates in the laboratory manual.
- Locate as many as possible of the parts of the digestive system given in boldface in this exercise on the charts and models which show the human digestive system.
- Locate the parts of the heart given in boldface in this exercise on the sheep heart and on the human torso model.
- Answer the review questions at the end of this exercise.

INTRODUCTION

In this exercise, the anatomy of the fetal pig, *Sus scrofa*, will be studied. This will aid us in obtaining some knowledge of mammalian anatomy and physiology. You will also examine certain comparative aspects of the anatomy and physiology of man.

The animals that you will use were removed from their mothers which were slaughtered for food. Farmers market pregnant females because hogs are sold by the pound.

In the pig, fertilization of the egg occurs in the oviduct; by the time the early embryo reaches what is equivalent to the blastula, it becomes buried in the uterine wall where subsequent development takes place. Here not only do the extraembryonic membranes form, but also a new organ, the **placenta**. This organ is made up of the **extraembryonic membranes**: amnion, chorion, and the allantois, which become intricately intermingled with the lining of the uterus. Remember that at no time does the blood of the fetus (unborn) mix with that of the mother. Gases and small molecules are able to pass across capillary walls in the placenta. The fetus lives like a parasite on the mother, absorbing all of its nourishment and oxygen from and excreting all of

237

its wastes into the blood of the mother, all by way of placental circulation. The connection between the placenta and the fetus is through the **umbilical cord**. The period or length of pregnancy (gestation) in pigs is approximately 17 weeks. The fetal pigs used in lab will be within one or two weeks of birth.

Below is a list of anatomical terms which are used frequently in dissection directions. Familiarize yourself with the definitions before continuing this exercise.

Dorsal — near or toward the back.

Ventral — near or toward the belly.

Medial — near or toward the middle.

Lateral — near or toward the sides.

Anterior — near or toward the head end.

Posterior — near or toward the tail end.

Caudal — referring to the tail or tail end.

Cephalic — referring to the head or head end.

Longitudinal — in the axis from head to tail.

Transverse — a thin section which cuts across the body at a right angle to the long axis.

Superficial — on or near the surface.

Pectoral — relating to the chest or shoulder region.

Pelvic — relating to the hip region.

Distal — free end of a limb or projection or toward this free end of a limb or projection.

Proximal — end attached to the body, or toward the end attached to the body.

The fetal pigs have been preserved in either formalin or isopropyl alcohol. These preservatives tend to dehydrate your fingertips; therefore, students may desire to lightly grease their fingertips with petroleum jelly.

The instructor will make available one fetal pig for each two students. The fetal pigs will be evenly distributed in reference to male and female. Upon receiving a fetal pig, wash it in running water and place it in a dissecting pan for observation. The aborted pigs will usually have a slash on one side of the neck. This marks the location where the blood was drained from the pig and where red and blue liquid latex were injected into a major artery or vein, respectively. The latex has become solid and rubbery in texture in the fetal pig. This strengthens the blood vessels and aids in their identification.

Measure your pig from snout to anus and refer to the following table to determine its approximate age. Remember that 1 cm. = 10 mm.

11 mm. — 21 days	40 mm. — 56 days
17 mm. — 35 days	220 mm. — 100 days
28 mm. — 49 days	300 mm. — 115 days (full term)

Please bear in mind, as you dissect your fetal pig, that there are very few differences between the anatomy of the pig and that of the human being. As you dissect out each system, make a mental comparison between yourself and the fetal pig concerning the arrangement of the internal organs.

EXTERNAL ANATOMY

Beginning at the anterior end, locate the **mouth**, which leads into the **oral cavity**, the **nose**, the **nostrils**, which lead into the **nasal cavity**, the **ears**, the **external ear canal**, which leads inward from the ears. Locate the **nictitating membrane** in the corner of the eye, the **eyeball**, **eyelashes**, and **eyelids**.

Now, lay the pig on its side and identify the major body divisions beginning anteriorly with a large **head**, a short thick **neck**, a cylindrical **trunk** with two pairs of **appendages**, and a short **tail**. The **anus** is located ventral to the tail.

Examine the forelimbs and locate the **shoulder**, **elbow**, and **foot**. On the hindlimbs find the **hip**, **knee**, **hock joint** (ankle), and the **foot**.

Turn the pig ventral side up and locate the large **umbilical cord** in the abdominal region. This cord connects the fetus to the mother at the placenta. Cut a half inch off the umbilical cord and observe the umbilical blood vessels. The blood vessels in the umbilical cord consist of two small **umbilical arteries** having relatively thick walls and an **umbilical vein**, considerably larger than the two arteries. The umbilical arteries carry blood from the fetal pig to the placenta, while the single umbilical vein returns blood to the fetal pig's body from the placenta. A fourth very small vessel, the **allantoic duct**, serves to carry some of the small amount of urine formed by the kidneys away from the fetus. Refer to Figure 28.1. Make a sketch below of a section through the umbilical cord, and label the four vessels.

CROSS SECTION OF UMBILICAL CORD

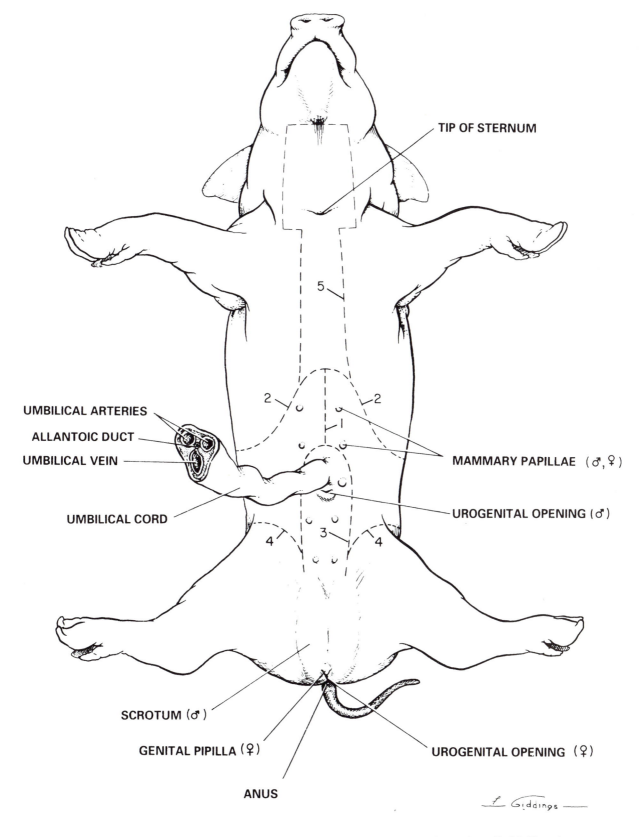

TIP OF STERNUM

UMBILICAL ARTERIES

ALLANTOIC DUCT

UMBILICAL VEIN

UMBILICAL CORD

MAMMARY PAPILLAE (♂, ♀)

UROGENITAL OPENING (♂)

SCROTUM (♂)

GENITAL PIPILLA (♀)

UROGENITAL OPENING (♀)

ANUS

_f_ Giddings

FIGURE 28.1 The Fetal Pig (External Features and Dissection Guidelines)

Also, on the ventral surface on either side of the umbilical cord, find the row of **nipples** or **teats**. The number of nipples indicates the number of mammary glands.

Next, determine the sex of your pig. You will be expected to recognize both male and female pigs, even though you have only one sex for dissection. In the male, the opening of the urogenital tract, which serves both reproductive and excretory functions, lies at the end of the **penis**, just posterior to the umbilical cord. The **scrotum**, which contains the **testes**, is situated ventrally with respect to the anus. The female pig has a single urogenital opening, located ventral to the anus, with a dorsal projection of tissue, **genital papilla** below it.

DISSECTION OF THE FETAL PIG

For this dissection you will need string, scissors, a sharp scalpel or a single edge razor blade, a blunt probe, forceps, and some dissecting pins. Place the pig on its dorsal surface in the dissecting pan. Referring to Figure 28.1, tie a string around one forelimb and place the string around and under the pan and tie it to the other forelimb. Make sure that the forelimbs are spread apart. Repeat this procedure for the hindlimbs.

NOTE

When preparing to put the pig back in the storage container for future use and reference, do not untie the string. Simply slide the pan from underneath the pig. Also, put a tag on the pig, identifying the members of the group. Use a pencil when writing on the tag to avoid fading.

Dissecting does not mean cutting up the specimen. Instead, it means exposing a specimen to view. Use the scalpel or razor blade carefully and sparingly. You may find that the most useful tool is the dull probe, which can be used to separate organs from membranes. Be very careful with your dissection. Any tissues or organs that are removed should be deposited in a trash can and never in the lab sink.

RESPIRATORY SYSTEM

This system functions in the exchange of gases between the internal and external environments of the organism. Air enters and leaves through the **nostrils** which lead, via the nasal cavity, into the **pharynx** or throat. The pharynx is where food and air passages cross and is located posterior to the **oral cavity**.

Expose the pharynx by inserting scissors into the corners of the mouth and cutting the jawbones. You may have to cut approximately one to one and a half inches. Separate the jaws further by pushing down on the tongue until you find a cartilaginous projection, the **epiglottis** at the base of the tongue. This flap aids in preventing food from entering the air passageway which leads into the lungs. The epiglottis covers an opening known as the **glottis**, the opening of the windpipe or trachea. Refer to Figure 28.3-A.

Figure 28.1 indicates the incisions required to open the thoracic and abdominal regions. A scalpel and forceps should be used during this part of the dissection. Make cuts as indicated by the dotted lines in Figure 28.1. To enter the abdominal cavity, you will pass through an outer layer of skin, several layers of muscle, and a tough inner glistening membrane, the **peritoneum**, which lines the body cavity and surrounds the internal organs.

Pick up the umbilical cord and pull lightly on it. Passing from the base of the umbilical cord to the liver, the large brown organ covering most of the anterior portion of the abdominal cavity, is the **umbilical vein**. After identifying it, cut it, leaving two stump ends that can be located later.

A certain amount of brown liquid may be present in the body cavity. This is clotted blood and it should be poured into the sink, after which you should completely rinse the body cavity with cold tap water. Now, pin back the skinflaps, and fold back between the hind legs the skin and muscle to which the umbilical cord is attached.

Locate the muscular **diaphragm** which separates the thoracic cavity from the abdominal cavity, and aids in the movement of gases into and out of the lungs.

To expose the viscera in the thoracic region, continue the initial incision from the abdominal cavity forward to the clump of hairs under the chin. Gradually deepen the incision until you have cut through the **sternum**, or **breastbone**. The muscle is particularly thick in the neck region. Avoid cutting the organs and blood vessels in the neck and chest regions. Sever the edges of the diaphragm where it is attached to the rib cage. Next, break the rib cage by applying pressure from your thumb on the sternum. The flap of tissue which contains the sternum can be carefully trimmed away and discarded. Some of the ribs may have to be cut with scissors. If so, be careful not to cut into the organs of the thorax.

Referring to Figure 28.2 locate the **larynx**, or voicebox, and slit it open along its midline to expose the small, paired lateral flaps internally known as

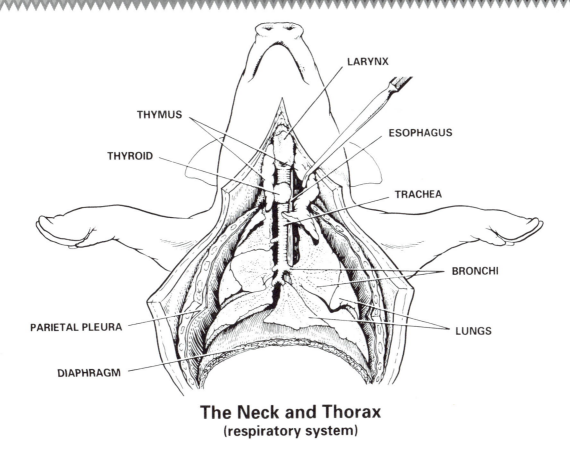

LARYNX

THYMUS

THYROID

ESOPHAGUS

TRACHEA

BRONCHI

PARIETAL PLEURA

LUNGS

DIAPHRAGM

The Neck and Thorax
(respiratory system)

FIGURE 28.2 The Fetal Pig (Respiratory System)

the **vocal folds**. In the fetal pig, these are not yet well-developed. Posterior to the larynx is the cartilaginous-ringed **trachea**. In the neck you will expose extensions of the **thymus gland**. The major portion of this gland is located on the ventral surface of the anterior portion of the heart. The extensions of the thymus gland may have to be removed. Also, in a midventral position, locate the small, dark, pea-shaped **thyroid gland** which lies on the surface of the trachea just anterior to the heart. Leave the thyroid gland in place.

Follow the trachea posteriorly until it branches into two **bronchi**, one leading to each **lung**. The left and right lungs are in separate cavities lined by a thin, fleshy **pleural membrane**. Trace the path of a bronchus into a lung by scraping away the lung tissue. The bronchus continues to branch into smaller tubes called **bronchioles** which lead to tiny air sacs or **alveoli** which are surrounded by a network of blood vessels. The alveolus is the site of gas exchange in the lungs.

In order to observe the nature of the lung tissue, remove the anterior lobe of the left lung. It should be noted that all dissections regarding the pig refer to the body of the pig. Therefore, if the directions identify an organ on the left side, this refers to the left side of the pig. Place the lung in a small dish of water. Holding it with your forceps, gently tease the lung apart with the blunt wooden base of a probe. Using a stereomicroscope, identify the branching network of bronchioles and blood vessels.

Having examined the respiratory system in the pig, consult the human torso model and the color plates of the human respiratory system in the laboratory manual. Identify the main structures of this system.

DIGESTIVE SYSTEM

Digestion begins in the **oral cavity** with the chewing of food. The **tongue** is attached at the back of the oral cavity. Observe the small, underdeveloped **teeth** in both the upper and lower jaws. The first set of teeth in mammals is called **milk teeth**. These are later replaced by the **permanent teeth**. Referring to Figure 28.3A note the ridged surface on the roof of the oral cavity. This is the **hard palate**. Posterior to the hard palate is the **soft palate**.

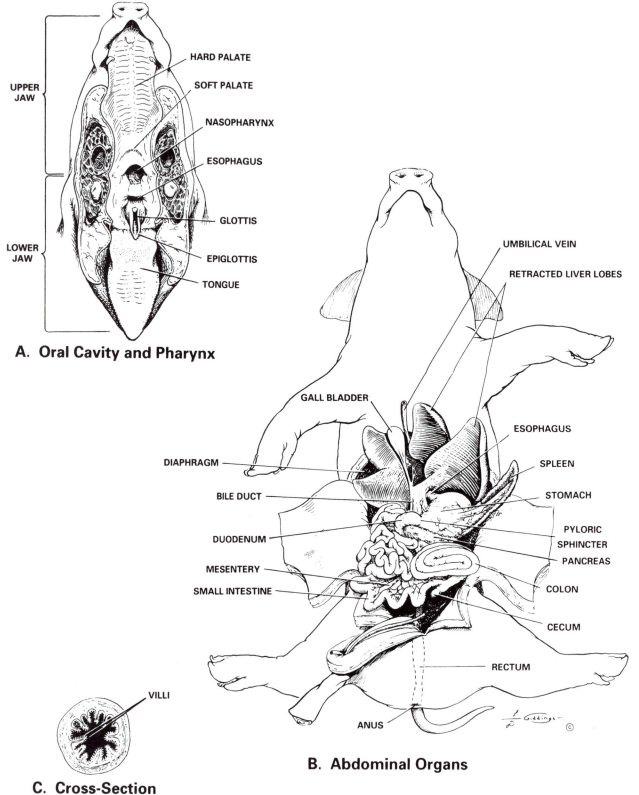

A. Oral Cavity and Pharynx

UPPER JAW

HARD PALATE
SOFT PALATE
NASOPHARYNX
ESOPHAGUS
GLOTTIS
EPIGLOTTIS
TONGUE

LOWER JAW

B. Abdominal Organs

UMBILICAL VEIN
RETRACTED LIVER LOBES
GALL BLADDER
ESOPHAGUS
SPLEEN
DIAPHRAGM
STOMACH
BILE DUCT
PYLORIC SPHINCTER
DUODENUM
PANCREAS
MESENTERY
COLON
SMALL INTESTINE
CECUM
RECTUM
ANUS

C. Cross-Section of Small Intestine

VILLI

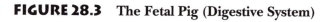

FIGURE 28.3   The Fetal Pig (Digestive System)

Locate the pharynx. Food leaves the pharynx via the **esophagus** on its way to the stomach. The opening into the esophagus is dorsal to the glottis. Insert a blunt probe into the opening and follow the esophagus posteriorly through the thoracic cavity. In the thorax, the esophagus lies just dorsal to the trachea. The esophagus penetrates the diaphragm and continues posteriorly to join the bag-like organ, the **stomach**. The **cardiac sphincter** is the muscle which forms the boundary between the esophagus and the stomach. Refer to Figure 28.3 in identifying the rest of the digestive system.

The stomach can be divided into the larger anterior **cardiac** portion and the lower tapering **pyloric** portion. This posterior segment of the stomach joins the small intestine. Open one side of the stomach and examine its interior surface. Does it appear smooth or rough? Locate the **liver**, the large lobed, reddish-brown organ which lies posterior to the diaphragm. Notice that the liver consists of several lobes which are attached only at the dorsal and anterior margins. Among other functions, the liver produces bile, which emulsifies fats in the small intestine. The bile is temporarily stored in a small sac, the **gall bladder**. This organ may be found by lifting up the extreme right lobe of the liver to which the gall bladder is attached. At the posterior end of the stomach, locate the hard ring of smooth muscle, the **pyloric sphincter**. This sphincter muscle forms the boundary between the stomach and the **small intestine**. Constriction of this muscle prevents food from escaping into the **small intestine** before the stomach has finished processing it.

The portion of the small intestine into which the stomach empties is the **duodenum**. Locate a long, whitish, cauliflower-like organ, the **pancreas**, lying dorsal to the duodenum and the stomach. The digestive enzymes produced by the pancreas pass into the duodenum. Try to locate the **pancreatic duct**. It may be difficult to find. Also locate the **bile duct** as it enters the duodenum from the gall bladder.

To the left of the stomach observe the **spleen**, a reddish brown tongue-shaped organ which functions in destroying old red blood cells in the adult. Notice that the small intestine is coiled, therefore providing an increased area for digestion and absorption of food. The small intestine is also held in place by a mass of sheetlike membranes, the **mesenteries**. Slit open a short portion of the small intestine and find the **villi** with a stereomicroscope. The villi are microscopic finger-like projections which serve to increase surface area also.

The small intestine continues posteriorly, merging with the first of three segments of the large intestine, the **colon**, which is a compact, rounded mass of intestine bound firmly by mesentery. Find the **cecum**, a blind pouch located posterior to the junction of the small and large intestines. The pig does not have an appendix. In man, the appendix is attached to the posterior end of the cecum. The straight most posterior section of the large intestine is the **rectum** which opens to the outside by way of the **anus**. The cecum, colon, and rectum make up the large intestine.

If time allows, sever the coiled small intestines just below the duodenum, and sever the colon at the point where it joins the rectum. Carefully cut the mesenteries holding the long intestinal sections in place so that it can be laid out in a straight line. How long are the intestines?

Locate as many of the structures as possible in the digestive system on the human torso model and color plates in the manual.

CIRCULATORY SYSTEM

The major **arteries** and **veins** are injected with latex. The arteries, which carry blood away from the heart, are injected with red latex, and the veins, which carry blood toward the heart, are injected with blue. Sometimes the pressure of injection causes the latex to cross capillary beds in some places. Therefore, veins occasionally contain some red latex, while arteries may have blue latex.

Arterial System (Refer to Figures 28.4, 28.5, 28.6, 28.7)

Locate the **heart** in the thoracic cavity and carefully trim away the **pericardial sac** that surrounds the heart (removed from Figures). The major portion of the heart is composed of the **right** and **left ventricles**. These are separated from one another on the ventral surface by the **coronary artery**. The left ventricle alone makes up the posterior tip, or **apex**, of the heart. Anteriorly and laterally are two dark projections, the right and left **auricles**, which are parts of the upper heart chambers, the **atria**. The term "auricle" comes from a root meaning "ear." The projections looked like ear flaps on each side of the heart. The auricles do not have any particular function.

Locate the **pulmonary trunk**. It is a large vessel which carries blood from the right ventricle of the heart. This large artery crosses from the upper right over the ventral surface of the heart and arches to the dorsal side of the heart, where it branches. Carefully remove the connective tissue surrounding the pulmonary trunk and trace it until it branches by displacing the heart to the right (the pig's right)

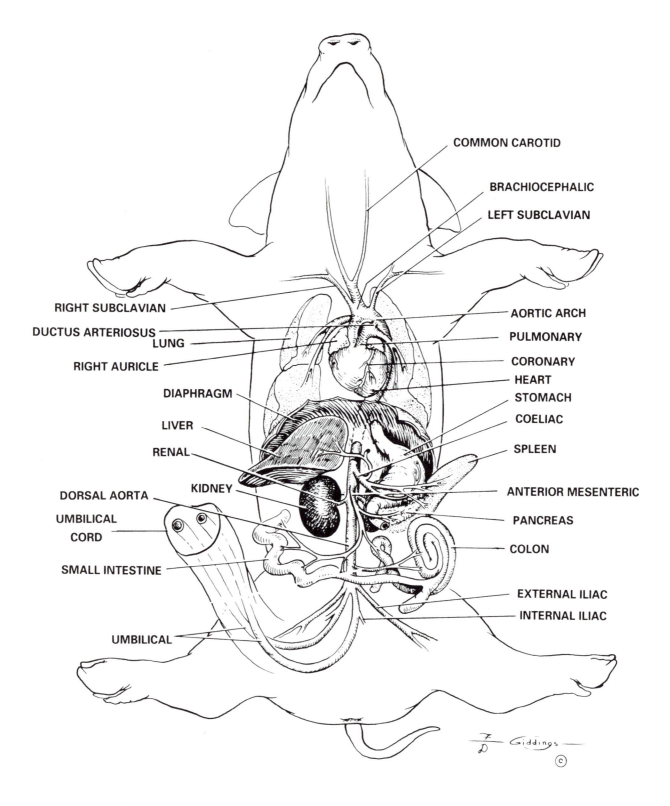

FIGURE 28.4   The Fetal Pig (Major Arteries)

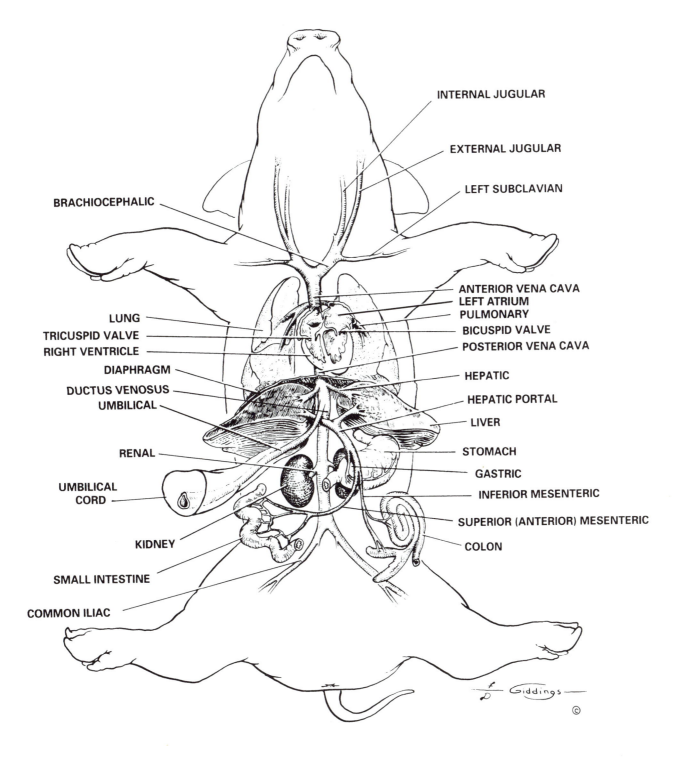

INTERNAL JUGULAR

EXTERNAL JUGULAR

LEFT SUBCLAVIAN

BRACHIOCEPHALIC

ANTERIOR VENA CAVA
LEFT ATRIUM
PULMONARY
BICUSPID VALVE
POSTERIOR VENA CAVA

LUNG
TRICUSPID VALVE
RIGHT VENTRICLE

DIAPHRAGM
DUCTUS VENOSUS
UMBILICAL

HEPATIC
HEPATIC PORTAL
LIVER

RENAL

STOMACH
GASTRIC
INFERIOR MESENTERIC

UMBILICAL
CORD

SUPERIOR (ANTERIOR) MESENTERIC

COLON

KIDNEY

SMALL INTESTINE

COMMON ILIAC

FIGURE 28.5   **The Fetal Pig (Major Veins)**

side. The **right pulmonary artery** goes to the right lung and the **left pulmonary artery** goes to the left lung. Next, locate the large, thick-walled **aortic arch**, or **aorta**, which carries blood from the left ventricle. It passes anteriorly for a short distance and makes a sweeping 180° left turn and comes to lie posteriorly along the dorsal wall of the thoracic and abdominal cavities.

Observe that the pulmonary trunk connects directly to the aorta on the left side of the heart. After the trunk branches to the right and left lungs, the short, thick interconnecting vessel is known as the **ductus arteriosus**. The ductus arteriosus shunts blood into the **dorsal aorta**, bypassing the lungs. The lungs are nonfunctional in the fetal pig. After birth, the arterial duct normally closes off with connective tissue.

Carefully dissect away the connective and muscle tissues in the neck to expose the arterial branches going to the head.

Two major arterial branches emerge from the aortic arch a short distance from the heart. The first is the **innominate artery** or **brachiocephalic artery** which supplies blood to the right forelimb and head. This branch continues toward the head, then divides into two branches. The first is the **right subclavian artery** which supplies blood to the right forelimb and right ventral chest wall. The second branch continues anteriorly, dividing into two **common carotid arteries**, which supply blood to the head. The second branch off of the aortic arch is the **left subclavian artery** which carries blood to the left forelimb and the left ventral chest wall.

Find the dorsal aorta in the abdominal cavity. The first major branch is the **coeliac artery** which supplies the stomach, spleen, and liver. The **anterior mesenteric artery** branches from the dorsal aorta just posterior to the coeliac artery. It supplies blood for the pancreas and the small intestine. Continue tracing the aorta posteriorly and locate the **renal arteries**, one leading to each kidney. Posterior to the renal arteries locate the paired **genital arteries** which are thread-like in size and lie on the extreme ventral surface of the dorsal aorta. The genital arteries lead and supply blood to the sex organs. At the posterior end of the body cavity, the dorsal aorta divides into two pairs of arteries that supply blood to the legs, the **external iliac arteries** and the **internal iliac arteries**. The two large **umbilical arteries** pass ventrally from the internal iliac arteries.

Venous System (Refer to Figure 28.5)

Find the large vein which enters the right atrium anteriorly. This is the **anterior vena cava** or **precaval vein**. There are four major veins which drain the anterior region of the body that unite to form this blood vessel. The two **external jugular veins** lie parallel to the carotid arteries. These are paralleled by two **internal jugular veins**. Also locate the **left** and **right subclavian veins** which drain each of the forelimbs, respectively.

Locate the **azygos vein** which lies to the left of and parallel to the dorsal aorta in the chest cavity. The azygos vein drains the muscles between the ribs and empties directly into the right atrium.

Locate the large **posterior vena cava** or **postcaval vein** which runs parallel to the dorsal aorta in the lower abdomen. Note where the two **renal veins** (one from each kidney) join the postcava at the kidneys. Posteriorly, the posterior vena cava will divide into the paired **common iliac veins**. Further subdivision of the common iliac veins does occur, but the veins are difficult to find.

Hepatic Portal Division of the Venous System
(Refer to Figure 28.5)

When you originally opened the abdomen, you cut the **umbilical vein** which leads from the umbilical cord into the liver. One of its branches leads into the hepatic portal system, and the other branch, the **ductus venosus**, leads directly into the posterior vena cava. The ductus venosus allows some blood rich in oxygen and nutrients to be pumped out to the body without passing through a capillary bed in the liver. Try to find the ductus venosus which may have been cut when you opened up the pig.

The organs of the gastrointestinal tract are drained by the **hepatic portal system**. This is a set of veins which collects blood from the digestive tract and filters it through the liver before the blood enters the heart.

In order to expose the hepatic portal vessels move the stomach, spleen, pancreas, and small and large intestines to the left. They may not be injected with blue latex in your pig; therefore, only the largest vessels can be identified.

Locate the **hepatic portal vein**, the main vein of the hepatic portal system. It carries blood from the intestine to the liver. In the liver the blood passes through a capillary bed, where toxic materials are removed from the blood and the nutrient content of the blood is regulated by cells of the liver. Blood leaving this capillary bed eventually enters the posterior vena cava via the **hepatic veins**.

Locate the **superior (anterior) mesenteric vein** which is the union of branches from the many coils of the small intestine.

Also locate the **gastric vein** which drains the pyloric region of the stomach before it joins the hepatic portal vein.

The Heart

Do not remove the pig's heart. For this dissection, you will use the sheep's heart.

Open the sheep's heart by making a midventral slit if this has not already been done. Use Figure 28.7 as a guide. Blood leaving the right atrium passes through a one-way valve, the **tricuspid valve**, into the right ventricle. Find the **bicuspid**, or **mitral valve**, in the same position in the left side of the heart. The tough cords holding the edges of the valves in place are called **chordae tendinae**. The chordae tendinae, which prevent the valves from flapping up into the atria when the ventricles contract, are attached posteriorly to **papillary muscles**, columns of muscle arising from the wall of the ventricle. In the fetal pig, the **foramen ovale**, an opening between the two atria, allows the lungs to be bypassed. It is reduced to a closed depression at birth. Locate this depression in the interatrial septum. Look down into the stub of the aortic arch

and pulmonary trunk and observe the flaps that compose the **semilunar valves**. These valves prevent backflow of blood from the arteries.

After finding the bold face parts given previously on the sheep heart, locate the same parts on the beef heart on display and the human torso model.

URINARY SYSTEM (Refer to Figure 28.8)

Locate the paired **kidneys**, lying dorsally against the abdominal wall. Notice that each kidney is covered by a thin membrane, the **peritoneum**, on its ventral surface. Next, find the **adrenal gland**, a narrow whitish body about half an inch long which lies medially along the anterior edge of the kidney. Remove the peritoneum from a kidney. On the medial side of the kidney is a concave depression, the **hilum**. At this point the **renal artery** and **renal vein** attach, carrying blood into and out of the kidney, respectively. The hilum is also the point where the **ureter** leaves the kidney, carrying urine to the **urinary bladder**. Expose one of the ureters by picking away the peritoneum covering it.

During fetal life, urine exits from the anterior end of the urinary bladder by way of the **allantoic**

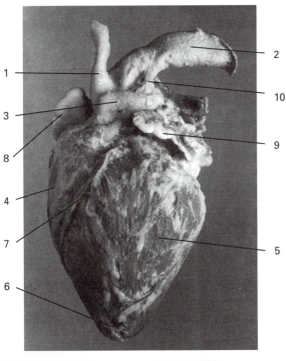

1. BRACHIOCEPHALIC ARTERY
2. AORTA
3. PULMONARY ARTERY
4. RIGHT VENTRICLE
5. LEFT VENTRICLE
6. APEX
7. ANTERIOR LONGITUDINAL SCILCUS
8. RIGHT AURICLE
9. LEFT AURICLE
10. DUCTUS ARTERIOSUS

FIGURE 28.6 Sheep Heart Lateral

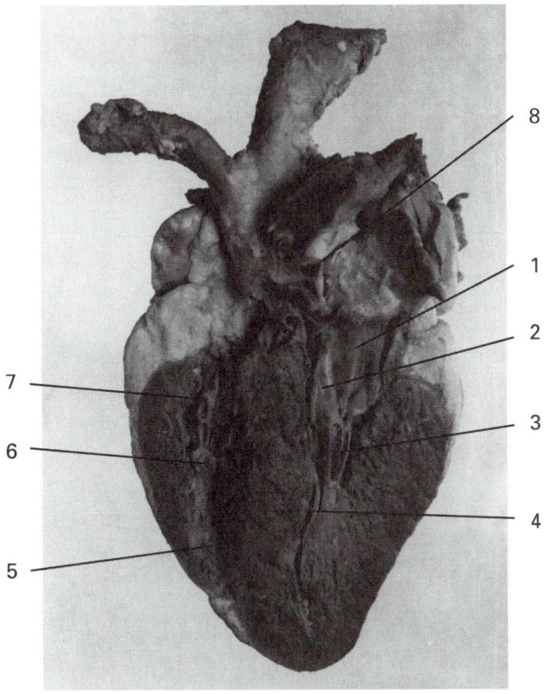

1. LEFT ATRIUM
2. BICUSPID VALVE
3. CHORDAE TENDINAE
4. LEFT VENTRICLE
5. RIGHT VENTRICLE
6. PAPILLARY MUSCLE
7. TRICUSPID VALVE
8. SEMILUNAR VALVE

FIGURE 28.7 Sheep Heart (L.S.)

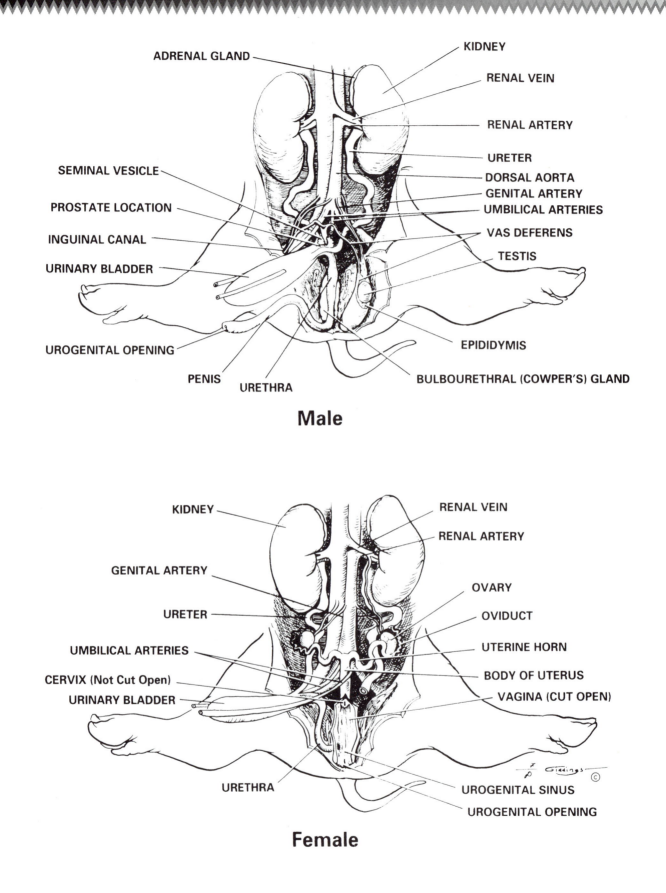

Male

Female

FIGURE 28.8   The Fetal Pig (Urogenital Organs)

duct, through the umbilical cord to the placenta. In the adult male, urine is voided to the outside of the body by the **urethra**. This is a tube-like structure which passes posteriorly from the urinary bladder for a short distance and then turns anteriorly and ventrally to enter the **penis**. In the adult female the urethra continues posteriorly to enter the **urogenital sinus**, or **vestibule**, which lies approximately one-half of an inch from the urogenital opening.

Using your razor blade or scalpel, section one of the kidneys in place, cutting from the lateral or the medial side on a plane parallel to the dorsal side of the animal. Note that at the center of the medial portion of the kidney is an irregular cavity, the **pelvis**. Here the urine is collected and through this region the branching blood vessels pass. The pelvis is continuous with the ureter.

The outermost portion of the kidney, the **cortex**, shows many small striations perpendicular to the outer surface. This region and the **medulla** region, which lies medially to it, are composed of thousands of minute excretory tubules called **nephrons** associated intimately with capillaries. Urine is formed in these regions and then drains to the pelvis.

REPRODUCTIVE SYSTEM

The reproductive system in mammals is closely associated with the urinary system. Together, these form the **urogenital system**. In this exercise, they will be discussed separately, but in Figure 28.8 they are combined.

Female Reproductive System

Pull the umbilical cord posteriorly and find the urinary bladder. The urinary bladder continues posteriorly joining the urogenital sinus via the urethra. Cut along one side of the urogenital sinus and through the cartilage of the pelvic girdle. Now you should be able to lay the legs out flat.

Referring to Figure 28.8, locate the pair of gonads the **ovaries**, about one half an inch posterior to each kidney. The ovary is a small, bean-shaped, light colored structure. The lateral and dorsal surface of each ovary is partially covered by the mouth of the **oviducts**, or **fallopian tubes**. The oviducts lead posteriorly to the **horns of the uterus**. The horns join along the midline to form the **body of the uterus**. Posterior to the body of the uterus, locate the **vagina**. The vagina joins with the **urethra** forming the **urogenital sinus**, or **vestibule**, an area shared by the urinary and reproductive systems. The **glans clitoris**, a genital papilla that is not visible in Figure

28.8, a small rounded papilla arises from the ventral floor of the urogenital sinus. This structure may not be apparent in your specimen. The **glans clitoris** (clitoris) is the female homolog to the penis in the male. The urogenital sinus opens to the outside via the **urogenital opening**, just ventral to the anus. Now locate the dorsally directed **genital papilla**.

Male Reproductive System

Referring to Figure 28.8, locate the pair of gonads, the **testes**, where sperm is produced. The testes begin their development in the body cavity, just posterior to the kidneys. Before birth they descend into the paired **scrotal sacs**, located between the hind legs and just ventral to the tail. In younger specimens, the testes have not yet descended and are located between the abdominal cavity and the scrotal sacs. Each scrotal sac is connected to the body cavity by the **inguinal canal**.

Locate the opening of the left inguinal canal. Then make an incision through the skin and muscle layers from a point over this opening to the left scrotal sac. In this way, the canal and whole sac will be exposed. Open the sac and find the testis. Note the much-coiled tubule, the **epididymis**, which lies along the surface of the testis. This is continuous with the sperm duct, or **vas deferens** which passes back toward the body cavity. In the body cavity, the sperm ducts from each testis loop over the umbilical arteries and ureters and unite dorsally at the posterior end of the urinary bladder with the **urethra**.

Near where the vasa deferentia (singular: vas deferens) join the urethra and dorsal to the urethra, locate a small pair of light colored glands, the **seminal vesicles**. A small rounded gland, the **prostate**, may be found in the dorsal surface of the urethra just posterior to the junction of the urinary bladder and the urethra. The prostate may be difficult to locate.

Cut through the cartilage of the pelvic girdle on the left side. Trace the urethra posteriorly until it makes a U-turn and proceeds anteriorly until it opens to the outside via the urogenital opening of the penis. Now locate a pair of white glands, the **bulbourethral** (**Cowper's**) **glands**, one on each side of the urethra near the U-turn. These three types of glands form a fluid, **seminal fluid**, which together with sperm constitutes the **semen**. Seminal fluid nourishes and provides transport for sperm from the epididymis.

Nervous System

You will examine the nervous system of the fetal pig in Exercise 30.

REVIEW QUESTIONS

1. The fetal pig is attached to the placenta by means of the _____ cord.

2. The sites of gas exchange in the lungs are tiny sacs called_____.

3. The flaplike projection which covers the glottis is the _____.

4. The _____ produces bile.

5. Bile is stored in the _____.

6. _____ are fingerlike projections inside the small intestine.

7. Blood enters the right atrium via the two major veins, _____ and

   _____ .

8. Blood is pumped from the right atrium to the right ventricle via the _____ valve.

9. Blood leaves the right ventricle via the _____ which branches to the lungs.

10. Blood is pumped out through the semilunar valves into the _____ which branches to distribute blood to all parts of the body.

11. Sperm is stored in the _____ until ejaculation.

12. The female equivalent of the penis in the male is the _____.

13. What is semen? _____

14. Describe the function of seminal fluid. _____

    _____

15. The thin membrane covering the kidneys is called the _____.

16. Describe the function of the placenta. _____

    _____

17. How old is your pig? _____

18. Are mammary glands present in both sexes? _____ How many? _____

19. How many lobes do you find for each lung?_____

20. Define emulsify. _____

    _____

21. What is the function of the foramen ovale? _____

    _____

22. What happens to this opening at birth? _____

    _____

Circulation

OBJECTIVES

After completion of this exercise, the student should be able to do each of the following:

- List the main components of blood.

- Determine by microscopic examination whether blood is that of a mammal or non-mammal.

- Upon microscopic examination, distinguish among each of the following: **erythrocyte, neutrophil, eosinophil, basophil, lymphocyte,** and **monocyte.**

- Identify the origins of the heart sounds heard through a stethoscope.

- Determine heart rate and pulse rate and tell how they vary with exercise.

- Use a sphygmomanometer to determine blood pressure and explain what the figures stand for.

- Locate a valve in an arm vein.

- Answer the questions throughout this exercise including the review questions at the end.

INTRODUCTION

In small organisms consisting of one or a relatively few cells, oxygen and dissolved food can diffuse directly into the protoplasm of the cell from the environment. Carbon dioxide and other wastes produced in the cell can diffuse into the environment. In larger, multicellular organisms, however, most cells are too far from the external environment for simple diffusion of nutrients and wastes to be adequate. In higher animals, well-developed circulatory systems transport materials throughout the body. Blood, the circulatory fluid of higher animals, is distributed throughout the body via a network of vessels (arteries, capillaries, and veins). The primary force behind its movement is the contraction of the heart.

Blood is composed of two basic parts: a fluid portion called **plasma**, and solid matter suspended in the plasma. The suspended matter consists of small cytoplasmic fragments called blood **platelets** and different kinds of **corpuscles**, or blood cells; these include the red blood corpuscles, or **erythrocytes**, and the white blood corpuscles, or **leukocytes**.

The erythrocytes, which are much more numerous than the leukocytes, transport oxygen. They contain a red pigment called **hemoglobin**, which combines readily with oxygen in a loose compound called **oxyhemoglobin**. Oxyhemoglobin is a brighter red than hemoglobin; this is why blood from an artery, which contains oxyhemoglobin, is a brighter red than blood from a vein, which contains oxygen-poor blood. The red blood cells are rounded disk-shaped

251

cells without nuclei when mature. They do not live very long (approximately 120 days) and are constantly replaced.

Leukocytes are usually larger than erythrocytes, and each one has a nucleus. They are motile and move with an amoeboid motion. White blood cells as a group, have several functions, one of which is to engulf bacteria and other foreign organisms which might invade the body. Certain of these cells are attracted to points of infection, where they accumulate and do battle with the invading bacteria. In the process, some of the white cells are killed and some of the surrounding tissue broken down. The combination of bacteria, broken-down tissue, and white blood cells forms the whitish semiliquid called **pus** that sometimes accumulates around a wound.

Blood platelets, cell fragments, are very small bodies that are important in the clotting of the blood.

This laboratory exercise may be said to have two purposes: first, to study blood cells and, second, to introduce certain laboratory techniques used in the study of blood and circulation.

COMPARISON OF MAMMALIAN AND NON-MAMMALIAN BLOOD

Obtain slides of blood smears from various types of organisms and make sketches of the blood cells seen in the space provided. Identify and label the red blood cells (RBC's) and white blood cells (WBC's) in each slide. The red blood cells will always be the most numerous cells.

What characteristic can be used to distinguish between mammalian and non-mammalian blood?

What selective advantage could this characteristic confer on mammals?

How does the presence or absence of a nucleus in a RBC affect (1) its life span, (2) its size, (3) its efficiency in transporting oxygen?

MORPHOLOGICAL CHARACTERISTICS OF BLOOD

Stained blood smears are customarily used to examine and identify the WBC's. Examine a stained smear of human blood under low power and high power of the microscope. Refer to Figure 29.1 and Figure 27.1 and to colored figures in a reference text. To get the best view of this slide use the oil immersion lens.

HUMAN BLOOD   **CHICKEN BLOOD**   **FROG BLOOD**   **RAT BLOOD**

ERYTHROCYTES (RBC's)

These are the most numerous objects seen in the stained smear. They are round, biconcave disks. The centers of the **erythrocytes** may appear light because the cells are thinnest there.

LEUKOCYTES (WBC'S)

Although there are a number of ways of classifying the **leukocytes** of man, the following outline of names and descriptions is generally accepted:

Granular Leukocytes (Granulocytes)

These are of several kinds, all of which have characteristically lobed nuclei and all of which have granular cytoplasm. Because of the lobed nature of the nuclei these cells are sometimes designated as **polymorphonucleocytes**, or "polymorphs." They are produced mainly in the red bone marrow. When granular leukocytes are stained with certain dyes, such as the mixture known as Wright's stain, the nucleus stains a dark purple color and the cytoplasmic granules do not all become colored the same way. On the basis of this differential staining, the granular leukocytes are further separated into groups, of which the following three kinds are usually listed.

Neutrophil

This kind of granular leukocyte has many very small dark blue granules distributed throughout the cytoplasm and is more abundant than all the other leukocytes combined. These comprise 54–62% of all leukocytes.

Eosinophil

This leukocyte has large distinctly red granules scattered throughout the cytoplasm. The nucleus most often shows only two to four lobes. Only 1–3% of the leukocytes are eosinophils.

Basophil

This kind of granular leukocyte has relatively few large lavender granules distributed somewhat irregularly throughout the cytoplasm. A few granules may overlie the nucleus which is likely to be less distinctly lobed than that of other granular leukocytes. Often the nucleus of the basophil seems only slightly indented or irregular in outline. Basophils comprise less than 1% of all leukocytes.

Agranulocytes

Lymphocytes

These are the second most numerous of the leukocytes. The cytoplasm of these leukocytes may

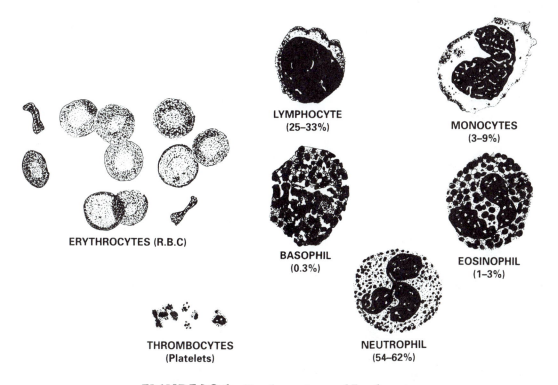

ERYTHROCYTES (R.B.C)

THROMBOCYTES
(Platelets)

LYMPHOCYTE
(25–33%)

BASOPHIL
(0.3%)

NEUTROPHIL
(54–62%)

MONOCYTES
(3–9%)

EOSINOPHIL
(1–3%)

FIGURE 29.1 Erythrocytes and Leukocytes

or may not contain small, either red or clear blue, granules, or a mixture of the two. Generally, however, **lymphocytes** are recognized by the pale to dark blue cytoplasm which surrounds a round, oval, or slightly indented purple or dark blue nucleus. Lymphocytes vary considerably in size. Smaller ones will have only a narrow band of cytoplasm around the nucleus, while larger lymphocytes will have a broader band of cytoplasm. Small lymphocytes have a diameter only slightly wider than that of an RBC. Large lymphocytes may have a size that is close to that of granulocytes. They are produced in lymphoid tissue such as that in the spleen and lymph nodes. Between 25–33% of all leukocytes are lymphocytes.

Monocytes

These cells are the largest of the leukocytes. They have a wide band of gray-blue cytoplasm containing very small clear blue granules which are generally apparent only under high magnification in a well-prepared and perfectly stained blood smear. The purple nucleus may be round, irregular in outline, or indented, but is so often "horseshoe" shaped that this characteristic is likely to be the best means of identifying them. Like the granular leukocytes, they are produced mainly in red bone marrow. From 3–9% of the leukocytes are **monocytes.**

BLOOD PLATELETS (THROMBOCYTES)

Find a cluster of blood **platelets**. These are not cells, but are particles of protoplasm separated from the cytoplasm of special giant cells in the bone marrow. They appear as tiny blue bodies, usually found in groups on stained smears.

What is the function of the platelets?

PHYSIOLOGICAL PROPERTIES OF BLOOD

BLOOD TYPE

There are four basic types: A, B, AB, and O. Blood type is determined by the presence of certain proteins on the RBC; these proteins are known as **antigens**. In addition to protein factors on the RBCs,

blood also contains certain proteins known as **antibodies** in the plasma. Some antibodies fight organisms entering the body. The disease-fighting antibodies usually develop after birth, but the blood antibodies are present at birth. A person never has the same type antibody as he does antigen. Why?

Blood Type	Antigen Type	Antibody Type	% Distribution (USA)
A	A	B	40
B	B	A	10
AB	A & B	None	5
O	none	A & B	45

In addition to the antigens mentioned above, a person may have still another blood antigen on his RBC's, the **Rh antigen**. If the antigen is present, the individual is considered Rh^+. If the antigen is absent, the individual is Rh^-. The Rh factor is of particular importance in the case of an Rh^- woman being pregnant with an Rh^+ child. If, for some reason, blood leakage occurs between the child and the mother and the Rh^+ blood comes into contact with the Rh^- blood of the mother, she may develop antibodies (Anti-Rh) against the Rh antigen of the child. These antibodies can enter the second child's bloodstream via the placenta and cause clumping of the RBC's, leading to anemia, jaundice, or even death.

The importance of blood typing is best demonstrated in the case of whole blood transfusions (i.e., both cells and plasma are given to another individual). Both the donor and recipient must be blood typed to ensure compatibility of their blood. If they are not compatible, clumping of RBC's (**agglutination**) will occur and may cause death. If a person with blood type A were to donate blood to another person with blood type B, what type of results would you expect? In this case, the recipient's A antibodies will cause a clumping of the RBC's (containing A antigens) he receives from the donor. These clumps can then cause blockage of tiny capillaries in the heart, brain, kidney, etc.

What blood types are compatible?

Would blood typing be critical if a person were to receive a transfusion of blood plasma only? Why?

CLOTTING TIME OF BLOOD

Rapid blood clotting is important in preventing excessive blood loss. This process is dependent on many different factors. Essentially what happens is that platelets, upon contact with collagen fibers exposed by the injury, release **thromboplastin**, which combines with calcium to convert **prothrombin** in the plasma to an active enzyme, **thrombin**. Thrombin acts upon another blood protein, **fibrinogen**, converting it to strands of **fibrin**. These strands form a mesh which traps cells and forms the framework for the development of a clot.

HEMOGLOBIN LEVEL IN THE BLOOD

There are many different types of anemia, the most common of which is characterized by a deficiency of hemoglobin in the RBC's. Because of this deficiency, the oxygen-carrying capacity of the blood is lowered. An easy, but not extremely accurate, test for hemoglobin level is the Tallquist method, in which the color of fresh blood is compared with standard colors on a printed chart. The darker red the blood, the higher the hemoglobin concentration.

Below are the accepted hemoglobin percentages for normal and anemic blood.

Normal	Anemia
above 85% — male	below 70% — male
above 80% — female	below 70% — female

HEMATOCRIT

The percentage of blood volume occupied by RBCs is known as the **hematocrit** or packed cell volume. It is used as a preliminary diagnostic test for anemias. If a tube of blood is properly centrifuged, the RBC's will pack into the bottom portion of the tube and the plasma will occupy the upper portion of the tube. From such a tube, it is easy to determine the percentage of volume occupied by RBC's.

Normal Values:

     Male — 40–54%   Female — 37–47%

CIRCULATORY PHYSIOLOGY

Heart rate, pulse rate, and blood pressure can be used to indicate whether the heart and blood vessels are functioning normally. Figure 29.2 shows the flow of blood through the human heart.

HEART SOUNDS

Beating of the heart is associated with a number of sounds, two of which can be heard readily with an ordinary stethoscope. The first sound is the result of the closing of the AV valves, caused by the contraction of the ventricles. It is a low-pitched "LUB" sound. The second sound is the result of the closing of the semilunar valves, due to the relaxation of the ventricles. When this occurs, blood falls back into the cuplike semilunar valves and causes them to "inflate" with a snap — like a parachute catching air. This produces a high-pitched "DUB" sound.

Procedure

1. To hear the first sound, place the stethoscope near the 5th or 6th rib, left of the center.

2. To hear the second sound, place the stethoscope near the 2nd rib, slightly to the left of center.

When one of these sounds is not distinct, but is more of a gurgle, a person is said to have a heart murmur. What could cause this? _____

_____

_____

HEART RATE

The speed at which the heart contracts is determined by many different factors, e.g., sex, age, size, physical condition, and mental state. The two heart sounds (LUB and DUB) make one complete heartbeat. The normal resting rate is 70–100 beats/min.

Procedure

1. Listen for the heart beat.

2. Count the number of beats per 15 seconds. Multiply by 4 to obtain heart rate per minute.

_____

3. Determine heart rate for each of the following conditions:

    At rest (lying) _____

    At rest (sitting) _____

    At rest (standing) _____

    After exercise (standing)_____

    Under which condition was the heart rate fastest?

    _____

    Slowest? _____

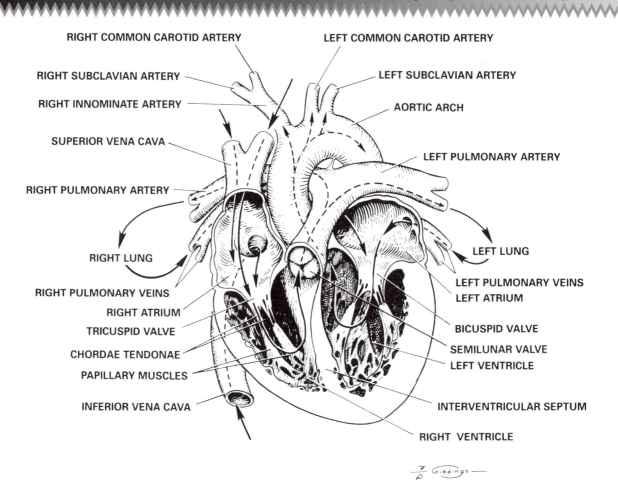

RIGHT COMMON CAROTID ARTERY

LEFT COMMON CAROTID ARTERY

RIGHT SUBCLAVIAN ARTERY

LEFT SUBCLAVIAN ARTERY

RIGHT INNOMINATE ARTERY

AORTIC ARCH

SUPERIOR VENA CAVA

LEFT PULMONARY ARTERY

RIGHT PULMONARY ARTERY

RIGHT LUNG

LEFT LUNG

RIGHT PULMONARY VEINS

LEFT PULMONARY VEINS
LEFT ATRIUM

RIGHT ATRIUM

BICUSPID VALVE

TRICUSPID VALVE

SEMILUNAR VALVE

CHORDAE TENDONAE

LEFT VENTRICLE

PAPILLARY MUSCLES

INTERVENTRICULAR SEPTUM

INFERIOR VENA CAVA

RIGHT VENTRICLE

FIGURE 29.2 Human Heart

PULSE RATE

One complete heartbeat corresponds to the pulse beat. The pulse beat is felt during the contraction of the ventricles. The most common place for the pulse to be taken is at the wrist. Average rate is 72 beats per minute (same as average heart beat). Pulse rate is also affected by such things as age, sex, activity, and body temperature.

Procedure

1. Place the fingertips gently but firmly over the radial artery along the inside of the wrist, below the base of the thumb.

2. Take a reading for 15 seconds, and multiply by 4 to obtain the pulse rate per minute. _____

_____

3. Determine the pulse rate for each of the following:

At rest (lying) _____

At rest (sitting) _____

At rest (standing) _____

After exercise (standing)_____

How long does it take for your pulse rate to get back to the value recorded at rest standing?

_____

(If it takes about 3 minutes or less, this indicates you are in good physical condition.)

How did your pulse rate differ going from sitting to standing? _____

BLOOD PRESSURE

A blood pressure reading is obtained by listening for the pulse sounds in the brachial artery. As the blood is pumped into the blood vessels by the heart, it causes pressure to be exerted against the inner walls of the arteries. During heart contraction (**systole**), the blood pressure in the arteries reaches its maximum. During heart relaxation (**diastole**), blood pressure reaches its minimum. This blood pressure is measured indirectly by the use of a **sphygmomanometer**. Factors having greatest influence on blood pressure are force of heartbeat, resistance of artery walls, elasticity of artery walls, volume of blood, and viscosity of blood.

Procedure

1. Place the cuff of the sphygmomanometer around the upper arm just above the elbow.

2. Shut off the valve near the rubber bulb and pump air into the rubber cuff by squeezing the bulb until the pressure reaches about 150 mm of mercury. This will stop the flow of blood through the brachial artery.

3. Place the stethoscope over the artery at the bend of the elbow; slowly release the thumb screw valve allowing air to leave the cuff.

4. Keep listening for blood seepage through the brachial artery. When the pressure in the cuff equals the pressure in the artery, the pulse will be heard distinctly. This thumping sound is produced by the contraction of the ventricles forcing blood into the arteries.

   This is the systolic pressure. _____(Normally this varies from 90-140 mm.)

5. Continue to listen as the pressure in the cuff decreases. The pulse will eventually become very soft and disappear. Record the pressure when the pulse sound disappears._____

   This is the diastolic pressure. (Normally this ranges from 60–90 mm.)

6. Measure blood pressure after exercising.

   Express blood pressure as $\dfrac{\text{systolic}}{\text{diastolic}}$

   How did **exercise** change the blood pressure?

   _____

   Is your pressure close to $120/_{80}$? _____

VALVES IN VEINS

The blood vessels taking blood back to the heart contain blood at a very low pressure. Therefore, to prevent backflow of blood, valves exist in veins. The location of these valves can be demonstrated as follows.

Procedure

1. Expose right forearm. Compress the vessels near the elbow until the veins in the forearm stand out prominently.

2. Lay index finger on the left hand on one of the veins near the wrist; with thumb press the blood in the vein from the finger toward the heart.

3. If the blood flows back to the finger, lay the finger at the point the thumb reaches and repeat the process. Continue until a point is found beyond which the blood will not return to the finger. This indicates the location of a valve.

4. With the finger just below the valve, attempt, by pushing downward over the vessel with the thumb, to force the blood past the valve.

Conclusions:_____

_____

_____

_____

_____

_____

CIRCULATORY PATTERN IN FROG'S FOOT

A live frog under anesthesia or with its brain removed has been positioned so that you can observe the web of the foot under low magnification. The web of thin skin between the toes affords an opportunity to observe the movement of RBC's through a capillary network. Note the rate of movement, the size of different vessels, the relative size of the RBC's and the distribution of the vessels. The black, irregular or star-shaped bodies are melanophores (pigment cells).

Does the direction of flow within the vessel reverse itself or does it always flow in one direction?

_____

Is the rate of flow correlated with size of vessel?

_____

Is it possible for blood in these vessels to exchange gases with the external environment at this point?

_____

REVIEW QUESTIONS

1. What is indicated by clotting of the blood in each of the following sera:

    Anti-A _____

    Anti-B _____

    Anti-Rh _____

2. Which type of blood agglutinates only with anti-Rh? _____

    _____

3. Name two ways of testing the blood for anemia. _____

4. Why are the nuclei of the human white blood cells blue in a prepared blood slide? _____

    _____

5. Why do the very numerous human red blood cells appear red? _____

6. Do the frog red blood cells appear red on the slide?_____

    Why? _____

7. Would a hematocrit of 30% be normal, or abnormal? _____

8. A person with low hemoglobin might be suffering from _____.

9. Platelets are important in the_____ process.

10. Explain the roles in blood clotting of thromboplastin, prothrombin, thrombin, fibrinogen, and fibrin.

    _____

    _____

    _____

11. State the function of white blood cells. _____

    _____

12. Is a neutrophil an erythrocyte or leukocyte?_____

13. Trace the flow of blood through the heart, list the chambers, valves, and vessels of the body through which you would journey. If you were a RBC, start your journey at the superior vena cava and end with the aorta. _____

    _____

    _____

    _____

14. What makes the "lub" heart sound?_____

    The "dub" heart sound?_____

Nervous System

OBJECTIVES

After completion of this exercise, the student should be able to do each of the following:

- Name the twelve pairs of **cranial nerves** and give the function and origin of each.
- Identify the major regions of the sheep, pig, and human brains, and know the major function of each.
- Diagram, label, and explain the processes involved in a **reflex arc**.
- Identify nervous tissue and its major components under the microscope.
- Answer the review questions at the end of this exercise.

INTRODUCTION TO THE NERVOUS SYSTEM

The nervous system of any animal, and especially that of man, is one of the most complex systems of the body from the standpoints of both structure and function. A complete understanding of the human nervous system is beyond the scope of this manual, but certain fundamental concepts, as outlined in the objectives, should be understood.

The nervous system functions to integrate the activities of the body by interpreting and responding to internal and external stimuli. Then, in response to the commands of the nervous system, continuous adjustments are made by practically all tissues of the body, especially the muscles and the glands.

NERVE TISSUE

The nervous system is composed of nerve tissue made up of nerve cells (**neurons**) and supportive tissue. In the brain and spinal cord, the supportive tissue is comprised of **glial cells**. In the cranial and spinal nerves, the supportive tissue is connective tissue.

Neurons are specialized for the reception of stimuli and the transmission, interpretation, and coordination of nervous impulses within the nervous system. Refer to Figure 27.1 for a diagram of the neuron.

Try to find a somewhat isolated neuron on a prepared slide. The nucleus is located in an enlarged portion of the cell called the **cell body**. Extending from the cell body are the nerve fibers which account for most of the cell length. Each vertebrate nerve fiber carries impulses in a specified direction. A **dendrite** picks up impulses and carries them toward the cell body and an **axon** carries impulses away from the cell body. It is not possible to distinguish dendrites from axons on your slide. Sketch a neuron from your slide and label the **nucleus**, the **cell body**, and a **nerve fiber**.

There are three functional types of neurons. A **sensory neuron** carries impulses toward, while a **motor neuron** carries impulses away from the brain or spinal cord. **Association neurons** (or **interneurons**) are located within the brain or spinal cord.

ORGANIZATION OF THE NERVOUS SYSTEM

All of the neurons and their supportive cells collectively make up the nervous system which can be divided into the central and peripheral parts. The **central nervous system** includes the **brain** and **spinal cord** and the **peripheral nervous system** includes the **cranial nerves** leading to and from the brain, and the **spinal nerves** leading to and from the spinal cord. Nerves, in general, are bundles of nerve fibers (axons, dendrites, or both) with the neuron cell bodies being located in the central nervous system or as "bumps" (**ganglia**) along the nerves.

BRAIN AND CRANIAL NERVES

Observe the brain of the dissected fetal pig on demonstration in the lab. The tough outer covering of the brain (especially around the cerebellum) is called the **dura mater**. The brain is composed of three primary parts: (1) the **cerebrum** (**cerebral hemispheres**) continuing forward as the **olfactory lobes**; (2) the **cerebellum** posterior to the cerebral

hemispheres and considerably smaller; and (3) the **medulla oblongata**, the enlarged anterior end of the spinal cord under the cerebellum (not easily seen from the view that you now have). Note the convolutions in the cerebrum — the folds are called **gyri** (singular — **gyrus**) and the depressions or grooves are called **sulci** (singular — **sulcus**).

Also available for study in the lab are preserved sheep brains. Use the labeled pictures in Figures 30.1–30.3 to find the parts of the brain. The cerebrum, cerebellum, and medulla are similar in appearance to the pig's brain. The two cerebral hemispheres are separated superiorly by the **longitudinal cerebral fissure** but connected inferiorly by the **corpus callosum** composed of nerve fibers connecting the two hemispheres. In a sagittal section of the sheep brain, the **pons**, **thalamus**, **hypothalamus**, **cerebral peduncles**, and the **corpora quadragemina** can be located. This view of the brain also shows that the interior of the brain has a number of **ventricles** (hollow chambers). Although the first and second ventricles are in the cerebral hemispheres and are not visible here, the third ventricle is housed within

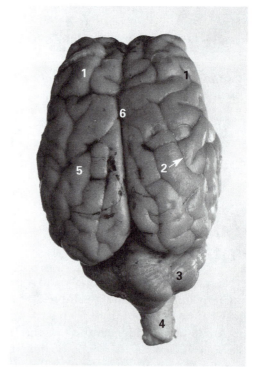

1. **CEREBRAL HEMISPHERES**
2. **SULCUS**
3. **CEREBELLUM**
4. **SPINAL CORD**
5. **GYRI**
6. **LONGITUDINAL CEREBRAL FISSURE**

FIGURE 30.1 Dorsal View of Sheep's Brain

the thalamus, and the fourth ventricle is in the area of the medulla. On the underside of the brain, one can see the **olfactory lobes** (smaller than in the pig) and the structures associated with the entry of the optic nerve into the brain (**optic nerve**, **optic chiasma**, and **optic tract**).

Using Figure 30.4 as a guide, locate the brain structures on the human brain model.

Refer to your textbook and in the space provided, give the basic function of the brain structures you have been locating.

Cerebrum:

Pons:

Medulla oblongata:

Cerebellum:

Thalamus:

Hypothalamus:

Ventricles:

Spinal cord:

Corpus callosum:

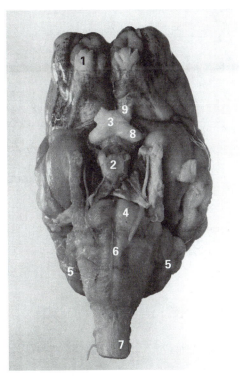

1. OLFACTORY BULB
2. PITUITARY GLAND
3. OPTIC CHIASMA
4. PONS
5. CEREBELLUM
6. MEDULLA OBLONGATA
7. SPINAL CORD
8. OPTIC TRACT
9. OPTIC NERVE

FIGURE 30.2   Ventral View of Sheep's Brain

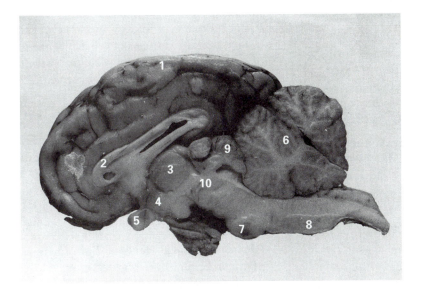

1. CEREBRAL CORTEX
2. CORPUS CALLOSUM
3. THALAMUS
4. HYPOTHALAMUS
5. OPTIC CHIASMA
6. CEREBELLUM
7. PONS
8. MEDULLA OBLONGATA
9. CORPORA QUADRIGEMINA
10. CEREBRAL PEDUNCLE

FIGURE 30.3   Sagittal View of Sheep's Brain

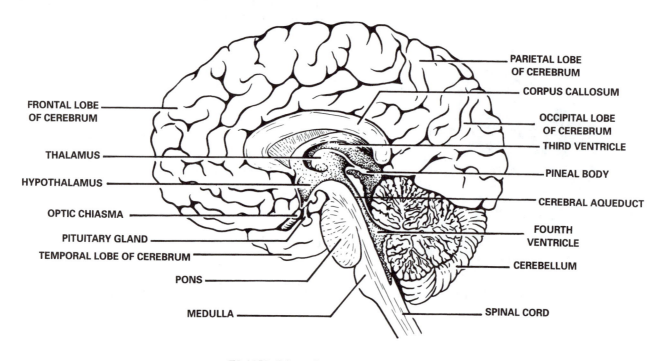

FIGURE 30.4   The Human Brain

In reptiles, birds, and mammals, there are 12 pairs of cranial nerves connected to the brain (fish and amphibians have 10 pairs). Following is a list of the cranial nerves, their origins, distribution, and their functions. They are in order from the most anterior to the most posterior. Their numbering begins with the most anterior and proceeds likewise.

I. **Olfactory Nerve**: The olfactory nerve arises from the olfactory lobe and is made up of many branches which extend through the skull into the nasal cavities to the olfactory epithelium. Function: sensory — smell.

II. **Optic Nerve**: Origin from the posterior portion of the forebrain and extends to the retina of the eye. The two optic nerves cross over to opposite sides of the brain in the **optic chiasma**. Function: sensory — sight.

III. **Oculomotor Nerve**: Origin from the ventral region of the midbrain and innervates four eye muscles (inferior oblique, superior rectus, inferior rectus, anterior rectus, also iris, lens, and upper eyelids). Function: motor — moves the eyeball, iris, lens, and upper eyelid.

IV. **Trochlear Nerve**: Origin from the dorsal region of the midbrain and innervates the eye muscle, superior oblique. Function: motor — moves eyeball.

V. **Trigeminal Nerve**: Origin from the side of the pons and innervates the skin of the face, the mouth, tongue, teeth, muscles of the jaw, and the top and sides of the head. This is the largest cranial nerve. Function: mixed, mostly sensory.

VI. **Abducens Nerve**: Origin from the ventral pons and innervates the eye muscle, posterior rectus. This is one of the smaller nerves. Function: motor — moves eyeball.

VII. **Facial Nerve**: Origin very near the trigeminal; innervating the muscles of the face, anterior 2/3 of tongue, muscles of mastication and muscles of the neck. This is a very large nerve. Function: mixed — motor: functions in the muscular changes in facial expression, chewing, and neck movement — sensory: taste.

VIII. **Statoacoustic Nerve (or Auditory)**: Origin from the side of the pons and passes to the inner ear to (1) Organ of Corti in cochlea; (2) semicircular canals. Function: sensory — (1) hearing, (2) equilibrium.

IX. **Glossopharyngeal Nerve**: Origin from the side of the medulla and goes to the muscles and membranes of the pharynx, and to the posterior 1/3 of the tongue. Function: mixed —

motor; controls motion in pharynx and tongue — sensory: receives sensations of taste and touch.

X. **Vagus Nerve**: Arises from the side and floor of the medulla and passes to the lungs, heart, stomach, intestine, esophagus, pharynx, and vocal cords. Function: mixed — motor; pharynx, vocal cords, lungs, esophagus, stomach, heart (inhibits heartbeat) — sensory: vocal cords and lungs.

XI. **Spinal Accessory Nerve**: Origin from the side of the medulla and innervates the muscles of the shoulder, palate, larynx, vocal cords, and neck. Function: motor — muscles of pharynx, larynx, and neck.

XII. **Hypoglossal Nerve**: Origin from the ventral region of the medulla and innervating the muscles of the tongue and neck. Function: motor — movement of tongue and neck.

SPINAL CORD AND SPINAL NERVES

Look at a slide of a cross section through the spinal cord under low power. The interior contains a butterfly-shaped area composed of **gray matter** (masses of neuron nerve fibers, interneurons, and motor neuron cell bodies). The gray matter is surrounded by **white matter** (masses of axons carrying impulses up and down the cord). The white matter in living material appears lighter in color than the gray matter because of the insulating **myelin sheaths** which surround the axons but not the dendrites and cell bodies. Observe the gray matter under high power and locate some of the neuron cell bodies. The small hole in the center of the cord is the **central canal** which is continuous with the ventricles of the brain.

Connected to the spinal cord are pairs of spinal nerves (33 pair in the pig, 31 pair in humans). Use Figure 30.5 and look again at the dissected fetal pig

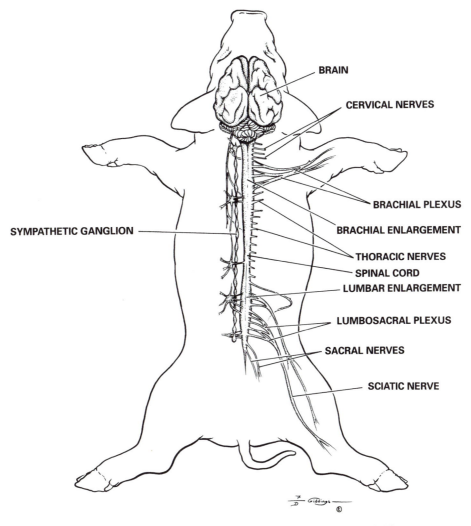

FIGURE 30.5 Fetal Pig Nervous System, Dorsal View

to locate the spinal cord and the spinal nerves coming from it. There are 8 Cervical, 14 Thoracic, 7 Lumbar, and 4 Sacral pairs of spinal nerves. Remove the skin, connective tissue, and muscles from the middoral line of the back to expose the anterior and posterior regions of the vertebral column. Next, carefully remove parts of vertebrae in the neck to expose the spinal cord. Note the tiny spherical **dorsal root ganglia** located on the dorsal root of the nerves just before they enter the spinal cord. These **may not** be visible on each nerve since they are easily destroyed during dissection. Note in Figure 30.5 the two enlargements (brachial and lumbar) on the spinal cord in the region of the appendages. Coming off of the brachial region of the spinal cord is an area composed of tough interconnected white nerves known as the **brachial plexus** which supply the forelimbs and pectoral muscles or chest muscles. Coming off the lumbosacral region of the spinal cord locate the **lumbosacral plexus** made up of branches from several of the spinal nerves. Locate the **sciatic nerve** extending down the hind leg by removing the skin, connective tissue, and muscles from the hip and upper leg.

HUMAN REFLEXES

REFLEX ARC

Some stimuli cause rapid responses that come about involuntarily, and often without the intervention of the higher centers of the brain, although in many cases, those centers are informed and may modify the response. These automatic responses are called **reflexes**. The sensory impulses that initiate a spinal reflex enter the spinal cord through the dorsal root; within the cord, they usually synapse with association neurons, which in turn transmit the impulses to motor neurons, and then to some effector, such as a muscle. The whole pathway, from sense organ through the CNS to effector, is called a **reflex arc**. Reflex activity may involve different levels of the central nervous system, depending on the complexity of the reflex. The complexity increases as more neurons become involved in the reflex.

Figure 30.6 is a diagram of a reflex arc with association neurons. The sensory neuron synapses with several association neurons in the gray matter of the spinal cord. Some of these association neurons may synapse directly with motor neurons on the same side, but some cross to the other side of the cord and there synapse with the other motor neurons

and with the additional association neurons that run in **ascending tracts** through the white matter of the cord to the brain.

Examination of specific reflexes such as the patellar reflex is used diagnostically as an indication of normal nervous system function.

For each of the reflexes listed, indicate which structure is the **receptor**, (i.e., receiving the stimulus) and which structure is the **effector** (i.e. skeletal muscle, causing the response). Be aware of the neural pathway between the two.

PATELLAR REFLEX (KNEE JERK)

Have the subject sit on a table so that his legs, from the knee down, hang freely from the table edge. Use a rubber mallet to tap the patellar ligament just below the knee. Note any movement of the lower leg. Test the opposite leg.

Repeat under each of the following conditions, noting how the reflex is affected in each case.

Subject willfully trying to stop the reflex:

Subject tightly clenching fists:

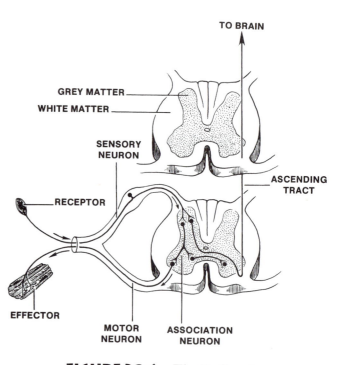

FIGURE 30.6 **The Reflex Arc**

Subject adds a column of figures:

Receptor:_____ Effector:_____

ACHILLES REFLEX (ANKLE JERK)

Have the subject sit on a chair, his feet hanging free over the edge of the chair. The tester should then place palm of hand under foot and hold it parallel to the floor. Then with the other hand, tap the Achilles tendon and note movement of the foot as it presses into hand of tester. See Figure 2-8 (Exercise 32) in the color plates for muscle identification.

Have the subject pull against the back of the chair and repeat the test. Result?

Receptor:_____ Effector:_____

PHOTO-PUPIL REFLEX

Have the subject sit facing a bright light (goose-neck lamp), with eyes closed and covered with hands. After a minute, have him uncover and open one eye. Note the size of the pupil.

Have the person open the other eye, but keep it shielded from the direct light with his hand in between. Observe both pupils. Explain the difference in size. **Refer to your textbook** for information on muscles controlling the constriction and dilation of the iris.

Receptor:_____ Effector:_____

CONVERGENCE REFLEX

Observe the position of the eyeballs while the subject is looking at an object 20 feet away.

Now have him focus on a pencil held 10 inches from his face. Does the position of the eyeball change as the gaze shifts from a far to a near object?

How?

Refer to your textbook for information on the muscles that control eye movement.

Receptor:_____ Effector:_____

FROG REFLEXES — OPTIONAL

As you examine some reflexes in a frog do not forget that spinal reflexes, though they do not require the intervention of the brain, may nevertheless be modified by it, and that most reflexes are not nearly as simple as we have outlined above.

REFLEX INVOLVING RELATIVELY SIMPLE RESPONSE

The instructor will prepare a frog in which the cerebral hemispheres have been destroyed by cutting off the cranium (upper part of the head) just posterior to the eyes. Such an animal is called a "decerebrated" frog and is incapable of sensations such as pain. Any response of muscles to external stimuli in such a frog are reflexes. Suspend the frog by the jaw from a clamp on a ring stand so the legs hang freely (Figure 30.7). Bring a small beaker containing 1% hydrochloric acid up under one foot so that one toe of the frog contacts the acid. Be sure that no other part of the foot or leg touches the sides of the beaker.

Describe what happens.

FIGURE 30.7   Frog Preparation

Submerge the foot in a large beaker of 2% sodium bicarbonate solution for a few seconds, to neutralize the acid, then rinse the foot in a large beaker of tap water.

REFLEX INVOLVING COMPLEX PURPOSEFUL RESPONSE

After the frog has rested for several minutes and the legs have relaxed, place, on the upper part of the thigh, a piece of filter paper (about 1/4 in. square) which has been moistened with 30% acetic acid.

Describe the frog's reaction.

Dip the entire leg of the frog in the 2% sodium bicarbonate solution and rinse with tap water. After the legs are relaxed, repeat the experiment, but this time hold the toes of the leg on which the paper is placed, so that the frog cannot move that leg. Describe what happens.

How did the nerve impulses reach the muscles in the leg which responded, but which was not stimulated by the acid?

EFFECT OF STRENGTH OF STIMULUS ON REFLEX TIME

After the frog has rested for several minutes and the legs are relaxed, suspend the preparation as before and perform the following experiment.

Submerge the long toe of the frog in 0.05% HCl solution. Carefully measure with a second hand of a watch or clock, the time between the application of the acid and the reflex response, if any. (It may take as long as 1½ min.)

Time: _____

After rinsing the toe in a beaker of 2% sodium bicarbonate, dry the toe and repeat the above procedure using 0.25% HCl.

Time: _____

Repeat using 0.5%, HCl.

Time: _____

Repeat using 1% HCl.

Time: _____

Plot your data on the graph below. Explain your results.

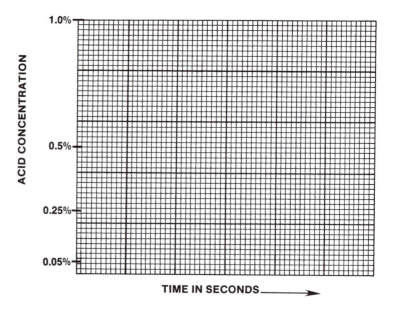

Does an increase in the strength of the stimulus cause the nerve impulses to travel with greater or lesser speed over the nerve fibers?

_____

_____

Why was reflex time different with the higher concentrations of acid?

_____

_____

REVIEW QUESTIONS

1. What region of the brain has, as one of its functions, the coordination of muscular activity? _____

_____

2. Define these terms:

Receptor _____

Effector _____

Reflex _____

Stimulus _____

Ascending tract _____

Gyri _____

Central nervous system _____

Peripheral nervous system _____

Dendrite _____

Axon _____

Ganglia _____

Glial Cell _____

3. Give the functions of the following:

Corpus callosum _____

Motor neuron _____

Myelin sheath _____

Optic nerve _____

Abducens nerve _____

Vagus nerve _____

Glossopharyngeal nerve _____

4. Describe the sequence of events in a simple reflex arc. _____
_____
_____

5. In the patellar reflex activity was the student able to control their response to the stimulus? _____

Explain: _____
_____
_____
_____

6. Compare the functions and locations of sensory neurons, motor neurons, and association neurons. _____
_____
_____

7. How do white matter and gray matter differ from each other in the spinal cord? _____
_____
_____

8. What is a plexus and name two within the peripheral nervous system._____
_____
_____

SENSORY MECHANISM

EXERCISE

31

OBJECTIVES

After completion of this exercise, the student should be able to do each of the following:

- Describe the effects of stimuli on those sense organs capable of responding to them.
- Label the major structures found in the various layers of the skin and know the function of each.
- Identify taste buds under the microscope.
- Describe the functions of the major parts of the human ear and eye.
- Label the parts of the human ear and eye.
- Be able to diagram, label, and explain the functioning of the retina.
- Define and explain the causes of **afterimage, nearsightedness, farsightedness, colorblindness**, and **astigmatism.**
- Answer the review questions at the end of this exercise.

INTRODUCTION TO SENSORY MECHANISMS

The nervous system functions in coordinating body activities in response to changes in internal and external stimuli. The particular part of the nervous system specialized in detecting these environmental changes (**stimuli**) is the **sensory receptor**, which initiates the transmission of neural impulses. Receptors vary from simply being "free" nerve endings (pain receptors) to being organized in complex sense organs (eye, ear). The actual sensation detected by the individual is not so much a result of the particular type of stimulus as it is a result of the specific region of the brain to which the impulses are sent.

Although it is said that man has five senses (**sight, taste, hearing, smell**, and **touch**), many more than five types of stimuli can be interpreted by the brain. Stimuli arising from the external environment are detected by the group of receptors known as **exteroceptors**. This group includes the eyes, the auditory part of the ear, the olfactory epithelium, and the receptors located in the skin. Stimuli arising within the body are detected by the group of receptors known as **interoceptors**. This group includes numerous receptors located in the visceral organs (giving rise to sensations of pain, hunger, thirst, nausea, etc.) and in the muscles, tendons, and joints (giving rise to sensations concerned with body position and movement).

Certain receptors display a phenomenon known as **adaptation**. Under continuous stimulation, these receptors respond less and less to a particular stimulus, thus enabling the individual to "get used to" a specific odor, temperature, pressure, etc.

In this exercise, you will learn the basic structure of certain sensory receptors. In addition, a number of experiments will be done on the sensory structures to help you understand their functioning more clearly.

CUTANEOUS SENSES

SKIN STRUCTURE

The skin (Figure 31.1) is an organ composed of three basic layers: epidermis, dermis, and subcutaneous layer. The **epidermis** is a region in which mitosis occurs continuously to produce more layers of squamous epithelium, as the outermost layer is being sloughed off continuously. The **dermis** is a region composed mainly of connective tissue. It is in this area that such structures as sweat glands, lymph and blood vessels, hairs (with associated arrector muscle and sebaceous glands), sensory receptors, and nerves are located. The **subcutaneous layer** is the region composed of loose connective tissue that binds the rest of the skin to the body. Here are located numerous fat cells, blood vessels, lymph vessels, the bases of hair follicles, nerves, and deep pressure receptors. Using your text and the model keys as a guide, find the parts of the skin indicated in the diagram on the model and in prepared slides.

There are structurally distinct cutaneous exteroceptors for **pain**, **touch**, **pressure**, **warmth**, and **cold** located in the skin. Stimulation of any one of these will generally give rise to a specific sensation; simultaneous stimulation of different types of receptors will give rise to sensations such as itching, crawling, burning, etc.

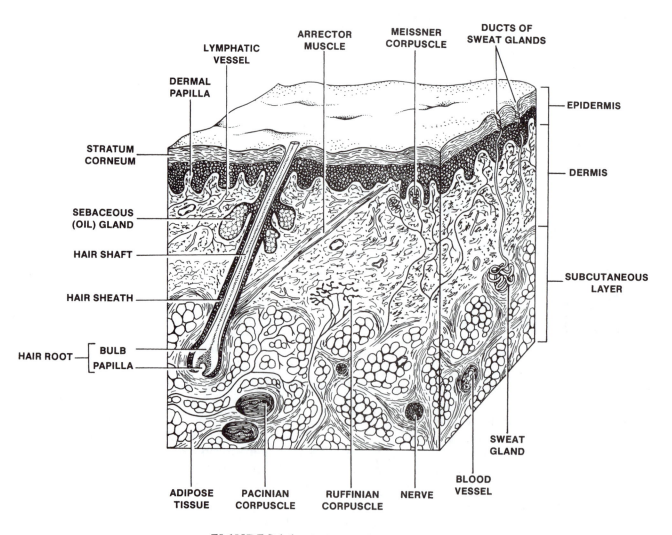

FIGURE 31.1 Section of Human Skin

SKIN SENSATIONS

Working with your lab partner, perform the following experiments and answer the associated questions.

Two-Point Discrimination Test

The distribution of **tactile** (touch) receptors varies over different areas of the body. With two clean toothpick points, touch the skin on the following suggested areas of the body. Start with the points very close together and increase the distance between the points until the person is able to discern two distinct points. Be sure to lift the points off the skin when increasing the distance between points. Record this distance in millimeters for each of the areas tested.

Tip of Nose _____

Back of Neck _____

Palm of Hand _____

Tongue _____

Back of Hand _____

Ball of Thumb _____

Lips _____

What can you conclude from the above data? _____

_____

_____

Adaptation of Touch Receptors

Note the time on a watch with a second hand, and place a coin, such as a penny or a dime, upon the skin of the inside of the forearm. How long does the sensation last?_____

   This disappearance of sensation is due to the fact that the receptors "adapt" to the particular stimulus and, therefore, initiate no more nerve impulses until a change takes place in the stimulus. Repeat the experiment on a new spot on the skin, but after the sensation disappears, add two more coins of the same denomination on the top of the coin. Did the sensation return? _____

How long did the sensation last when 3 coins were used? _____

Adaptation of Heat and Cold Receptors

Set up 3 beakers, one with cold water, one with water at room temperature, and one with tolerably hot water. Place one index finger in the cold water and the other index finger in the hot water.

What are your initial sensations? _____

_____

What are your sensations after several minutes

(fingers still in the beakers)? _____

_____

   IMMEDIATELY after removing fingers from hot and cold beakers, plunge both fingers simultaneously into the beaker of water at room temperature. Even though both fingers are supposedly being stimulated by the same temperature of water, are the

sensations the same? _____

Are the stimuli really identical for each finger?

_____

What are the receptors actually responding to?

_____

TASTE (GUSTATORY SENSE) AND SMELL (OLFACTORY SENSE)

TASTE BUD STRUCTURE

The **taste buds** are barrel-shaped structures that open through the surface of epithelial cells by way of pores. Within the taste bud are the actual receptor cells with hairlike extensions that detect the materials that are dissolved in saliva. The taste buds are most numerous on the tongue, specifically on some of the **papillae** ("bumps") of the tongue surface. The papillae are rounded mounds, each surrounded by a deep groove. It is into these grooves that the pores of the taste buds open.

   Examine a slide of a taste bud. Make a diagram of what you see and label all the parts you can identify. Use the space provided for your diagram.

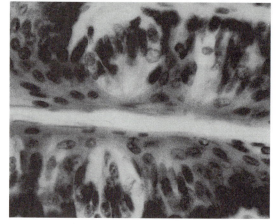

Photomicrograph of human taste bud (430X)
Courtesy of Turtox, Inc.

Stimulation of Taste Buds

Dry the upper surface of your tongue with a clean tissue and have your lab partner place a few granules of sugar on it. Can you taste the sugar immediately?

Why?_____

_____

Combined Effects of Taste and Smell

Obtain several pieces of distinctly flavored chewing gum. Dry the surface of your tongue, hold your nostrils shut and **close your eyes**. Have your partner place a piece of gum (flavor unknown to you) on

your tongue. Can you identify the flavor?_____

Now, still holding your nostrils shut, chew the gum. Record your sensations. Open your nostrils and record the changes in sensation.

Nostrils closed: _____

Nostrils opened: _____

Dry the surface of your tongue and **shut your eyes**, but do not pinch your nostrils shut. Have your lab partner place one flavor of gum on your tongue while holding a piece of different flavor under your open nostrils. Record your sensations. _____

_____

_____

DIAGRAM OF HUMAN TASTE BUD

OLFACTORY RECEPTORS

The **olfactory receptors** are specialized nasal epithelial cells covering about a postage stamp-sized area in the upper region of each nasal cavity. The receptors themselves are cells containing hair-like extensions stimulated by molecules dissolved in the mucous covering the nasal epithelium.

DISTINGUISHING BETWEEN SENSATIONS

Sensations arising from stimulations of taste and smell receptors are often difficult to distinguish from each other. Some of the following experiments are designed to illustrate the relationship between gustatory and olfactory sensations.

Combined Effects of Taste, Smell, and Touch

Obtain small 1/4 inch cubes of carrot, onion, potato, and apple. Dry your tongue, shut your eyes, and pinch the nostrils shut. Have your lab partner place one of the materials on your tongue. Attempt to identify (1) immediately, (2) after chewing (nostrils closed), and (3) after opening the nostrils. Record your observations on the table below.

Substance	Immediately	Observations After Chewing	Open Nostrils
Carrot			
Onion			
Potato			
Apple			

AUDITORY AND EQUILIBRIUM SENSES

EAR STRUCTURE

The ear is a complex sense organ that can be divided into three basic regions: the **external ear**, the **middle ear**, and the **inner ear**. The external ear composed of the **pinna** (**auricle**) and the **external auditory canal** which function in funneling sound waves into the ear. The middle ear is composed of the **tympanic membrane** (eardrum) and the **middle ear bones** (**malleus, incus, stapes**). These four structures function in transmitting sound vibrations across the **middle ear cavity** to the inner ear. The middle ear cavity and the pharynx are connected by the **Eustachian tube**. The inner ear is actually a double organ. The exteroceptor portion (**cochlea**) is stimulated by vibrations coming from the stapes. The stapes vibrates against the **oval window** membrane of the cochlea. The oval window membrane is at the base of the upper canal. As the oval window vibrates, it in turn causes wave formation in the fluid of the upper canal. The membrane of the **round window** moves outward as the membrane of the oval window moves inward, allowing the fluid to oscillate back and forth. This produces movement of fibers in the **basilar membrane** of the central canal, which in turn causes the compression of individual sensory hairs in the **organ of Corti**. The hair cells connect with the **cochlear branch** of the **statoacoustic nerve** which carries the impulses to the brain. It is in the temporal lobe of the cerebrum where the sound discriminations are made.

In addition to hearing, parts of the inner ear are concerned with the sense of balance and equilibrium. Fluid in the **semicircular canals** moves as the position of the head changes, simulating tufts of sensory hairs at the bases of the canals. The overall position of the head when still is detected by hair cells in the two chambers of the **vestibule**, at the junction of the semicircular canals and the cochlea. Small crystals of calcium carbonate called **otoliths** entangled in these hairs exert unequal pull on the hairs in response to gravitational pull. Nerve impulses from the semicircular canals and vestibule travel to the brain through the **vestibular branch** of the statoacoustic nerve. Locate each of the parts labeled in Figure 31.2 on the model of the human ear. Refer to Figure 31.3 while identifying the parts of the organ of Corti. Identify the following: **vestibular membrane**,

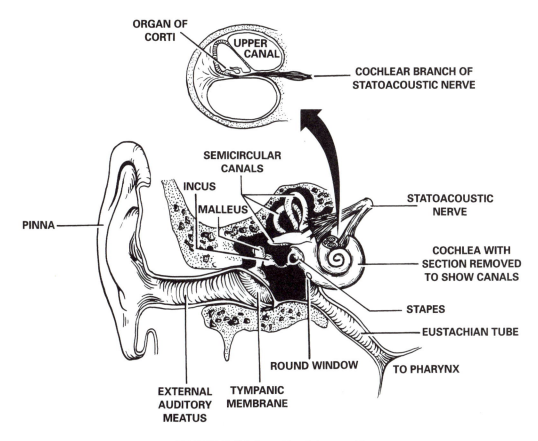

FIGURE 31.2   **The Human Ear**

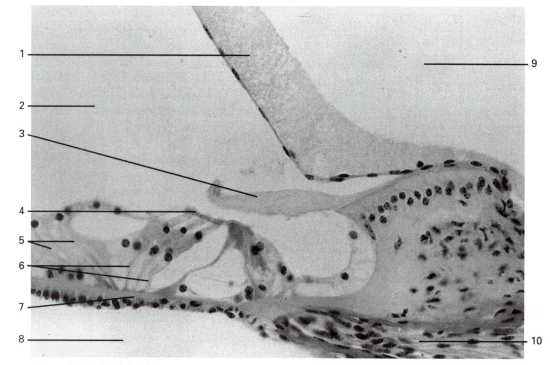

FIGURE 31.3. Organ of Corti, Within the Cochlea of the Inner Ear

1. Vestibular membrane
2. Cochlear duct
3. Tectorial membrane
4. Dendritic endings of hair cells
5. Supporting cells
6. Nerve fibers
7. Basilar membrane
8. Scala tympani
9. Scala vestibuli
10. Cochlear nerve

cochlear duct, tectorial membrane, basilar membrane, and cochlear nerve, while observing a slide of the Organ of Corti.

EXPERIMENTS ON THE EAR

Visual Sensations in Equilibrium

Normal maintenance of equilibrium is dependent on sensations from a number of receptors, including those of the semicircular canals, those of muscles and joints, and those associated with vision. The role of vision can be demonstrated by performing the following experiment. With eyes open, stand up straight and raise one foot about 12 inches off the floor. To check the stability of your stance, have a lab partner determine approximately how many inches you sway from side to side. Close your eyes and repeat the procedure. Again have a lab partner check the amount of sway. What differences are noted? ___

_____

_____

Air Pressure in the Middle Ear

Close your nostrils and mouth and swallow hard. Describe what sensations you experience in your ears.

What is the function of the Eustachian tubes?

_____

_____

Why should there be equal pressure on both sides of

the tympanic membrane? _____

_____

_____

Bone Conduction of Sound Waves

Wash and dry the handle of a tuning fork. Cause it to vibrate by tapping the prongs against the palm of the hand. Hold the fork close to the ear until the sound is nearly inaudible, then place the handle firmly between the teeth. Can you hear the sound now? Explain. _____

_____

_____

_____

Again set the tuning fork to vibrating and place the handle on top of the head. Where does the sound seem to originate? Now close one ear. Where does the sound seem to come from now? _____

_____

_____

Auditory Adaptation or Fatigue

Place the earpieces of a stethoscope in your ears. By pinching the tube, close the passage that leads to your left ear. Sound a tuning fork close to the open end of the stethoscope. When the sound has become almost inaudible to the right ear (through the open tube), open the pinched tube leading to your left ear. What happens? Explain. _____

_____

_____

VISUAL SENSE

EYE STRUCTURE

The optical system of the eye consists of a lens system (cornea and a crystalline lens) and a light-sensitive surface (retina). The function of the lens system is to focus light from objects so that the light-sensitive surface is stimulated by the pattern of light and dark areas of the image. The light sensitive surface converts the image pattern into nerve impulses to be transmitted to the brain by way of the optic nerve. Locate each of the parts labeled on Figure 31.4 on the model of the human eye.

DISSECTION OF SHEEP EYE

Obtain a preserved sheep's eye and find the stumps of the optic nerve and the eye muscles. Then examine the eye and identify the **sclera**, the white cartilaginous coat covering the hole outside. Note the cloudy area in the front, which is the **cornea** (a specialized area of the sclera that is transparent when living). Identify the **pupillary opening** in the pigmented **iris**, and the **lens**, partially visible through the pupil. Carefully bisect the eye with a sharp scalpel *in a plane parallel to the front surface*, so that all

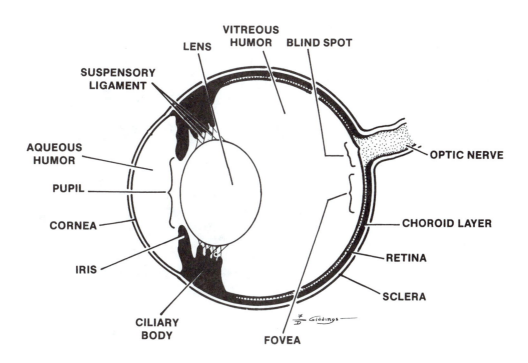

FIGURE 31.4 **The Human Eye**

of the cornea, iris, and lens will be in one half and the back portions of the eye in the other half. The jellylike material filling the main cavity of the eye is the **vitreous humor**, which aids in maintaining eye shape.

Examine the back half from the inside. The pale, loose tissue is the **retina**. Note the central **blind spot,** where the nerve fibers from the retina join the **optic nerve** and where the **retinal blood vessels** pass into the eye. Lift the retina and examine the underlying **choroid coat**, pigmented black. Note the shiny, greenish material covering part of the choroid coat. This is the **tapetum lucidum**, which acts as a mirror reflecting incident light back into the retina. A tapetum is not present in man. Describe the position of the tapetum. Scrape off some of the choroid layer and note that the choroid lies to the inside of the sclera, the outer eye covering already identified above.

Examine the front half of the eye from the inside. Identify the **lens** and note its attachment to the **suspensory ligaments**. The latter are contained in the fine, radially striated membrane which surrounds the lens. The suspensory ligaments are attached peripherally to the **ciliary body**, a ring-shaped muscle pigmented black on its surface. Remove the lens and look through it. Is the curvature of the lens surface the same in front and back? Follow the choroid coat toward the front and note that it continues forward as the **iris**. Probe through the pupil into the anterior chamber containing the **aqueous humor**, a space bounded in front by the cornea.

STRUCTURE OF THE RETINA

As was mentioned above, the retina is the layer of the eye which contains the receptor cells responsible for initiation of neural impulses stimulated by light. Looking at Figure 31.5, note that the retina consists of two layers: an outer thin **pigment layer** adjacent to the choroid layer and an inner **neural layer** adjacent to the vitreous humor.

The pigment layer is composed of the tips of the receptor cells, the **rods** and **cones**. The remainder of these cells lie in the neural layer and synapse with **bipolar cells**, which in turn synapse with **ganglion**

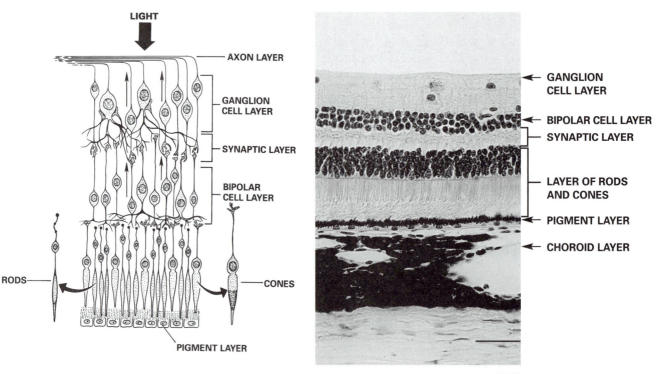

250X

FIGURE 31.5   **The Human Retina**

cells. The axons of the ganglion cells come together at the back of the eyeball to form the optic nerve. Because of this arrangement, with the pigment layer beneath the neural layer, light must pass through the ganglion and bipolar cells to reach the pigment layer and generate an impulse.

EXPERIMENTS ON THE FUNCTIONING OF THE EYE

Blind Spot

With the left eye closed, hold the following black cross and dot about 20 inches from your face with the cross directly in front of the right eye. You should be able to see both figures even though you are looking at the cross. Keep the left eye closed and slowly bring the page closer to the face while continuously looking at the cross with the right eye. At a certain distance the dot will disappear from view, because its image is falling on the blind spot of the eye. Continue to move the page toward the eye. Does the dot reappear? Explain.

Repeat the experiment using the left eye; this time concentrate on the dot instead of the cross. Are results similar to those for the right eye? _____

_____

What is the blind spot of the eye? _____

_____

Visual Acuity Test

Using the Snellen Chart determine which line of letters you can easily recognize from the standard distance of 20 feet from the chart. Have your lab partner work with you to note which line you can read accurately. In writing the results of this test, set up the following equation:

$$\frac{20 \text{ (standard distance from the chart)}}{\underline{\quad} \text{ (the line you can read from 20 ft. away)}}$$

For example, if, with your right eye, you can read the line labeled as 40 ft., your vision would be 20/40 for your right eye (i.e., at 20 ft. your right eye can see what the normal-visioned individual can see at 40 ft., indicating poorer than average vision). Test each eye separately (without glasses first and then with, if you wear them). Record your results below.

Right Eye _____   Left Eye _____

Most cases in which the vision is poorer than average result from structural defects in the eye. If the eyeball is too short, or if the lens is too flat, the incoming light rays will focus at a point behind the eyeball. This is known as **hyperopia**, or **farsightedness**, because the person can see distant objects more clearly than those which are close up. If the eyeball is too long, or if the lens is too convex, **myopia**, or **nearsightedness**, results because the focal point falls in front of the retina. Another common defect, **astigmatism**, occurs as a result of unequal curvature of the cornea or lens. In this condition light rays focus in a line rather than in a point.

Colorblindness and Astigmatism

Examine the charts available to test for astigmatism and colorblindness. Do you have either of these conditions? _____

What type of colorblindness do you have? _____

_____

Demonstration of Negative Afterimage

Color with bright green, bright yellow, and black crayons or felt pens the American flag pictured below. Place the colored flag at a distance of approximately one foot in front of you in a strong light and concentrate your gaze at the small target "X" in the center of the flag. Keep your eyes focused on the target for at least ten full seconds. Now look at a blank piece of white paper or at the white ceiling of the room. Describe, in the space below, the afterimage you see.

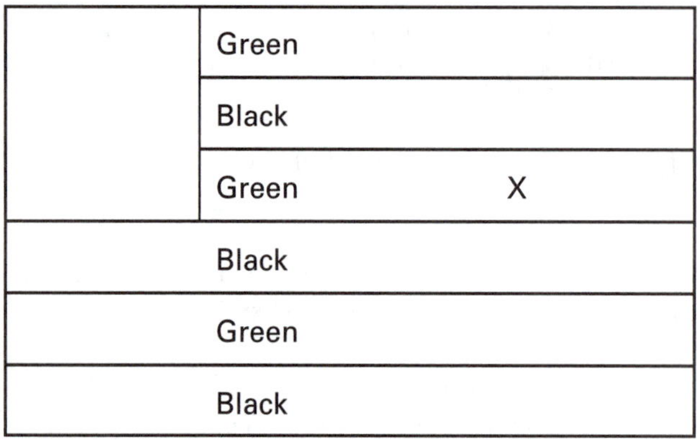

Green	
Black	
Green	X
Black	
Green	
Black	

Afterimages appear to form as a result of fatigue in the receptor cells, and possibly in the bipolar and ganglion cells as well. It is thought that the cells increase their rate of firing in response to one color, but decrease their rate of firing in response to its complementary color. When one stares intently at a uniform green field for a period of time, the cells which fire at an increased rate for green (green on-red off cells) become fatigued, so that if one looks at a uniform white field, the reduced firing of these cells is interpreted as redness.

REVIEW QUESTIONS

1. List the five cutaneous senses and the receptors that monitor them.

   a. _____

   b. _____

   c. _____

   d. _____

   e. _____

2. In the two point discrimination test, which area had the highest number of exteroceptors? _____

   _____

3. Describe the steps involved in the passage of sound through the ear._____

   _____

   _____

   _____

   _____

   _____

4. From your tests with olfactory and gustatory receptors, what are your conclusions regarding the relationship of these two senses? _____

   _____

   _____

5. What is auditory adaptation or fatigue? _____

   _____

6. What structure helps to regulate the air pressure on either side of the tympanic membrane? _____

7. Trace the steps that would be involved in the passage of a beam of light through the eye until the impulse leaves through the optic nerve._____

   _____

   _____

8. Describe the functions of the following:

Vestibular branch of the statoacustic nerve _____

Round window _____

Oval window _____

Cochlear branch of the statoacustic nerve _____

Organ of Corti _____

Semicircular canals _____

Retina _____

Crystalline lens _____

9. What kind of lens would be prescribed to correct farsightedness? _____

Nearsightedness?_____

Astigmatism? _____

THE HUMAN SYSTEMS

INDEX

THE HUMAN SYSTEMS

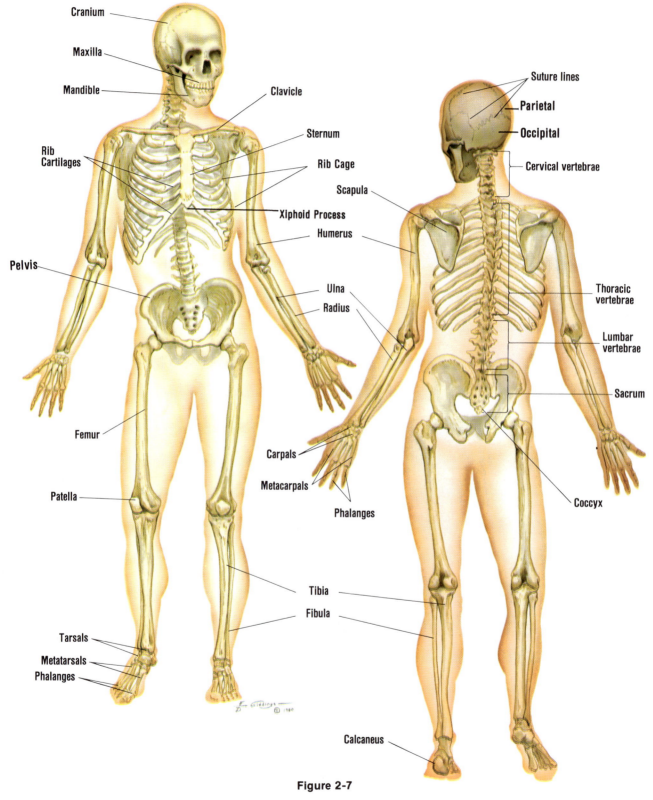

Cranium

Maxilla

Mandible

Clavicle

Sternum

Rib Cartilages

Rib Cage

Scapula

Xiphoid Process

Humerus

Pelvis

Ulna

Radius

Femur

Carpals

Metacarpals

Patella

Phalanges

Tarsals

Metatarsals

Phalanges

Tibia

Fibula

Calcaneus

Suture lines

Parietal

Occipital

Cervical vertebrae

Thoracic vertebrae

Lumbar vertebrae

Sacrum

Coccyx

Figure 2-7

THE SKELETAL SYSTEM

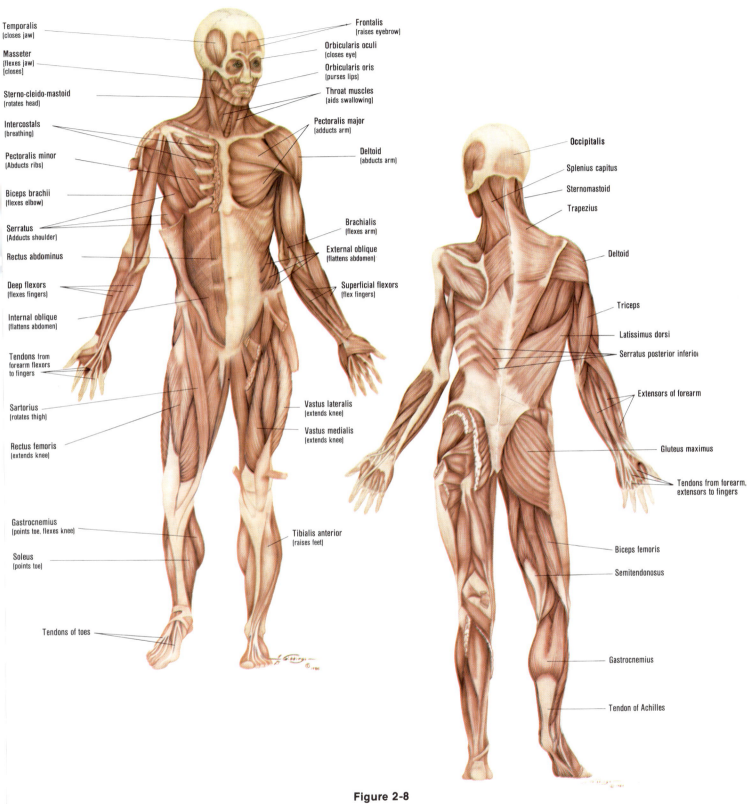

Temporalis
(closes jaw)

Masseter
(flexes jaw)
[closes]

Sterno-cleido-mastoid
(rotates head)

Intercostals
(breathing)

Pectoralis minor
(Abducts ribs)

Biceps brachii
(flexes elbow)

Serratus
(Adducts shoulder)

Rectus abdominus

Deep flexors
(flexes fingers)

Internal oblique
(flattens abdomen)

Tendons from
forearm flexors
to fingers

Sartorius
(rotates thigh)

Rectus femoris
(extends knee)

Gastrocnemius
(points toe, flexes knee)

Soleus
(points toe)

Tendons of toes

Frontalis
(raises eyebrow)

Orbicularis oculi
(closes eye)

Orbicularis oris
(purses lips)

Throat muscles
(aids swallowing)

Pectoralis major
(adducts arm)

Deltoid
(abducts arm)

Brachialis
(flexes arm)

External oblique
(flattens abdomen)

Superficial flexors
(flex fingers)

Vastus lateralis
(extends knee)

Vastus medialis
(extends knee)

Tibialis anterior
(raises feet)

Occipitalis

Splenius capitus

Sternomastoid

Trapezius

Deltoid

Triceps

Latissimus dorsi

Serratus posterior inferior

Extensors of forearm

Gluteus maximus

Tendons from forearm,
extensors to fingers

Biceps femoris

Semitendonosus

Gastrocnemius

Tendon of Achilles

Figure 2-8

THE MUSCULAR SYSTEM

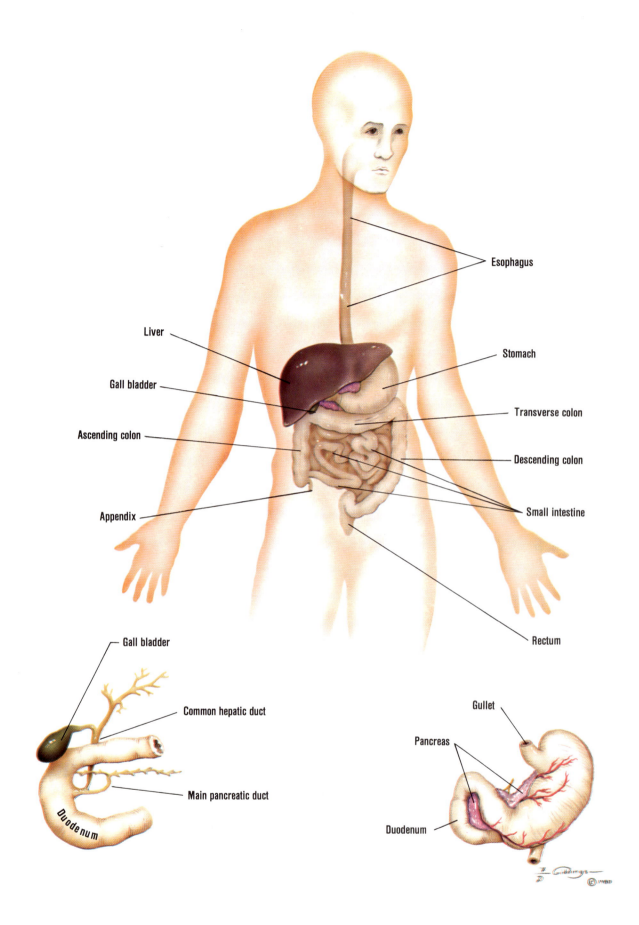

Esophagus

Liver

Stomach

Gall bladder

Transverse colon

Ascending colon

Descending colon

Appendix

Small intestine

Rectum

Gall bladder

Common hepatic duct

Main pancreatic duct

Duodenum

Gullet

Pancreas

Duodenum

Figure 2-11

THE DIGESTIVE SYSTEM

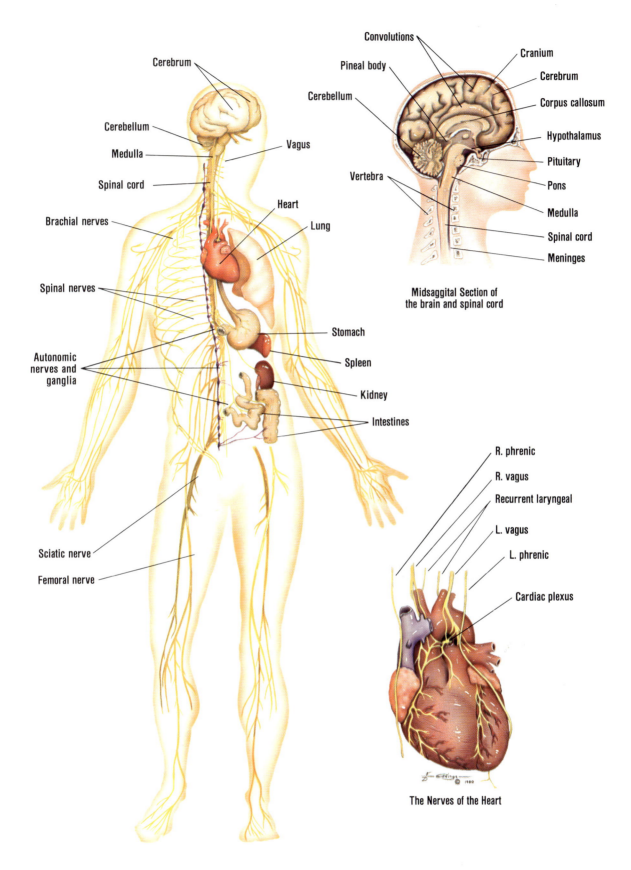

Cerebrum

Cerebellum

Medulla

Spinal cord

Brachial nerves

Spinal nerves

Autonomic
nerves and
ganglia

Sciatic nerve

Femoral nerve

Vagus

Heart

Lung

Stomach

Spleen

Kidney

Intestines

Convolutions

Pineal body

Cerebellum

Vertebra

Cranium

Cerebrum

Corpus callosum

Hypothalamus

Pituitary

Pons

Medulla

Spinal cord

Meninges

Midsaggital Section of
the brain and spinal cord

R. phrenic

R. vagus

Recurrent laryngeal

L. vagus

L. phrenic

Cardiac plexus

The Nerves of the Heart

Figure 2-12

THE NERVOUS SYSTEM

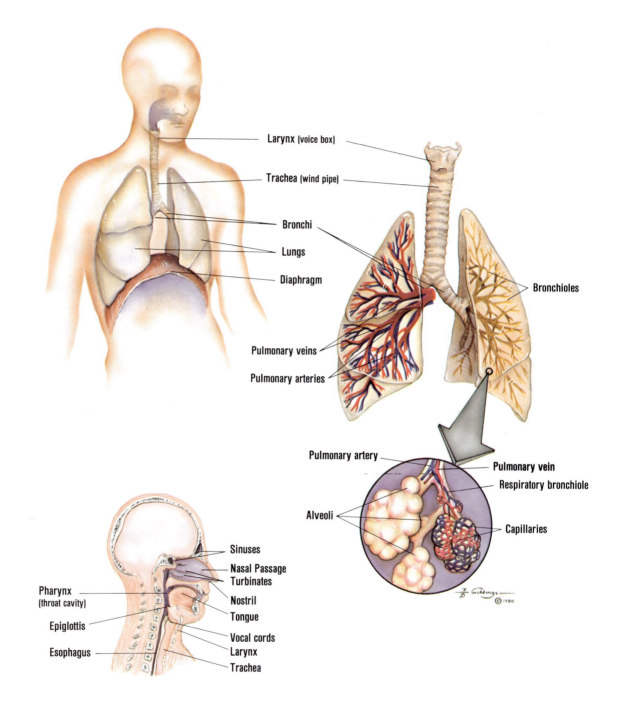

Larynx (voice box)

Trachea (wind pipe)

Bronchi

Lungs

Diaphragm

Pulmonary veins

Pulmonary arteries

Bronchioles

Pulmonary artery

Pulmonary vein

Respiratory bronchiole

Alveoli

Capillaries

Sinuses

Nasal Passage

Turbinates

Pharynx
(throat cavity)

Nostril

Tongue

Epiglottis

Vocal cords

Esophagus

Larynx

Trachea

Figure 2-13

THE RESPIRATORY SYSTEM

Skeletal System

OBJECTIVES

After completion of this exercise, the student should be able to do each of the following:

- Diagram and label the parts of a bone in longitudinal and cross sections.
- Identify the major parts of bone tissue under the microscope.
- Identify the major bones of the human skeleton (both articulated and disarticulated) as well as in diagrams.
- Name and describe the three major types of joints and give examples of each.
- Name the various types of freely movable joints and give examples of each.
- Identify the various types of teeth in the human.
- Diagram and label a longitudinal section through a human tooth.
- Describe the functions of the various types of teeth and the layers of the individual tooth structure.
- Answer the review questions at the end of this exercise.

INTRODUCTION TO THE SKELETAL SYSTEM

The skeletal system in the vertebrate animal is comprised of **bone, cartilage**, and **ligaments.** It functions in such things as body support and protection of internal organs, and also provides sites for skeletal muscle attachments, thus functioning in locomotion. In addition, it is involved in the formation of blood cells and serves as a storage site for calcium and phosphorus salts.

In this exercise we will examine the structure of the skeletal system, beginning with bone and cartilage and its microscopic structure and ending with the gross anatomy of the entire system. Various types of joints (sites where two bones meet) and their movements will also be studied.

BONE STRUCTURE

PARTS OF A BONE

Bones are comprised of a type of connective tissue in which the **osteocytes** (individual cells) are separated by an intercellular matrix hardened by the deposition of tricalcium phosphate $(Ca_3(PO_4)_2)$ and calcium carbonate $(CaCO_3)$. This bony tissue exists as two types: a solid **compact** type and a **cancellous** type (porous spongy type). In long bones, the **diaphysis** (shaft portion) is primarily a cylinder of compact bone surrounding a central cavity containing **yellow bone marrow.** The outer surface of the diaphysis is covered by a tough fibrous membrane known as the **periosteum.** The **epiphysis** (enlarged end) of long bones is primarily spongy bone whose spaces contain red bone marrow. The spongy bone is covered by a thin layer of compact bone, which is covered in turn by **articular cartilage.** In young animals there is a thin layer of cartilage (**epiphyseal disc**) between the diaphysis and epiphysis; this disc is the site of bone elongation.

SLIDE OF GROUND BONE

The structure of spongy bone is such that the osteocytes can easily acquire their nutrients and eliminate their wastes by the blood capillaries in the numerous marrow-filled spaces around the bony network. In compact bone, because of its density, blood vessels

from the periosteum are distributed through the bony tissue in passageways known as **Haversian canals.** These canals contain blood and lymph vessels, as well as nerve fibers. The osteocyte in its **lacuna** (cavity) communicates with the Haversian canal and neighboring osteocytes by even smaller canals known as **canaliculi,** into which protoplasmic extensions of the bone cell extend. The osteocytes are distributed in concentric rings around a central Haversian canal; the concentric layers of matrix surrounding the canal are known as **lamellae.** The entire structure made up of the Haversian canal, lamellae, and osteocytes is known as an **Haversian system,** a microscopic structure extending lengthwise in compact bone. This structure is continuously being broken down and rebuilt by the body.

Examine a prepared slide of ground bone. Note that the actual osteocytes are no longer present in the preparation. Locate the structures printed in boldface in the previous paragraphs with the aid of Figure 32.1 and Figure 27.1. Is there more than one Haversian system present? _____

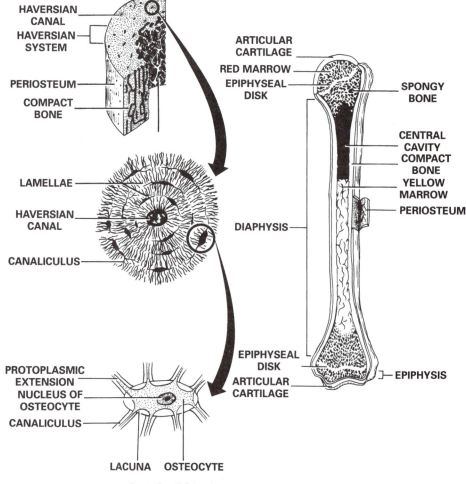

FIGURE 32.1 Bone Structure

CARTILAGE AND LIGAMENTS

Cartilage is a type of connective tissue in which the cells (**chondrocytes**) are separated by a matrix of collagen or elastic fibers imbedded in a firm gel. This matrix allows cartilage to function as a supportive tissue while still being somewhat flexible. Similar to osteocytes, the chondrocytes are found in spaces called **lacunae,** but they are not arranged in rings connected by canaliculi as in bone.

Cartilage is found at the ends of bones in articulations, the nose, the pinna of the ear, the larynx. the trachea, the intervertebral discs (pads between vertebrae), and in the connections between the ribs and the sternum.

1. Observe prepared slides of hyaline and elastic cartilage and fibrocartilage and identify the parts in boldface type from the description above. Refer to Figure 27.1.

    Ligaments are composed of fibrous connective tissue and function to connect bones together. Fibrous connective tissue is composed of cells sandwiched between layers of tightly packed **collagen** and **elastic fibers.** Thus ligaments have a long, stringy texture. **Tendons** are also composed of fibrous connective tissue, but connect muscle to bone instead of bone to bone.

2. Locate the cells and the fibers in a prepared slide of dense (fibrous) connective tissue. Refer to Figure 27.1.

SKELETAL STRUCTURE

In vertebrates, most of the skeletal system is initially cartilaginous during embryonic development. Eventually, this cartilaginous framework is replaced by bone. At birth, man has about 270 bones making up his skeleton; in adulthood this number is reduced to about 206 due to fusion of individual bones.

In the following listing of skeletal bones, the skeleton has been divided into two portions: the **axial portion** (skull, backbone, and rib cage) consisting of about 80 bones, and the **appendicular portion** (arms, legs, and part of shoulder and hip girdles, all bones occurring in pairs).

AXIAL SKELETON
Skull

Bones of the Cranium (Brain Case)

Frontal — forms the forehead region; separated from the parietal bones by the **coronal suture** (unpaired)

Occipital — forms the back of the skull where it rests on the vertebral column, and through which the spinal cord of the nervous system passes as it connects with the brain; it is separated from parietal bones by the **lambdoidal suture** (unpaired)

Sphenoid — an irregular "butterfly" shaped bone forming part of the floor of the cranium, sides of skull, and backs of eye sockets (unpaired)

Ethmoid — at the top of the nasal chamber, between the eye sockets, and forms part of the wall of each eye socket, on the side toward the nose (unpaired)

Parietals — at the top and upper sides of the skull, behind the frontal bone; separated from each other by the **saggital suture**

Temporals — at the sides of the skull, spreading out around the opening of the outer ear canal, and containing the deeper parts of the hearing apparatus; separated from the parietals by the **squamosal suture**

Bones of the Face

Mandible — forms the lower jaw (unpaired)

Vomer — forms part of the thin, bony partition between the nasal cavities, the remainder of the partition being cartilage

Maxillae — together forming the upper jaw

Zygomatics — form the upper, bony prominences of the cheeks and the sides and lower borders of the eye sockets; each joins a zygomatic process of the temporal bone to form a zygomatic arch under and over which pass the strong chewing muscles

Lacrimals — form a small part of the medial walls of the eye sockets

Nasals — form the hard upper part or bridge of the nose (most of the nasal support is cartilage)

Inferior nasal conchae — curved, shell-like bones sloping inwardly, like shelves, form the lateral walls of the nasal chambers, below the curved **superior** and **middle conchae** which are parts of the ethmoid bone. The conchae are very brittle, thin, delicate bones and, therefore, are often damaged in preparation of skulls; they may be hard to locate if partly destroyed

Palatines — (not shown on figure) constitute the back part of the hard palate of the roof of the mouth

Ossicles or ear bones — (not visible on surface) paired bones in the middle ear cavities (6 bones)

Malleus or "hammer" — attached to the inside of the tympanic membrane or eardrum

Incus or "anvil" — in between the malleus and stapes

Stapes or "stirrup" — in contact with the oval window membrane

Hyoid bone — (not shown) a U-shaped bone suspended by ligaments from the pointed styloid processes on the lower part of the temporal bones. (This is not truly a bone of the skull, but it is included here as a matter of convenience.)

Use the diagrams of the skull (Figures 32.2 and 32.3) to locate the bones on the skulls available in lab.

NOTE

Coloring each bone of the diagram a different color will help in distinguishing them.)

Vertebral Column (26–33 Vertebrae; 24 are moveable; singular: vertebra). See Figure 32.4.

Cervical vertebrae — in the neck region (7 bones). The first cervical vertebra is the **atlas** and supports the head. The second vertebra is the **axis,** which has an **odontoid process** which projects through the ring of the atlas to allow the head to pivot

Thoracic vertebrae — in the upper back region. There are 12 of these vertebrae, 10 of which are attached to ribs

Lumbar vertebrae — in the region of the "small of the back" (5 bones)

Sacral vertebrae — the "sacrum" which consists of fused vertebrae, usually 5, and forms a triangular wedge of bone in the back part of the pelvic girdle (counted as 1 bone in the adult skeleton)

Coccygeal vertebrae (Coccyx) — 4 or 5 vertebrae, sometimes each separated from the others, but usually fused in various combinations to form the "tail" bone. (May be any number from 1 to 5, inclusive. There is a tendency for more fusion of these bones in the skeleton of the male than in the female.)

Thorax (25-27 bones)

Sternum (breastbone) — consisting of 3 parts (**manubrium, gladiolus** or body, and **xiphoid**) which become fused in elderly individuals (usually counted as one bone, but may be 3 in younger adults): the manubrium and gladiolus are named for their resemblance to the handle of a sword

Ribs or **Costae** (12 pairs)

"**true ribs**" — 7 pairs attached directly to the sternum by means of their cartilages

"**false ribs**" — 5 pairs: the first 3 pairs are attached to the cartilages of adjacent ribs rather than to the sternum directly; the last 2 pairs are termed "floating ribs" because they are not attached to the sternum either directly or indirectly.

APPENDICULAR SKELETON

Shoulder Girdle

Scapulae (singular, scapula) — shoulder blades

Clavicles — collar bones

Upper Appendages (30 bones, each)

Humerus — upper arm bone

Radius — lower arm bone, on the thumb side

Ulna — lower arm bone, on the little finger side; the longer of the two lower arm bones

Carpal bones: —in the wrist (8 bones)

Metacarpal bones — in the hand (5 bones)

Phalanges — finger bones; each finger has 3; the thumb has 2 (14 bones)

Hip Girdle

Innominate bones — consisting of the fused **ilium** (uppermost and largest portion), **ischium** (strongest portion — directed slightly posteriorly), and **pubis** (superior and anterior to ischium). counted as 1 bone on each side after the fusion has occurred; the joint between the two pubic bones is called the **pubic symphysis**

Lower Appendages (30 bones, each)

Femur — thighbone

Patella — kneecap

Tibia — larger, medial lower leg bone

Fibula — smaller, lateral lower leg bone

Tarsals — form the ankle (7 bones)

Metatarsals — foot bones (5 bones)

Phalanges — toe bones; big toe has 2, other toes have 3 (14 bones)

In Figure 32.5 (the human skeleton anterior and posterior views), label the bones indicated by label lines.

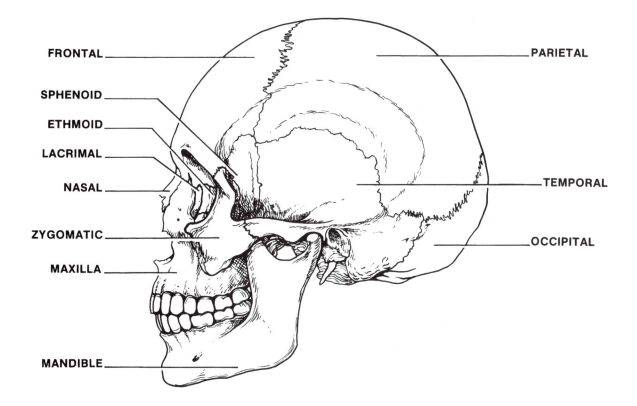

FIGURE 32.2   Human Skull (Side View)

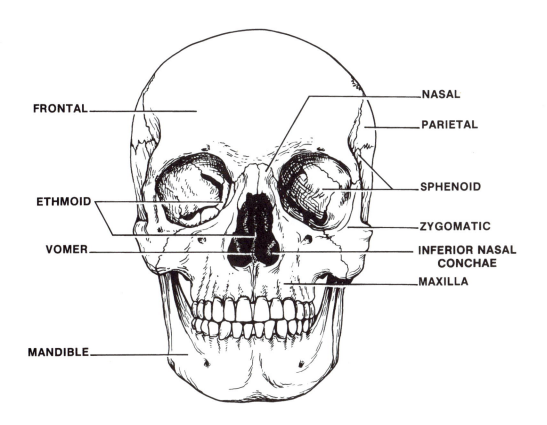

FIGURE 32.3   Human Skull (Front View)

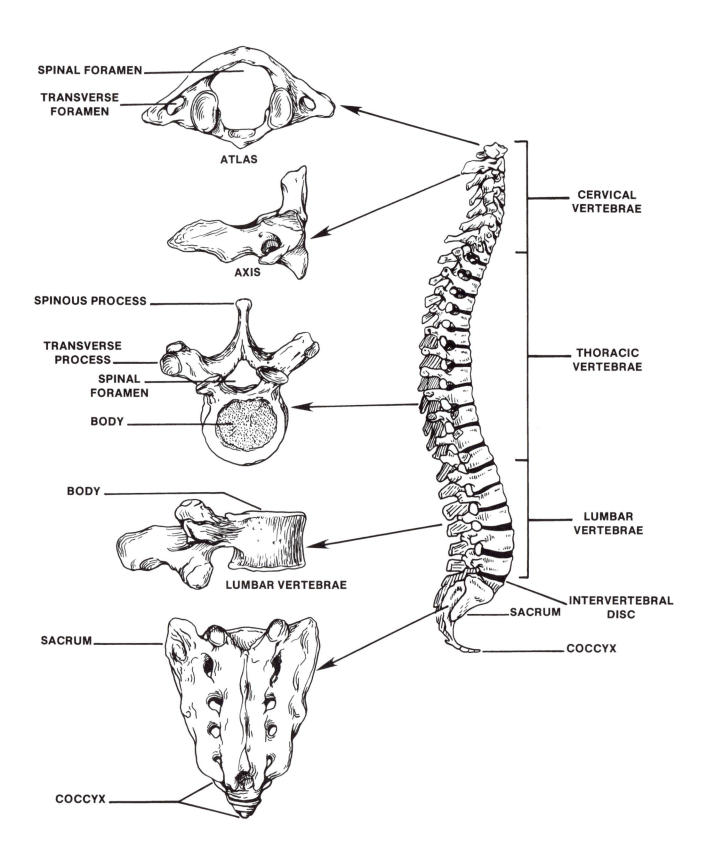

FIGURE 32.4 The Vertebral Column

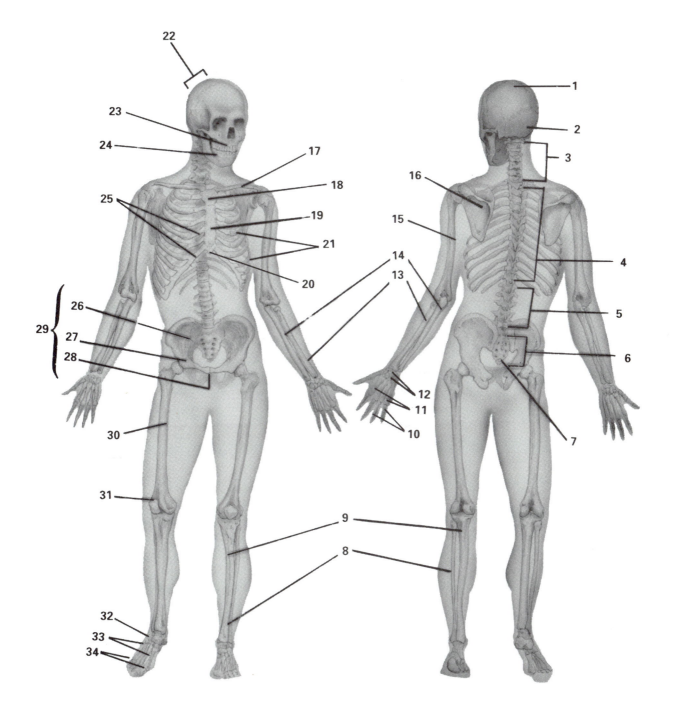

FIGURE 32.5 Human Skeleton — Anterior and Posterior Views

JOINTS (ARTICULATIONS)

Joints are considered to be the junctions between two bones and are classified on the basis of the amount of movement allowed. There are three basic categories:

IMMOVABLE JOINTS (SYNARTHROSES)

These are joints in which the bones are connected to each other by fibrous connective tissue. The sutures of the skull and the epiphyseal discs of the long bones are typical examples.

Diagram A of Figure 32.6 is a section through a suture. Note that the two bones are separated by **fibrous connective tissue** which is continuous with the **periosteum** on the external surface and with the **dura mater** on the inner surface.

SLIGHTLY MOVABLE JOINTS (AMPHIARTHROSES)

These are joints in which the ends of the bones are covered with **articular cartilage** and have a disc of **fibrocartilage** between the two bones which has a cushioning effect. The bones are held together by **ligaments.** A typical example of this type of joint is the junction between vertebrae, as shown in Diagram B of Figure 32.6.

FREELY MOVABLE JOINTS (DIARTHROSES)

These are joints in which the ends of the bones are covered with **articular cartilage** and the bones held together with **ligaments** as above; the distinguishing feature, however, is the presence of a **synovial membrane** on the inner surface of the ligament. This membrane produces a slippery viscous fluid which functions in lubricating the joints. This is the most common type of joint in the body. Some various types of freely movable joints are: (1) **ball and socket joint** (example — shoulder and hip): (2) **hinge joint** (example — elbow, knee, and ankle); (3) **gliding joint** (example — wrist); (4) **pivot joint** (example — between axis and atlas). Diagram C in Figure 32.6 is of the knee joint, one of six different types of diarthrotic joints. Working with the skeleton or your own body, determine the different types of movements allowed by various freely movable joints.

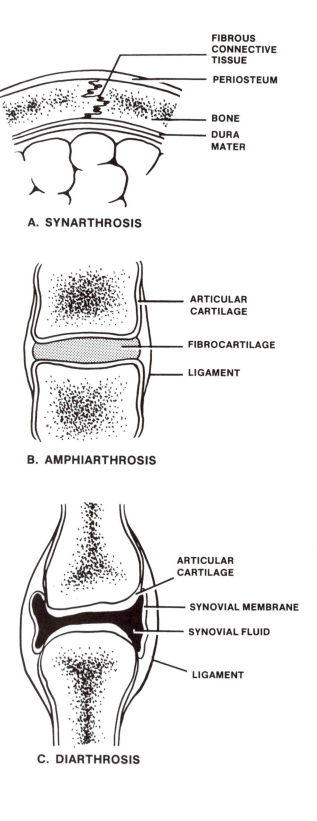

A. SYNARTHROSIS

FIBROUS CONNECTIVE TISSUE
PERIOSTEUM
BONE
DURA MATER

B. AMPHIARTHROSIS

ARTICULAR CARTILAGE
FIBROCARTILAGE
LIGAMENT

C. DIARTHROSIS

ARTICULAR CARTILAGE
SYNOVIAL MEMBRANE
SYNOVIAL FLUID
LIGAMENT

FIGURE 32.6 Types of Joints

TOOTH ANATOMY

TOOTH STRUCTURE

Associated with the skeletal system because of their chemical makeup, are the **teeth.** However, embryonically they are derived from ectoderm rather than from mesoderm as is the skeleton. Internally they have the same structure as scales. Examine the lower surface of the maxilla and upper surface of the mandible of a skull and notice the sockets in which the roots of the teeth are located. Usually by the time an individual is 2 years old he has acquired all 20 of his **deciduous** ("baby") **teeth.** At about 6 years of age, the **permanent teeth** begin erupting. The loss of the deciduous teeth arises from a reabsorption of the roots accompanying the growth of the permanent teeth underneath.

Figure 32.7 shows the permanent teeth of the human upper jaw. Locate the parts indicated and be able to identify these structures on the jaw models, head models, and skulls available in the lab.

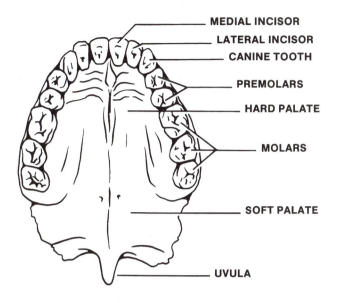

FIGURE 32.7 Human Upper Jaw

LONGITUDINAL SECTION OF A TOOTH

Refer to Figure 32.8 to locate the structures given below in boldface type.

The tooth is comprised of three main regions: the **crown** (above the gum); the **root** (embedded in the **bony socket** of the jaw, which is covered by the fleshy **gum**); and the **neck** (a slightly constricted area between the root and the crown). The hard parts of the tooth include the **enamel, dentine,** and **cementum**; even though these are like bone in the structure, they are much harder. (Why?) **Enamel** is the outer covering of the crown; **cementum,** continuous with the enamel, covers the root; **dentine** makes up most of the tooth (underneath the enamel and cementum).

The soft parts of the tooth are the **pulp** and the **peridontal membrane.** The pulp is the innermost part of the tooth, consisting of connective tissue plus capillaries, lymph vessels, and nerve fibers which enter the pulp through an opening (**foramen**) in the root. The peridontal membrane is a fibrous membrane covering the root and lining the bony socket (here it functions as the periosteum); its function is to anchor the tooth to the socket and to serve as a shock absorber.

Obtain a slide of tooth development. On this slide, the tooth has not yet erupted through the gum.

In the space provided on the following page make a diagram of the tooth on the slide and label the **enamel, dentine, pulp,** and **gum**.

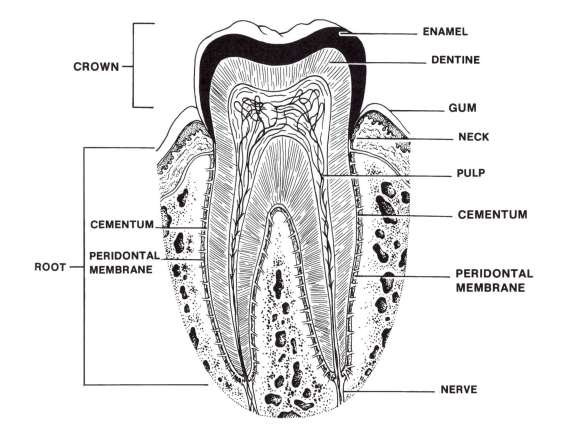

ENAMEL

DENTINE

CROWN

GUM

NECK

PULP

CEMENTUM

CEMENTUM

PERIDONTAL
MEMBRANE

ROOT

PERIDONTAL
MEMBRANE

NERVE

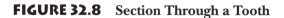

FIGURE 32.8 Section Through a Tooth

REVIEW QUESTIONS

1. What general functions does the skeletal system perform? _____

    _____

    _____

2. Is spongy bone found only in a long bone? _____ Explain. _____

    _____

    _____

    _____

3. What does the shape of different teeth indicate about their particular function? _____

    _____

4. Which teeth are considered to be the "wisdom" teeth? _____

5. How many permanent teeth are there in a complete set? _____

6. List the 3 major types of joints and give an example of each.

    _____

    _____

    _____

7. Identify the following terms:

    Osteocyte _____

    $Ca_3(PO_4)_2$ and $CaCO_3$ _____

    Epiphyseal disc _____

    Canaliculi _____

    Ossicles _____

    Axial skeleton _____

    Appendicular skeleton _____

    Ball and socket joint _____

    Atlas and Axis _____

8. What is the function of ligaments?_____

    tendons? _____

9. Name the five places in the body in which cartilage can be found. _____

_____

_____

Muscular System

EXERCISE
33
EXERCISE EXERCISE EXERCISE

OBJECTIVES

After completion of this exercise, the student should be able to do each of the following:

- Identify the three different types of muscle tissue under the microscope.
- Identify the type of muscle in the structures named and the function of the muscle in that structure.
- Define the terms **flexion, extension, abduction, adduction,** and **rotation** and give an example of each.
- Recognize on models or diagrams, the fourteen major human muscles mentioned in this exercise.
- Name the muscles involved in some of the simple activities outlined in this exercise.
- Define muscle **fatigue, threshold,** and **oxygen debt.**
- Define the following, and identify each on a physiograph or kymograph tracing: **tetanus, contraction, latent period, refractory period,** and **relaxation.**
- Answer the review questions at the end of this exercise.

INTRODUCTION

Approximately one half of the body is composed of muscle. This tissue is specialized for contraction, thus allowing for a movement of body parts, whether it be in the movement of an arm, leg, emptying of the urinary bladder, churning action of the stomach, etc. Within the vertebrates, muscle tissue has differentiated along three lines; these three resulting types of muscle will be described below. In addition to a study of muscle anatomy or structure, this exercise will include simple demonstrations of muscle physiology or function.

TYPES OF MUSCLE TISSUE

Obtain a slide of each of three muscle types. Examine each slide carefully and make a diagram in the spaces provided on the next page. Refer to Figure 27.1.

SKELETAL MUSCLE

Characteristics

1. Makes up most of the body muscle mass
2. Attached to bone (usually)

3. Under conscious control

4. Involved in locomotion, movement of bones, and maintaining posture

5. Each muscle fiber has many nuclei (along the edge of the fiber) and many cross striations (lines)

VISCERAL MUSCLE (Smooth Muscle)

Characteristics

1. Found in walls of internal organs

2. Not usually under conscious control

3. Involved in contraction of stomach, intestine, blood vessels, lungs, etc.

4. Each muscle cell has one nucleus, located in the center of the fiber; striations are lacking

CARDIAC MUSCLE

Characteristics

1. Located only in the heart

2. Not usually under conscious control

3. Branched muscle fibers; one nucleus per cell; cross striations present

4. Intercalated discs (dark lines at junction of 2 cells) present (function in rapid conduction of impulses from one cell to another)

OPERATION OF SPECIFIC ORGANS

Considering the following structures, indicate on the table at the top of the next page the type of muscle found in each structure and the particular role of the muscle tissue in each case.

MUSCLE ACTION

Contraction is the only action a muscle can perform. When a muscle relaxes, no work is done. Skeletal muscles do not work singly, but rather in groups — with one group being antagonistic to the other (i.e., having an opposite effect). Skeletal muscles can be classified on the basis of the type of movement they produce. There are a large number of different types of movement, but in this exercise we will restrict ourselves to the five listed below.

Flexion — this action is a movement of a body part such that the angle of a joint is decreased (e.g.,

Structure	Muscle Type	Role
Tongue & Jaw		
Pyloric sphincter		
Diaphragm & muscles between ribs		
Intestinal Wall		
Wall of Urinary Bladder		
Internal Anal Sphincter		
External Anal Sphincter		
Ventricle of Heart		
Walls of Arteries		
Walls of Uterus		

bending the elbow). A flexor muscle is one which bends a limb

Extension — extension involves a movement of a body part such that the angle of a joint is increased (e.g., straightening of the arm)

Abduction — this is movement of a body part away from the median axis of the body (e.g., lifting the arm straight out laterally)

Adduction — this constitutes movement of a body part toward the median axis of the body (e.g., lowering the arm to the side of the body)

Rotation — this constitutes twisting about an axis (e.g., turning the head from side to side)

MUSCLE IDENTIFICATION

Each skeletal muscle has two points of attachment: the more fixed parts being the **origin** and the more movable part being the **insertion.** Usually the origin is closer to the median axis of the body, while the insertion is more peripherally located. The two ends of the muscle are attached to different bones, otherwise contraction would not produce movement. When a muscle contracts, the origin usually remains stationary and the insertion is usually moved toward the origin. A muscle can be attached directly to a bone, to another muscle, or to a tendon.

MUSCLE LOCATION

Listed below are some of the major muscles of the human body. Using the charts available plus the descriptions provided, locate and label each muscle on the following diagram. To better distinguish one muscle from another, color each muscle a different color. The numbers 1-15 listed below do not correspond to the numbers on Figure 33.1.

Deltoid — shoulder muscle. Origin: clavicle and scapula. Insertion: upper part of humerus. Action: abduction of humerus

Biceps brachii — muscle of the anterior portion of the upper arm. Origin: scapula. Insertion: radius. Action: flexion of the forearm

Sternocleidomastoid — long muscle on the side of the neck. Origin: clavicle and sternum. Insertion: mastoid process of the skull. Action: contraction of one causes head to move toward the side; contraction of both simultaneously flexes head forward and downward on the chest

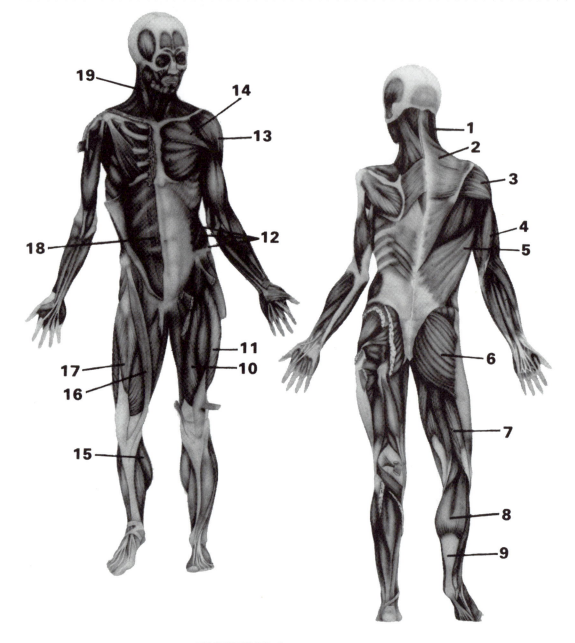

FIGURE 33.1   Human Musculature

Pectoralis major — triangular muscle in the upper chest. Origin: clavicle, sternum, and cartilages of the 6 upper ribs. Insertion: upper anterior portion of the humerus. Action: adduction and rotation of humerus

Rectus abdominis — long flat muscle covering the abdominal region. Origin: pubic bone. Insertion: sternum and cartilages of 5th, 6th, and 7th ribs. Action: flexion of the body at the lumbar region plus compression of the abdominal organs.

External oblique — large flat muscle on the side of the body (between rib cage and hip). Origin: external surface of the lower 8 ribs. Insertion: ilium crest and flat band of connective tissue in midline of abdomen. Action: flexion of body plus compression of abdominal organs

Sartorius — a long slender muscle running diagonally across the thigh. Origin: ilium. Insertion: inner surface of the upper part of the tibia. Action: flexion of the leg on thigh and of thigh on the pelvis and rotation of thigh laterally

Quadriceps femoris — very large muscle of the anterior part of the thigh; it consists of 4 parts:

Rectus femoris — located on the anterior part of the thigh

Vastus lateralis — located on the lateral part of the thigh

Vastus medialis — located on the medial surface of the thigh

Vastus intermedius — located underneath the rectus femoris (not visible on the diagram)

Origin: ilium (only on the rectus femoris) and femur (the other 3). Insertion: tendon passing over knee joint to attach to tibia. Action: extend leg (the entire quadriceps) and flex thigh (only rectus femoris).

Trapezius — triangular muscle in the upper back region. Origin: occipital bone of skull and cervical and thoracic vertebrae. Insertion: scapula and clavicle. Action: adduction of scapula, raising of scapula ("shrugging"), and extension of head backward

Latissimus dorsi — back muscle in the lumbar area. Origin: lower 6 thoracic vertebrae, all the lumbar vertebrae, sacrum, and ilium. Insertion: upper part of the humerus. Action: extends, adducts, and rotates arm medially and draws shoulder down and backward

Triceps brachii — muscle of the posterior surface of the upper arm. Origin: scapula, and humerus. Insertion: ulna. Action: extends and adducts forearm

Gluteus maximus — muscle of the buttocks region. Origin: ilium, sacrum, and coccyx. Insertion: femur. Action: extends, abducts, and rotates the thigh laterally

Biceps femoris — muscle of the posterior lateral surface of thigh. Origin: ischium and femur. Insertion: fibula and tibia. Action: flexes leg and rotates laterally after flexion

Semitendinosus — muscle of the posterior medial surface of thigh. Origin: ischium. Insertion: tibia. Action: extension and rotation of thigh and flexes leg. The biceps femoris and semitendinosus are called the hamstrings

Gastrocnemius — muscle making up the external part of the calf of the leg. Origin: femur.

Insertion: calcaneus (heel bone) via tendon of Achilles. Action: extension of foot (points toes) and flexes leg at the knee

ACTION OF MUSCLES DURING MOVEMENT

Perform each of the following actions. When doing so, observe in yourself or your lab partner, the contraction of the muscle involved. By palpitation, feel the muscle harden during contraction. Name the primary muscle responsible for the action, based on the labeled diagrams filled in earlier in this exercise.

Turning head to the side —

Rising on tiptoes —

Bending the knee —

Crossing one leg over the other —

Raising the arm outward—

Bending the forearm at the elbow —

Forcible expiration as in coughing —

Which muscles are usually used for intramuscular injections? _____

_____

ANALYZING MOVEMENTS

As your lab partner goes through the movements indicated by the following diagrams, Figure 33.2, analyze the movement, then draw in the muscle on each figure which causes the movement indicated. Be sure to show where the insertion and the origin are. Also, name the muscle and indicate the type of movement (i.e., adduction, flexion, etc.) produced.

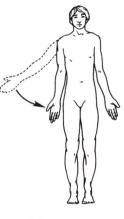

MUSCLE _____

ACTION _____

MUSCLE _____

ACTION _____

MUSCLE _____

ACTION _____

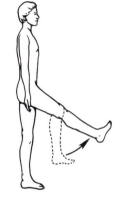

MUSCLE _____

ACTION _____

MUSCLE _____

ACTION _____

MUSCLE _____

ACTION _____

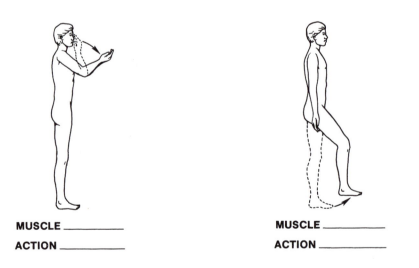

MUSCLE _____

ACTION _____

MUSCLE _____

ACTION _____

FIGURE 33.2   **Muscle Actions**

MUSCLE PHYSIOLOGY (OPTIONAL EXPERIMENT)

CONTRACTION OF CARDIAC MUSCLE

Your instructor will pith a frog which is a procedure that destroys the brain and spinal cord. The frog's heart will be exposed by dissection.

Is the heart still beating? _____

The heart will now be removed from the frog's body and placed in a solution of frog saline (Ringer's solution).

Does the heart continue to beat? Why? _____

_____

What control then, does a frog have over the contraction of its heart? _____

_____

PREPARATION OF THE GASTROCNEMIUS MUSCLE

The gastrocnemius muscle will now be prepared by: (1) removing the skin from the leg; (2) exposing the sciatic nerve on the dorsal side of the thigh near the femur; and (3) clipping the Achilles tendon from its attachment at the foot after tying a thread around the tendon near the body of the muscle. The origin of the gastrocnemius is anchored by a clamp or pinned down to a dissecting pan. The thread that is tied to the tendon is attached to some kind of recording device (the lever of a kymograph or a transducer of an electronic physiograph machine). The recording device will make a tracing on a piece of paper and when the muscle contracts, it creates a pull on the thread which causes a deflection on the tracing. The muscle is stimulated by an electric current applied to the sciatic nerve. The muscle must be kept moist at all times with Ringer's solution!

TWITCH THRESHOLD

With the stimulator set for single shocks of about 10 milliseconds duration and a voltage of zero, give a single shock to the muscle. No response should occur since the stimulus is below the threshold required for contraction. Increase the voltage a bit and deliver another stimulus to the muscle. Repeat, increasing the voltage a bit at a time, until a response is seen and recorded on the tracing. Record this voltage.

The muscle **twitch** is the response of a muscle (contraction, then relaxation) to a single applied stimulus. The **threshold** is the minimum stimulus that will elicit a response.

Increase the voltage and continue stimulation until the maximum muscular response is seen on the tracing.

How do you account for the increasing muscular response (to some maximum point) as stimulus strength is increased? _____

_____

_____

DURATION OF THE TWITCH

With the recorder set for maximum speed and the stimulator set for maximum stimulus, deliver single shocks to the muscle. Place one-second marks on the tracing. Measure the length, in millimeters, between the one-second marks. Determine the length of the **latent, contraction,** and **relaxation phases** of the muscle twitch. The latent phase is the fraction of a second in which a muscle does not respond after the stimulus is applied. Refer to Figure 33.3.

TETANY

Because of its short refractory period, skeletal muscle is susceptible to being thrown into a sustained contraction. This will occur if the stimuli are applied so often that the muscle does not have a chance to go into the relaxation phase after contraction. This sustained contraction is called **tetany.** Refer to Figure 33.4.

With the recorder set for medium speed and the stimulator on maximal stimulus, deliver stimuli of 1, 2, 3, 5, 7, 10, and 20 per second in succession to the muscle. Note the fusion of twitches into a tetanic contraction.

Switch off the stimulator, leaving the frequency at 20 stimuli per second. Switch on again and note the smooth contraction which results. Leave on until contraction strength begins to diminish. What is happening to cause this diminished response? ___

_____

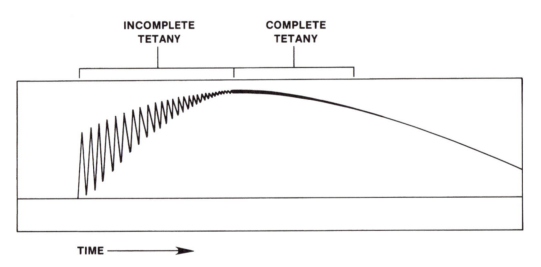

FIGURE 33.3 The Muscle Twitch

FIGURE 33.4 Tetany

FATIGUE

Repeated contraction of a muscle produces **muscle fatigue.** Fatigue is caused by the accumulation of waste products such as lactic acid, or by a deficiency in oxygen, glucose, and other raw materials.

To experience the sensation of fatigue, lay your left forearm on the table top, palm up. Flex and extend your fingers rapidly at the third joint (where the fingers join the hand) until they are fatigued.

Record the time needed to cause fatigue: _____

Rest the fingers for 10 min., then attach a weight to one finger. Again, record the time needed to cause fatigue: _____

Repeat the experiment after a 10 minute rest period, except this time put the cuff of a sphygmomanometer on the left upper arm and inflate it until no pulse can be felt at the wrist. At a pressure of about 130 mm it should disappear. Repeat the flexing of the fingers for a given weight as above.

Record the time needed to cause fatigue: _____

Conclusions:

REVIEW QUESTIONS

1. List the three major types of muscle and describe where they are generally located.

   _____

   _____

   _____

2. What type of muscle would be found in the following structures?

   Diaphragm _____

   Uterine walls _____

   Walls of urinary bladder _____

   Ventricle of heart _____

   Anal sphincter _____

   Biceps brachii _____

3. Define the following terms:

   Abduction _____

   Extension _____

   Intercalated disc _____

   Adduction _____

   Involuntary _____

4. Which muscle has its origin on the clavicle and scapula, insertion on the upper part of the humerus and

   has the function of abducting the humerus? _____

5. Name the four parts of the quadriceps femoris

   a. _____

   b. _____

   c. _____

   d. _____

6. Which type of muscle fits the following descriptions:

usually attached to bone, under conscious control, and has muscle fibers with many nuclei and cross striations?

_____

located only in the heart, not usually under conscious control, and has muscle fibers with single nuclei, cross striations, and intercalated discs between the cells. _____

found in walls of hollow internal organs, not usually under conscious control, and has muscle fibers with a single nucleus and lacking striations? _____

7. Recognize on a model the following major human muscles, also be able to tell their action:

MUSCLE   **ACTION**

Deltoid _____

Biceps brachii _____

Sternocleidomastoid _____

Pectoralis major _____

Rectus abdominis _____

External oblique _____

Sartorius _____

Quadriceps femoris _____

Rectus femoris _____

Vastus lateralis _____

Vastus medialis _____

Vastus intermedius _____

Trapezius _____

Latissimus dorsi _____

Triceps brachii _____

Gluteus maximus _____

Biceps femoris _____

Semitendinosus _____

Gastrocnemius _____

REPRODUCTION IN ANIMALS AND GAMETOGENESIS

OBJECTIVES

After completion of this exercise, the student should be able to do each of the following:

- Compare the advantages and disadvantages of **asexual** and **sexual** forms of **reproduction**.
- List three types of asexual reproduction.
- List the steps involved in **spermatogenesis** and **oogenesis**, and identify the points at which mitosis and meiosis occur in the process.
- Explain why conjugation in *Paramecium* is a type of sexual reproduction.
- Describe how **sperm** are formed in the **seminiferous tubules**.
- Identify the parts of the male and female reproductive systems and describe the function of each part.
- Describe the changes that occur as a **follicle** develops to maturity in the **ovary** and what happens to the follicle after **ovulation**.
- Describe the changes in the **endometrium** during the **menstrual cycle**.
- Explain how the levels of **FSH, LH, estrogen**, and **progesterone** are regulated during the menstrual cycle.
- Describe the points in the menstrual cycle at which each of the hormones listed above reaches a peak, and describe the effects produced by each.
- Answer the review questions at the end of this exercise.

INTRODUCTION

The creation of new cells (or organisms) from previously existing cells (or organisms) has developed along two basic lines: **asexual reproduction** in which no genetic variability exists between parent and offspring and **sexual reproduction** in which genetic variability does exist.

In this lab we will examine various methods of reproduction. The major portion of the lab, however, will be concerned with mammalian reproduction.

303

ASEXUAL REPRODUCTION

Asexual reproduction is common among lower plants and animals and has several advantages: (1) it involves only mitosis; (2) only one parent is necessary, therefore, no complex mating behavior is required; (3) large numbers of offspring can be produced at one time; (4) it is a means of dispersal of offspring; and (5) in a stable environment, it allows for non-variable "copies" of the parent. The major disadvantage is that, because of the genetic stability, there is reduced ability for populations to adapt to changes in the environment. The three main categories of asexual reproduction, which you have already studied, include: **fission** (in protozoa, bacteria, yeast, and some algae); **budding** (as in *Hydra*, sponges, yeast, and some jellyfish); and **fragmentation** or **regeneration** (in flatworms, segmented worms, starfish, and among filamentous forms of algae).

Many organisms reproduce both asexually and sexually. This is particularly true of internal parasites such as malarial organisms and liver flukes, which reproduce asexually during certain parts of their life cycles, thus increasing their numbers significantly, and improving their chances of dispersal. Thus, we find that asexual reproduction is widespread, even persisting, among forms that also reproduce sexually. This is another way of stating that the adaptive advantages of asexual reproduction may outweigh the disadvantages of reduced variation in genetic makeup.

SEXUAL REPRODUCTION

FUNCTION AND METHOD

Sexual reproduction is found in the whole range of plant and animal life — from the one-celled level to the multicellular mammalian level. Its primary advantage is that through the process of meiosis and fertilization it allows for genetic variations among organisms, i.e., offspring that are not identical to the parent. This variable is important in that it allows populations to adapt to environmental changes — it is the raw material upon which natural selection acts.

The major disadvantages would be the complex processes of sperm and egg development and the necessity for anatomical and behavioral features to bring sperm and egg together.

In the process of **conjugation** in *Paramecium*, the fusion of nuclei, which is the main feature of sexual reproduction, takes place. The micronuclei of the two organisms divide meiotically. When they unite side by side, they exchange one of their micronuclei. Then the two nuclei fuse in each cell. The two organisms each contain some nuclear material from each other and are different genetically than they were before conjugation. Mitosis follows, resulting in the production of four different offspring from each cell.

Fertilization in animals involves the fusion of nuclei in specialized gametes, one large and non-motile, the other small and motile. In some organisms, such as hydra, earthworms, and tapeworms, one individual may produce both sperm and eggs, although they usually do not fertilize their own eggs. In other cases (among invertebrates especially, e.g., aphids, and bees), the unfertilized egg may develop into an adult; a process called **parthenogenesis**. In those organisms in which there are separate sexes, producing unique gametes, there are numerous variations on the means of fertilization. In some cases (e.g., frogs) the gametes are shed into the water and fertilization occurs there. In other cases, development of special structures is necessary for fertilization to take place internally. In the rest of this lab we will deal with sexual reproduction in the latter group of animals with an emphasis on mammals.

GAMETOGENESIS

Gametogenesis is the process involving the development of gametes or sex cells. Gametogenesis in the male is called **spermatogenesis** (development of sperm); in the female it is called **oogenesis** (development of the ovum). Both meiosis and mitosis are involved in gametogenesis. During embryonic development certain cells become different from the other body cells; these are the primordial germ cells which will give rise to the sex cells. Each primordial germ cell undergoes many mitotic divisions, producing a large number of cells. These cells are diploid, therefore it is necessary for them to undergo meiosis in order to develop into functional gametes.

Figure 34.1 shows the general scheme of spermatogenesis and oogenesis. Be sure that you understand why each functional sex cell contains only one-half the chromosomes that are in the primordial germ cells.

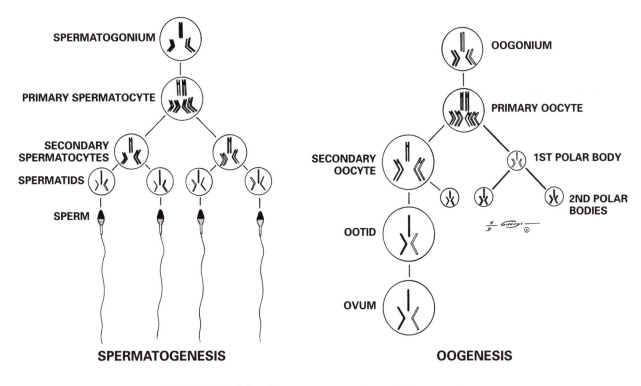

FIGURE 34.1 Spermatogenesis and Oogenesis

MALE REPRODUCTIVE SYSTEM

Testis

Under low power, examine a prepared slide of monkey testis. Note the round or oval cross sections of the **seminiferous tubules** in which the sperms are produced. Tubules are held together by loose connective tissue, which contains blood vessels, nerves, and **interstitial cells**. The latter are compact masses in the spaces between adjacent tubules, not particularly distinct from connective tissue. Such interstitial cells are endocrine; they are the main producers of the male sex hormones, or **androgens**.

Study a few seminiferous tubules under high power. The wall of each tubule contains an outer **basement membrane** which forms the boundary of several layers of stratified epithelial cells. Most of the cells of this epithelium are **spermatogonia**, the precursors of sperms.

The spermatogonia represent the diploid **generative epithelium** of the testis. The cells divide mitotically and mitotic stages may actually be visible.

As a result of the continuing divisions, some of the daughter cells, called **spermatocytes**, become displaced toward the **lumen** (central cavity) of the tubule. Spermatocytes then undergo meiosis and become **spermatids**, haploid spherical cells. The

spermatids differentiate into mature sperms, with their tails projecting into the lumen of the tubule. Subsequently, the sperms come to lie free in the lumen. New sperms continuously form, eventually displacing the older ones, and the latter are then gradually pushed out of the testicular tubules into the sperm duct.

Scattered among the developing sex cells are relatively large cells attached to the basement membrane but extending inward to the lumen of the seminiferous tubule. These cells are known as **Sertoli cells** and are believed to function in the nutritive support of the developing sperm.

Examine the slide and try to correlate the diagram in Figure 34.2 and Figure 34.3 with what you see. On the diagram, note the interstitial cells, Sertoli cells, spermatogonia, spermatocytes, spermatids, and sperm (developing and complete).

Sperm Production

Hormonal production in the male is not cyclical as in the female. Once secretion begins with puberty, it usually continues throughout life. The anterior pituitary hormones involved are **FSH** (stimulating spermatogenesis, i.e., development of primary spermatocytes into sperm) and **LH** (stimulating the

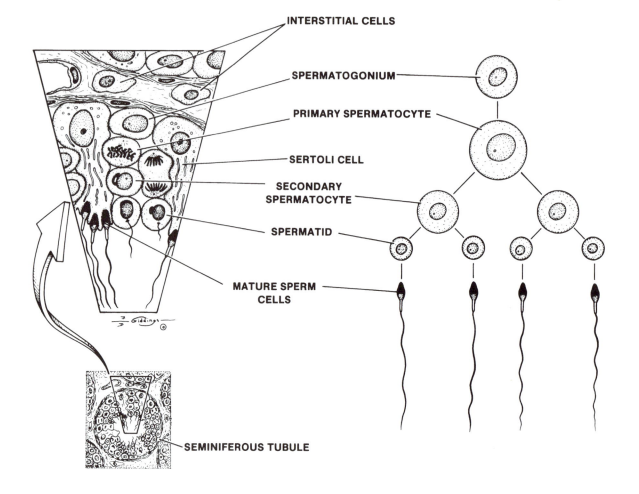

INTERSTITIAL CELLS

SPERMATOGONIUM

PRIMARY SPERMATOCYTE

SERTOLI CELL

SECONDARY SPERMATOCYTE

SPERMATID

MATURE SPERM CELLS

SEMINIFEROUS TUBULE

FIGURE 34.2   **Cross-section of a Mammalian Seminiferous Tubule**

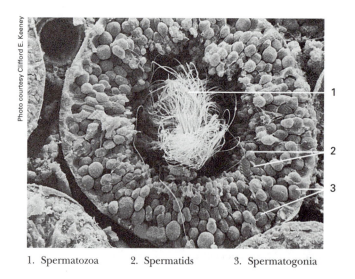

Photo courtesy Clifford E. Keeney

1. Spermatozoa   2. Spermatids   3. Spermatogonia

FIGURE 34.3   **Electron Micrograph of a Seminiferous Tubule**

interstitial cells to produce **testosteron**e). Testosterone promotes a variety of effects concerned with male sexual characteristics. Although there is no cyclic pattern in spermatogenesis as in oogenesis, there does appear to be a tendency for a decrease in the rate of sperm production after about 40 years of age due to degeneration of the seminiferous tubules.

Examine slides of different types of sperm (including living human sperm if available) and make drawings of them.

Anatomy

The main function of the reproductive system is to ensure continuation of species. The organs of this system are designed so that the sex cells (eggs and sperm) are brought together in the reproductive system of the female, so that fertilization may occur internally.

The male reproductive system consists of **testes**, **epididymides** (sing. = **epididymis**), **vasa deferentia** (sing. = **vas deferens**), and their associated glands (**prostate**, **bulbourethral glands**, and **seminal vesicles**), and **external genitalia** (the **penis**, and **scrotum**).

Using Figure 34.4 as a guide, locate the following structures on the model of the male pelvis:

ejaculatory duct	ureter
colon	prostate gland
scrotum	penis
vas deferens	urinary bladder
bulbourethral gland	urethra
inguinal canal	glans penis
testis	epididymis
corpus cavernosum	foreskin
seminal vesicle	pubic bone
ampulla of vas deferens	corpus spongiosum

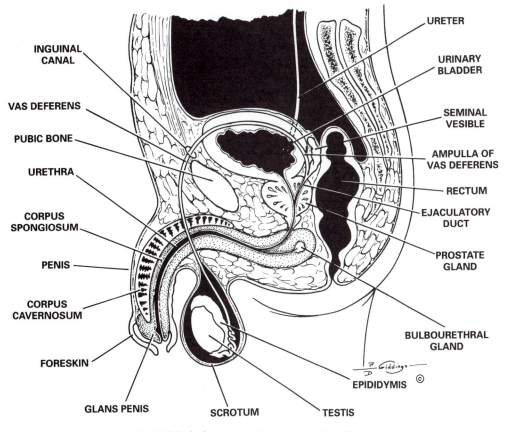

FIGURE 34.4 Male Reproductive System

FEMALE REPRODUCTIVE SYSTEM

Anatomy

The structures of the female reproductive system are homologous to those of the male reproductive system. They include **ovaries**, **oviducts**, **uterus**, **vagina**, and **external genitalia** (**clitoris**, **labial folds**, and **mons pubis**). The **mammary glands** of the female are also included in the reproductive system, but actually they should be included in the study of the skin, since they are modified sweat glands.

Find all of the parts listed below on the model of the female pelvis using Figure 34.5 as a guide:

ovary	oviduct	cervix
urinary bladder	clitoris	uterus
rectum	vagina	mons pubis
ligament	labium major	pubic bone
labium minor	urethra	

Ovary

As you read the rest of this exercise, refer frequently to Figure 34.6. In a section through a cat ovary, the outer tissues form the cortex, a layer not marked off sharply from the core tissues, or medulla. The surface layer of the cortex represents the **generative epithelium** from which strands of cells are budded off.

Below the cortical surface, note the presence of a distinct zone containing **developing follicles**. Each such follicle contains an **oocyte**, or immature egg, substantially larger than other cells in the vicinity. Directly surrounding the oocyte is a layer of more or less cuboidal **follicle cells**. At this stage the oocytes have usually completed the first meiotic division; the second division will occur after the egg leaves the ovary. In humans the second division will take place only if the egg is penetrated by a sperm cell.

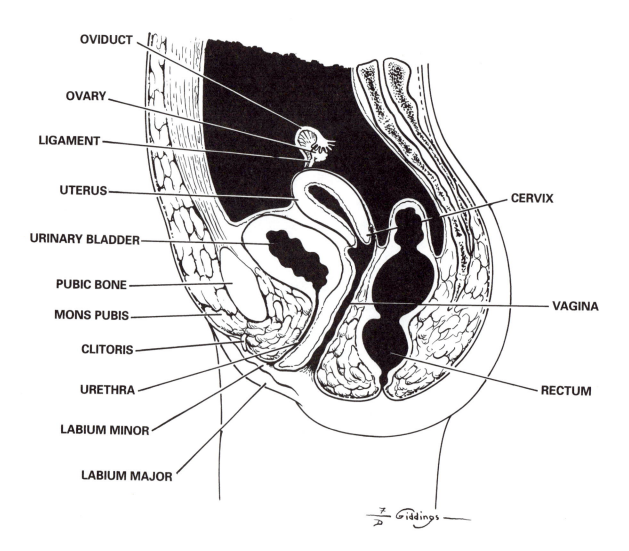

FIGURE 34.5 Female Reproductive System

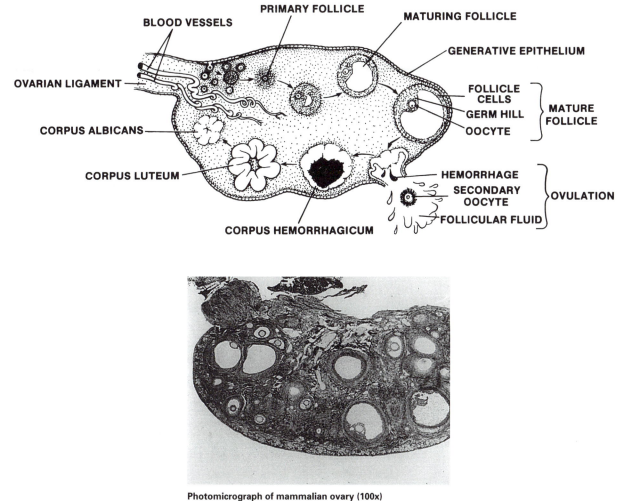

Photomicrograph of mammalian ovary (100x)
Courtesy of Turtox, Inc.

FIGURE 34.6 Diagrammatic Representation of an Ovary
Showing Follicle Development, Ovulation, and Formation of a Corpus Luteum

As the follicle (oocyte plus follicle cells) matures and grows, it ultimately comes to occupy the entire width of the ovarian cortex. Follicle cells divide (mitotically) and produce the female sex hormones, or **estrogens**. Follicle growth is accompanied also by the appearance of an enlarging fluid-filled cavity. In one region of the follicular wall a **germ hill** exists, which supports and surrounds the oocyte. When ovulation occurs, the oocyte escapes through ruptures in the follicular wall and the cortical tissue of the ovary.

Examine a prepared slide showing a mammalian ovary with a mature follicle. Make a diagram of it in the space to the right, labeling the mature follicle, oocyte, follicle cells, and primary follicle.

MAMMALIAN OVARY

MENSTRUAL CYCLE

Phenomena Occurring During the Cycle

The menstrual cycle involves cyclic changes occurring in the anterior lobe of the pituitary, ovaries, and uterus of the female. These changes deal with the preparation of the egg for, and its release at, the time of ovulation. In addition, the uterus undergoes changes in preparation to receive the developing embryo.

The textbook "average" for the length of the cycle is 28 days, with ovulation generally occurring about 14 days before the onset of the next menstrual flow period.

The steps involved in this cycle are (refer to Figures 34.7 through 34.9):

1. The high concentration of **FSH (Follicle Stimulating Hormone** from the anterior pituitary) in the bloodstream during the beginning of the cycle stimulates mitotic division in some of the follicles during the first half of the cycle.

2. The follicle cells begin producing **estrogen**. When the estrogen level gets above a certain threshold, it inhibits the production and release of FSH from the pituitary. This hormone is re-

sponsible for causing an increase in the thickness of the endometrium (the inner lining of the uterus where the fertilized ovum would implant). Later in the cycle, when the estrogen level drops, the level of FSH will begin to increase. This FSH-estrogen interaction is an example of a **negative feedback mechanism**, in which one substance stimulates the production of a second substance and the second substance inhibits the first.

3. Estrogen in the bloodstream stimulates the secretion of **LH (Luteinizing Hormone)** from the anterior pituitary. Therefore, as the estrogen level increases, the increasing level of LH stimulates further development of the follicle and causes **ovulation** (the release of the secondary oocyte from the mature follicle) at around day 14 of the cycle.

4. At the time of ovulation, some of the capillaries that had been nourishing the developing follicle rupture and release blood into the cavity of the follicle. What remains of the follicle together with its blood clot is known as the **corpus hemorrhagicum**. The clot will be removed by white blood cells, and with the removal of the clot and the thickening of the cell layers, the corpus

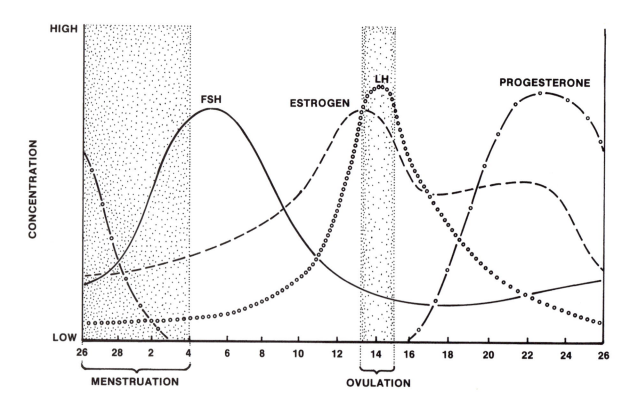

FIGURE 34.7 Plasma Hormone Levels During Menstrual Cycle

hemorrhagicum becomes the **corpus luteum**, capable of producing progesterone as well as small amounts of estrogen. While the secondary oocyte moves down the oviduct, the progesterone stimulates the endometrium to undergo preparations for the implantation of an embryo, should fertilization occur. Once the level of progesterone in the bloodstream increases, however, another negative feedback mechanism goes into effect, in which the secretion of LH is inhibited.

5. In the non-pregnant woman, the corpus luteum lasts for about 10–12 days and begins to degenerate on about day 20 or 21, due to the drop in the level of LH in the blood. The corpus luteum degenerates to form the **corpus albicans**, which gradually shrinks in size until only a microscopic scar remains on the ovary to indicate its location. This degeneration causes a decrease in the level of progesterone and estrogen (from the corpus luteum) arriving at the uterus via the bloodstream. As a result, the arteries supplying blood to the endometrium contract, causing the endometrial tissue to start degeneration and sloughing off (**menstruation**) due to an inadequate blood supply.

6. As soon as the level of estrogen drops due to the degeneration of the corpus luteum, the level of FSH increases as a result of the feedback mechanism mentioned in step 1. FSH then begins the cycle over again.

7. If pregnancy does occur, steps 5 and 6 are eliminated. The egg is usually fertilized in the upper 1/3 of the oviduct. The embryo, specifically the trophoblast cells of the blastocyst, and later the developing chorion, produces **HCG** (**human chorionic gonadotropin**). This hormone functions to maintain the corpus luteum and thus maintain the level of progesterone necessary to maintain the endometrium and sustain pregnancy.

Study demonstration slides of ovaries containing corpora hemorrhagica, corpora lutea and corpora albicans, diagramming each in the space below.

Examine demonstration slides of the endometrium at different stages of thickening. Make sketches in the space provided and compare these to Figure 34.8.

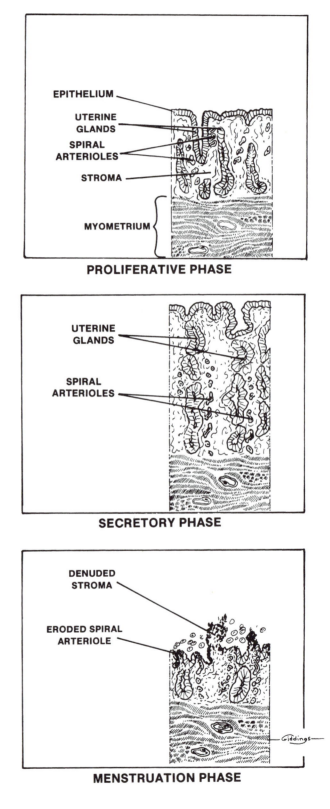

FIGURE 34.8   **Diagrammatic Representation of the Uterine Endometrium**

In Figure 34.9 showing changes in the ovary and endometrium during the menstrual cycle, identify:

follicle
corpus luteum
menstrual flow days
proliferative phase (estrogen stimulated
    thickening of the endometrium)
secretory phase (progesterone stimulated
    glandular development of the endometrium)
corpus albicans
ovulation time
corpus hemorrhagicum
secondary oocyte

For the four hormones important in this cycle, indicate their sources, peak levels during the cycle, and their functions.

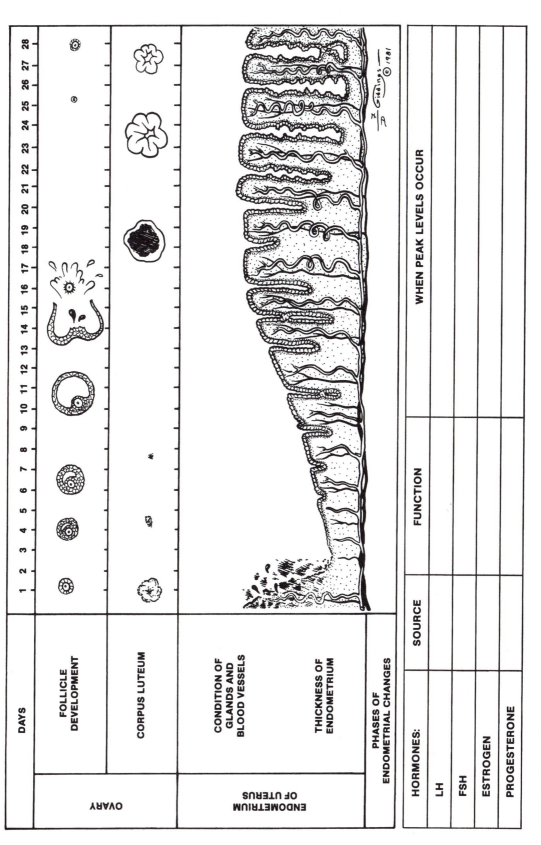

FIGURE 34.9 Diagrammatic Representation of Interrelationships of Reproductive Hormones and the Ovarian and Uterine Cycles

CONTRACEPTION

Various types of methods and devices are currently used to prevent pregnancy. Each of them varies in its effectiveness in preventing conception and in safety for the individual using it. Below is a list of the common contraceptive methods used today. Some of the common methods of contraception are shown in the lab. Observe the film in the lab entitled "Contraception" which explains these methods fully and discusses their effectiveness.

ORAL CONTRACEPTIVES

These pills contain synthetic estrogen and progesterone. The hormones function in setting up a negative feedback relationship with the anterior pituitary, thus preventing the release of LH and FSH. Consequently, ovulation does not occur. Some are believed to affect the cervical mucus and thus prevent the sperm from entering the uterus.

There are two general categories of contraceptive pills. In one type ("combination pills"), each pill contains both estrogen and progesterone, and the woman would take them for 20 days. Menstrual flow would begin approximately 3 days after taking of the last pill in the series. With the other type ("sequential pills"), a more normal hormonal level is established by the taking of pills containing only estrogen for the first 15 days and for the next 5 days a pill containing both estrogen and progesterone. Again, after these 20 pills are taken, menstrual flow would begin within a couple of days and on day 5 of the menstruation, a new series of pills would be begun.

The advantage of the pill is that it is the most effective contraceptive (other than sterilization), and it does not require any special preparation for birth control prior to intercourse. Disadvantages are the pill's potentially hazardous side effects; included among these are their link to certain vitamin deficiencies, increased blood pressure, and blood clot formation. The pill may also increase the risk of minor vaginal infections. Smokers and women over 35 should not use the pill. Conception rate is less than 2% and probably 0%, if a woman never forgets to take the pills as prescribed.

INTRAUTERINE DEVICES (IUD'S)

These small plastic devices are placed within the cavity of the uterus by a physician. They can be removed at any time the woman desires. Exactly how they work as birth control devices is not known; it is believed that they modify the environment of the cervix, uterus, and/or oviducts such that pregnancy is prevented either by preventing the sperm from reaching the egg or by preventing the implantation of the fertilized egg.

The advantage of the IUD is that it is extremely convenient since once it is inserted it requires little further attention. In addition, it does not alter the woman's natural hormonal cycle. As far as effectiveness is concerned. this varies with the particular "style" of the IUD; the newer types are more than 99% effective. Disadvantages are that in a few women they can be expelled spontaneously or may cause some discomfort or mild bleeding. On rare occasions pregnancy can occur and, depending on the type of IUD, either an apparently unaffected child plus IUD are expelled at the end of 9 months or the fetus may be incompletely aborted spontaneously, causing possible uterine infections and, in a few rare cases, death of the woman. Conception rate is 3% if the device is not expelled accidentally.

STERILIZATION

Male

This is the only foolproof method of birth control. What it involves is the surgical blocking or cutting of the vasa deferentia or tubes. **Vasectomy** is a relatively simple, rapid, inexpensive procedure that does not usually require hospitalization. Small incisions are made in the top of the scrotum and the vasa deferentia are either cut (with ends tied, or cauterized shut), clamped shut, or blocked with a plastic plug. Thus, even though sperm are still being produced, they are not transported to the urethra; instead, they are reabsorbed by the body. Hormonal production continues as normal, as well as secretory activity of the seminal vesicles, prostate and Cowper's glands. Conception rate is almost 0%. On occasion the male tubes have rejoined.

Female

In the female, the sterilization procedure can be done one of two ways; tubal ligation or laparoscopy. In **tubal ligation**, the oviducts are tied off and cut, preventing the sperm from reaching the egg. It is considered major surgery since an abdominal incision is required. **Laparoscopy**, on the other hand, is comparable to a vasectomy in regard to time and recovery rate. Essentially what this procedure involves is the insertion of a laparoscope (a tube through which the doctor can look into the abdominal cavity of the woman and locate the oviducts) and of an electric cauterizing instrument (to cauterize the severed ends of the oviducts) through tiny incisions in or near the navel (accounting for the common reference to this type of sterilization as being "belly button surgery" or "Band-Aid surgery"). It requires little or no hospitalization and most of the patients have either no reaction to the surgery or have a cramping response similar to menstrual cramps. The advantage of sterilization in the female is that it does not affect her natural hormonal production and the egg that is released simply disintegrates. Conception rate is almost 0%. Once in a great while the oviducts manage to grow back together.

MECHANICAL METHODS

The **condom** is a thin latex sac that fits over the erect penis. It functions to trap ejaculated semen in the sealed end of the sac so that the sperm cannot actually enter the female system. The advantage to this device is that it has no harmful effect and it does offer some protection against contracting venereal disease. The disadvantages are that it can rupture and that it can loosen as the penis becomes flaccid, allowing sperm to enter the vagina. Conception rate is 7%.

The **diaphragm** is a thin rubber cup with a flexible spring in the rim. It is coated with spermicidal cream or jelly, inserted into the vagina, and placed so that it covers the cervical opening. When it is properly positioned, the woman cannot feel it. It has to be inserted prior to having intercourse and should be left in place for a certain length of time afterwards to ensure that all sperm have died. Since there is variation among women as to the size of the cervical area. the correct size or fit of the diaphragm has to be determined by a doctor. It has no

harmful effects and, when it does fail as a birth control device, this is usually due to improper insertion or failure to use the spermicidal cream or jelly. Conception rate is 10%.

VAGINAL SPERMICIDES

These are creams, jellies, and foams that are inserted into the vagina to serve as a chemical barrier at the cervical area. The appropriate dosage is inserted with an applicator just prior to intercourse and is "good" for just one intercourse. Apparently, these spermicides work in affecting the pH of the vagina, making it intolerable for the sperm. Conception rate is 13%.

RHYTHM METHOD

In this type of birth control method, the couple uses the basic idea that fertilization can only occur around ovulation time, taking into consideration also the life span of the sperm (2–3 days) and the life span of the egg (1–2 days). The couple would abstain from intercourse for about 4 days before the **expected** ovulation and about 3 days after (to ensure that viable sperm and egg are not present at the same time). The big difficulty with this as a contraceptive technique is the inability to determine **exactly** when ovulation will occur, since the exact time of ovulation can fluctuate depending on physiological and psychological stresses the woman may be under. Hints about the timing of ovulation are sometimes obtained by noting an increase in body temperature and changes in stickiness of cervical mucus. Conception rate is roughly 25%.

REVIEW QUESTIONS

1. If a menstrual cycle is regularly 34 days, on about what day would you expect ovulation to occur? _____

   Why? _____

2. If it is correct to say that lack of one hormone is primarily responsible for the onset of menstruation, which one might it be? Why? _____

   _____

3. In what organs of the reproductive system (male and female) would there be a high rate of mitosis? _____

   _____

4. Which structures contribute to the formation of semen? _____

   _____

5. Where is the corpus luteum located during its functional life? _____

6. What does the corpus luteum develop from? _____

7. What happens to the corpus luteum when it is no longer functional? _____

8. Where is an IUD placed? _____

9. What structure in the female would have the same embryological origin as the penis in the male? _____

   _____

10. List in sequence the structures which sperm must travel through in the male reproductive system during an ejaculation. _____

    _____

    _____

COMPARATIVE EMBRYOLOGY

INTRODUCTION TO EMBRYOLOGY

In the process of sexual reproduction in animals, two gametes — sperm and egg — fuse to make one new cell. These two gametes are usually donated by two different parents. The new zygote resulting from this fusion contains material from both parents and is "new" in the sense that it contains new potentialities resulting from the mixture of the genetic material from the two parents. Embryology is the study of the development of the zygote, the product of sexual reproduction.

The zygote develops as a result of three kinds of activity: **mitosis** and subsequent growth, **differentiation of cells**, and **movement of cells**. Thus, as a result of embryonic activity, the one-celled fertilized egg (zygote) changes or develops into the adult form eventually. In observing the embryonic development of any animal we can divide it into the following stages:

Fertilization — the fusion of nuclei and other events associated with the union of sperm and egg. The result is the zygote.

Cleavage — a series of mitotic divisions undergone by the zygote. No growth in size of the entire structure occurs at this time. The actual pattern of cleavage — whether the whole zygote (**holoblastic cleavage**) or only a part of it divides (**meroblastic cleavage**) — varies from one species of animal to another, depending on the amount of yolk in the egg and its distribution. The end result of cleavage is the formation of a hollow ball of cells known as a **blastula**, the central cavity of which is known as the **blastocoel**.

317

Gastrulation — the migration of cells of the blastula resulting in the formation of a new cavity known as the **gastrocoel** (archenteron, or "primitive gut"). Further development leads to the formation of three distinct **germ layers** (embryonic cell layers) known as the **ectoderm**, **mesoderm**, and **endoderm**, from which all the organs of the new organism develop. At this stage, when the three germ layers are present, the structure is technically termed an embryo.

The development of internal shape characteristics of the animal, known as **morphogenesis**, results from the shaping of the germ layers due to differential growth, movement, and association of cells in the germ layers.

Neurulation — the development of the notochord, neural tube, and coelom. This only occurs in chordates.

Organogenesis — the differentiation and association of cells to form organ systems.

In this exercise, we will make a brief survey of four different animals: starfish, frog, chick, and human. This survey should, first of all, illustrate the basic similarities in the embryonic development of these entirely different animals. Secondly, even though there are similarities, there are also variations, based primarily on the amount of yolk present in the cytoplasm of the egg.

STARFISH DEVELOPMENT

Obtain one slide of the starfish development. On this slide, you will find all of the stages indicated in Figures 35.1–35.7. The stages on the slides are whole mounts, not sections. Identify each of the stages indicated on the slide.

NOTE

Draw representative stages of starfish development as indicated on Figure 35.8.

UNFERTILIZED EGG

In this nearly spherical cell a large nucleus and a nucleolus are clearly visible. A small amount of yolk (stored food) is present in the form of many small particles. Are these yolk particles present in any particular area of the cytoplasm, or are they scattered about?

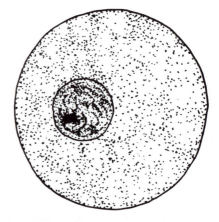

FIGURE 35.1   Unfertilized Egg

ZYGOTE (FERTILIZED EGG)

This is a single celled structure very much like the unfertilized egg in appearance. In contrast to the unfertilized egg, the zygote's nucleus is inconspicuous. Would this structure be 1N or 2N?

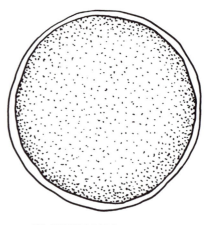

FIGURE 35.2   Zygote

EARLY CLEAVAGE

The 2-, 4-, and 8-celled stages are included in early cleavage (Figure 35.3). Find an example of each of these and note that the cells remain attached to each other. Is there any growth in size?

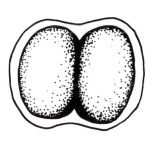 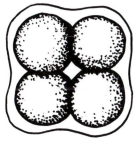 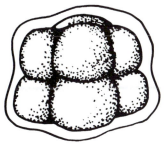

2-Cell Stage   **4-Cell Stage**   **8-Cell Stage**

FIGURE 35.3   Early Cleavage

How does the cell size in each of these stages compare with the size of the zygote? _____

_____

Are all the cells within a single stage the same size? _____

_____

LATER CLEAVAGE

The 16-, 32-, and 64-celled stages are included in later cleavage. The 64-celled stage is hollow and is called the **blastula**.

What happened to the individual cell size? _____

_____

BLASTULA

As cell division continues, the increasing number of cells become arranged around an enlarging central cavity known as the **blastocoel** (Figure 35.5). In the starfish, the walls of the blastula are usually one cell layer thick. Why would you expect your specimen to appear dark around the edges and light in the middle? _____

_____

Can you see any differences among the cells?

_____

How does the size of the blastula compare to earlier stages? _____

_____

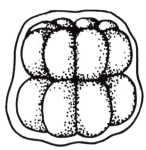 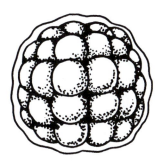 

16-CELL STAGE   **32-CELL STAGE**   **64-CELL STAGE**
(MORULA)   **(EARLY BLASTULA)**   **(BLASTULA)**

FIGURE 35.4   Later Cleavage

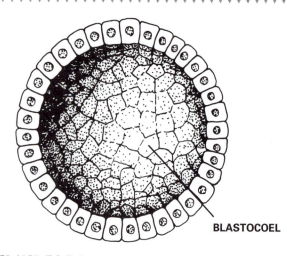

FIGURE 35.5   Late Blastula Cross-Section

GASTRULA

Shortly after the formation of the blastula, a small depression begins to appear at one end of the embryo; the appearance of this depression marks the beginning of gastrulation. As gastrulation proceeds, the depression invaginates (folds inward) more and more. Which of the three basic embryonic activities would this be? _____

_____

Find the embryos at various levels of gastrulation on your slide. The latest stages on your slide are those in which the inner end of the invagination is beginning to expand.

As a result of gastrulation in the starfish, the embryo produces two primary cell layers: an outer **ectoderm**, and an inner **endoderm**. The third layer, the **mesoderm**, develops later, between these two. Gastrulation, then, eventually results in an embryo with 3 primary germ layers, a mere remnant of the blastocoel, and a new cavity known as the **gastrocoel** or

archenteron. The gastrocoel is continuous with the outside through the **blastopore**. The gastrocoel will become the digestive cavity of the digestive tract. In the deuterostomes, the blastopore will eventually become the anus.

LARVAL STAGE

The gastrula stage is reached within a day or two after fertilization. Within another day or two, this stage undergoes some alterations to give rise to the **larval stage**, which is free-swimming. During a period of from several weeks to several months, the larva grows. After this time, it settles to the bottom and becomes a small starfish. Refer to Figure 35.7.

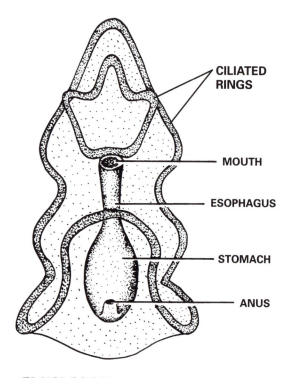

FIGURE 35.7   Bilateral Starfish Larva

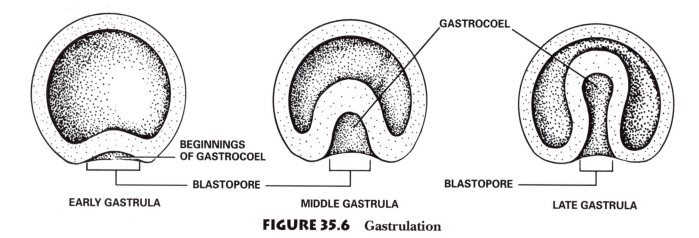

FIGURE 35.6   Gastrulation

UNFERTILIZED EGG   FERTILIZED EGG   TWO-CELL STAGE

FOUR-CELL STAGE   EIGHT-CELL STAGE   MORULA

BLASTULA, SECTION   EARLY GASTRULA, SECTION   MIDDLE GASTRULA, SECTION

LATE GASTRULA

FIGURE 35.8   Stages of Starfish Development

FROG DEVELOPMENT

Examine the charts, models, and whole specimens available in lab. Obtain slides of sectioned embryonic stages. Using Figures 35.9 through 35.16, locate the structures printed in boldface in this part of the exercise.

UNFERTILIZED EGG

The egg consists of two portions: a darkly pigmented portion, the **animal pole**, and a lightly colored, yolk-filled portion, the **vegetal pole**. In nature, when the egg is released into the water, the **gelatinous covering** produced by the oviduct absorbs water and swells, causing the eggs to be equidistant from each other. What is the importance of this swelling? _____

_____

ZYGOTE

Sperm have to penetrate the eggs before swelling of the gelatinous covering takes place. Once fertilization has occurred, the cell rotates in such a way that the heavier portion of the cell, the yolk-filled vegetal pole, is downward.

Suppose you had a dish of fertilized frog eggs; how would you know if they were all fertilized (i.e., what color would be facing upward)? _____

_____

Another indication that fertilization has occurred is the appearance of a pigmented area (**gray crescent**) between the yolk-filled and black portions.

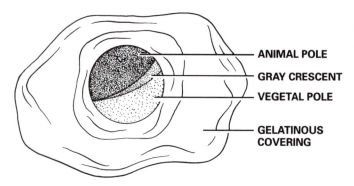

— ANIMAL POLE

— GRAY CRESCENT

— VEGETAL POLE

— GELATINOUS COVERING

FIGURE 35.9 Frog Zygote

CLEAVAGE

The beginning of the first cleavage is marked by the appearance of a groove on the animal pole end of the egg; this **cleavage furrow** gradually extends toward the opposite side of the zygote, dividing it into two cells. The next cleavage occurs at right angles to the first and produces the **4-cell stage**. The third cleavage occurs parallel to, but a little above the equator of the developing embryo. What would you suspect was the cause of this unequal division?

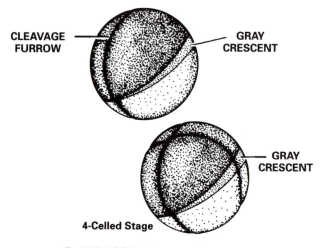

CLEAVAGE FURROW

GRAY CRESCENT

GRAY CRESCENT

4-Celled Stage

FIGURE 35.10 Early Cleavage

In which region of the developing embryo would you expect a more rapid rate of cell division? _____

_____

What effect would this have on cell size in the two regions of the developing embryo? _____

_____

Figures 35.11 and 35.12 are two diagrams of early and late cleavage. Note that the jelly layer is not shown in these and subsequent stages. In the space below each of the drawings, make a diagram of the section through these stages as seen in the prepared slides. Label the **animal pole**, **vegetal pole**, and **cleavage furrow** on your drawings. Can you find the yolk particles? _____

Give two ways in which the animal pole can be distinguished from the vegetal pole. _____

_____

Is the pigment localized in any particular region of the cell cytoplasm or is it distributed equally throughout? _____

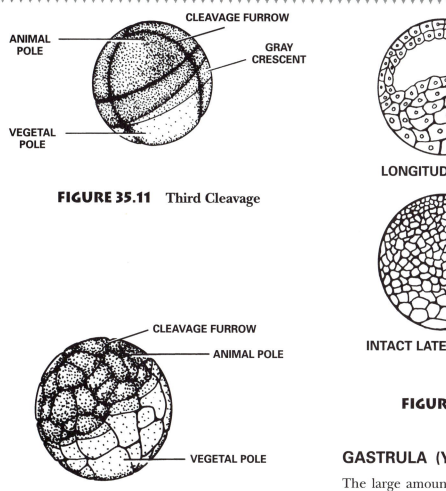

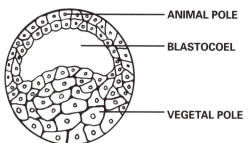

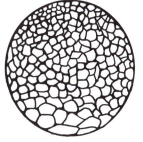

FIGURE 35.11 Third Cleavage

FIGURE 35.12 Late Cleavage

LONGITUDINAL SECTION, BLASTULA

INTACT LATE BLASTULA

FIGURE 35.13 Frog Blastula

What do you think is the function of the **membrane** (blue-purple color) around the entire structure on the slides? _____

_____

BLASTULA

The stage in which the **blastocoel** (internal cavity) is formed is called the blastula. Notice that, instead of being centrally located as in the starfish, the blastocoel is off center toward one pole of the developing embryo.

In which hemisphere is it located? Why? _____

_____

GASTRULA (Yolk Plug Stage)

The large amount of yolk in the frog egg prevents the type of invagination at one end of the blastula which is seen in the starfish. Instead, the more rapidly dividing cells from the animal pole grow downward over the yolk-filled cells, gradually enclosing them. **Gastrulation** begins with the pushing inward (or **invagination**) of these cells. This forms a crescent-shaped "line" on the surface known as the **dorsal lip**. The opening marked by the dorsal lip is known as the **blastopore**. This marks the posterior end of the embryo. In the prepared slide of the early gastrula, you can actually see a depression forming here as the surface cells move inward. The depression gradually enlarges to form the **gastrocoel** or archenteron, as seen in the slide of the late gastrula. Note the difference between this gastrocoel and the one in the starfish.

What will the gastrocoel give rise to? _____

_____

With continued cell division and the migration of cells inward, the animal pole cells encircle, and eventually line, the entire gastrocoel except for a small circular plug of yolk cells within the blastopore opening, the **yolk plug**. Eventually, even the yolk plug disappears as it is covered by the migrating

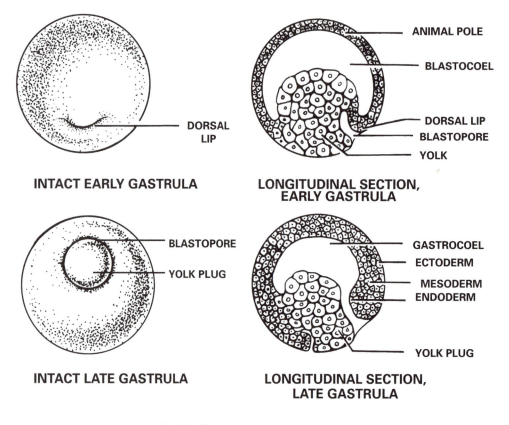

DORSAL LIP

INTACT EARLY GASTRULA

ANIMAL POLE

BLASTOCOEL

DORSAL LIP
BLASTOPORE
YOLK

LONGITUDINAL SECTION,
EARLY GASTRULA

BLASTOPORE

YOLK PLUG

INTACT LATE GASTRULA

GASTROCOEL
ECTODERM

MESODERM
ENDODERM

YOLK PLUG

LONGITUDINAL SECTION,
LATE GASTRULA

FIGURE 35.14   Frog Gastrulation

cells. As all of this is occurring, the three developing **germ layers** are becoming arranged in such a way that there is an outside layer of cells (**ectoderm**), an inner lining of the gastrocoel (**endoderm**), and a layer between the two (**mesoderm**).

Can you detect any visible differences between the cells of these three layers? _____

_____

Is there any difference in the overall shape of the structure? _____

_____

NEURULA

Near the end of gastrulation, the ectodermal cells in the mid-dorsal region of the embryo thicken to form a flattened area on the surface known as the **neural plate**. The sides of the neural plate, the **neural folds**, gradually fold upward forming a depression, the

neural groove, between them. Eventually the folds fuse, forming a closed tube, the **neural tube**, which will develop into the brain and spinal cord. Refer to Figure 35.15.

In the midline of the mesodermal layer, the cells develop into a cylindrical rod, the **notochord**.

Although the notochord is a rod extending along the length of the animal, how would you expect it to appear in your slide? _____

_____

What is the cavity in the area below the notochord? _____

_____

What is the function of the large cells that may be (depending upon where the animal was sectioned) located in this region? _____

_____

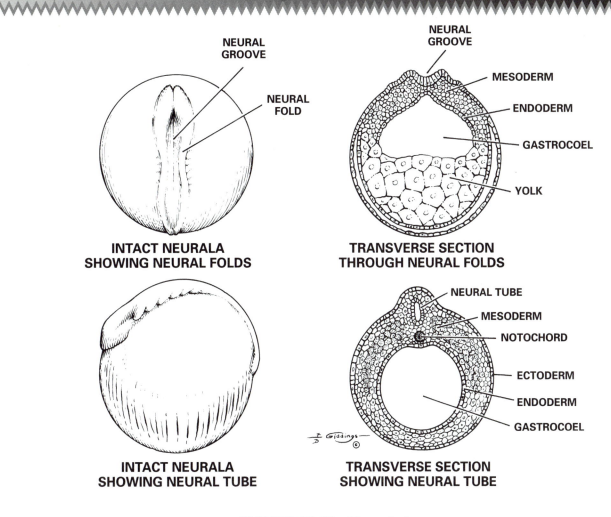

NEURAL GROOVE

NEURAL FOLD

INTACT NEURALA SHOWING NEURAL FOLDS

NEURAL GROOVE

MESODERM

ENDODERM

GASTROCOEL

YOLK

TRANSVERSE SECTION THROUGH NEURAL FOLDS

INTACT NEURALA SHOWING NEURAL TUBE

NEURAL TUBE

MESODERM

NOTOCHORD

ECTODERM

ENDODERM

GASTROCOEL

TRANSVERSE SECTION SHOWING NEURAL TUBE

FIGURE 35.15 Neurulation

LARVAL STAGE

With the formation of the neural tube, the embryo elongates. The anterior end becomes slightly enlarged, forming the **head** (Figure 35.16). A **tail** develops from the posterior end, and **gills** develop for gas exchange. The mouth opens so that feeding can occur. Over a period of growth of a few months, this **tadpole** will metamorphose into an adult frog capable of survival both on land and in water.

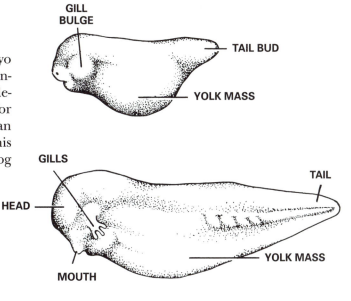

GILL BULGE

TAIL BUD

YOLK MASS

GILLS

HEAD

MOUTH

TAIL

YOLK MASS

FIGURE 35.16 Frog Larval Stages (Tadpole)

CHICK DEVELOPMENT

In both the starfish and the frog, all of the cells derived from the fertilized egg are used in making the new individual. In land animals, however, a number of cells are not used in the immediate makeup of the embryo body; instead, they give rise to certain temporary structures necessary for embryonic development on land. Figure 35.17 demonstrates the four membranes that develop in the chick to ensure survival:

Amnion — contains water to shield the embryo

Yolk sac — contains yolk as a food source

Allantois — stores waste materials produced by the developing embryo plus transporting respiratory gases between the embryo and its environment

Chorion — also functions in exchange of gases

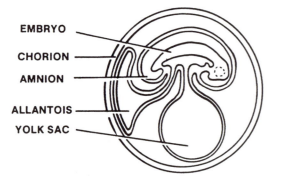

FIGURE 35.17 **Chick Embryonic Membranes**

Thus, the egg of the chick is a "self-contained" capsule simulating the watery environment of the lower vertebrates and yet showing remarkable adaptations to land development.

UNFERTILIZED EGG

Examine the egg on demonstration in lab. Figure 35.18 shows the internal structure of the chicken egg. Notice that in this case the everyday term of "egg" refers to more than simply the egg cell, **ovum**. The circular yellow mass that we would refer to as "yolk" as we are sitting at the breakfast table is actually the ovum. The yolk of the egg is in fact the single egg cell. It is composed of a large amount of **yolk granules** plus a small amount of yolk-free cytoplasm, the **germinal disc**, on the surface of the yolk. It is the germinal disc that contains the nucleus to be fertilized and that will undergo cell division to form the embryo. If the egg becomes fertilized, the germinal disc is referred to as the **blastoderm**. The yolk supplies nourishment to the developing embryo.

After the embryo has been ovulated, it will pass down the oviduct acquiring a number of other structures designed to aid survival in the land environment. Immediately around the ovum is the egg "white"; this is known as **albumen**, which functions as a water reservoir for the young embryo and as a food source for later development. Attached to the yolk and extending out into the albumen are dense cordlike structures, the **chalazae**, that serve to suspend the yolk in the albumen. In another region

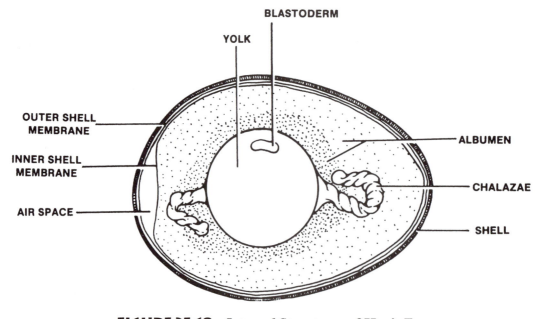

FIGURE 35.18 **Internal Structures of Hen's Egg**

of the oviduct, two thin shell membranes are deposited around the albumen; these function in decreasing water loss. The next section of the oviduct produces the **shell** and molds the egg into its customary shape. The hard protective shell is porous, allowing for gas exchange.

EARLY DEVELOPMENT

By the time the egg is laid, the stages of development through gastrulation have usually occurred, and the embryo is undergoing neurulation. Notice that only the blastoderm undergoes cleavage. Note also that, because of the tremendous amount of yolk, the blastula is not a "hollow ball of cells"; instead, it is a hollowed, flattened disc on the yolk surface. A third difference is with regard to gastrulation. Notice that the blastopore (the site where surface cells migrate inward to establish the three germ layers) is not a circular opening as was in the case of the starfish and the frog. In the chick, the blastopore is elongate and is referred to as a **primitive streak**.

LATER DEVELOPMENT

Obtain plastomounts of different stages of chick development and examine them under a stereomicroscope for the structures in Figure 35.19. Notice particularly the characteristic curvature of the embryo body as it is being separated from the underlying yolk. The circulatory system within the embryo is quite extensively developed; the heart would definitely be beating if this were alive. Why do you suppose there is such an extensive blood vessel network outside the embryo body? The 72-hour

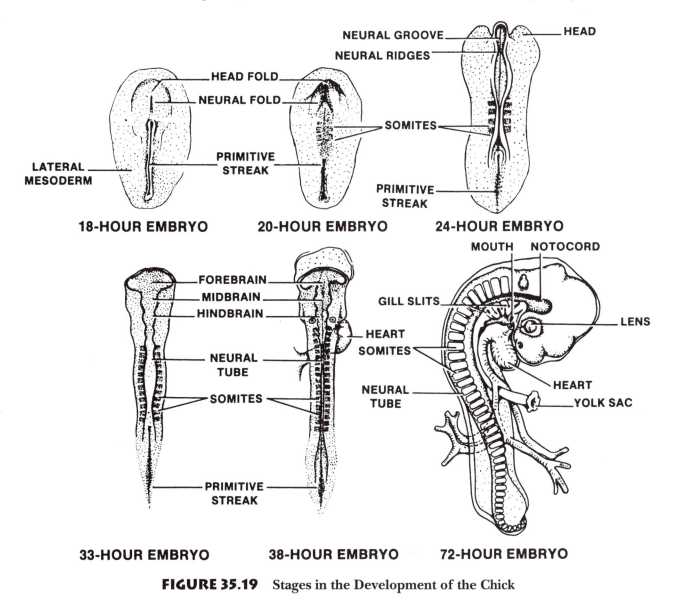

FIGURE 35.19 Stages in the Development of the Chick

embryo has two extreme bends which make it curve back on itself. Identify the **anterior limb bud** which will give rise to the wing.

Notice the bump-like structures on either side of the spinal cord (mid-dorsal region). These structures are the **somites**, blocks of mesodermal cells that will later give rise to muscles and vertebrae. They are also used as a means of determining how old the embryo is. How many pairs are there? _____

In looking at the embryo, can you determine if there is a difference in the rate of development in comparing the anterior and posterior regions of the body? Explain your answer. _____

_____

Chick development continues for 21 days, until hatching. Note that this animal does not go through a larval stage as did the starfish and frog.

EXTRAEMBRYONIC MEMBRANES OF THE CHICKEN AND MAMMALS

Human development parallels that of the chick even though the egg of humans does not contain an overabundance of yolk. Why? _____

_____

The reptiles were the first to lay eggs on land. Their eggs contained extraembryonic membranes by which the embryo carried out gas exchange, excretion of wastes, and consumption of stored food (yolk). These same membranes develop in the human, but are put to different uses since the human develops internally. Figure 35.20 shows the membranes in a chick egg and compares them to those in the human.

HUMAN DEVELOPMENT

Referring to Figure 35.21, note that fertilization normally occurs in the upper 1/3 of the oviduct, and early development of the embryo occurs during its movement down the oviduct. By the time the developing embryo reaches the uterine cavity (5–7 days after ovulation), it is in the **blastocyst** stage.

The blastocyst is equivalent to the blastula stage; the blastocyst cavity does not have a wall of uniform thickness: a cluster of cells occurs at one end. The wall of the blastocyst will give rise to the **chorion**. This membrane has two important functions: (1) during early development (first 3 months) it produces **HCG**, the hormone responsible for keeping the corpus luteum functioning; and (2) it initially absorbs nutrients from the endometrium through villi and later gives rise to the fetal portion of the **placenta**. The cluster of cells is known as the **inner cell mass**, which will give rise to 4 structures: **embryo**, **amnion**, which forms a protective, fluid-filled cavity around the developing embryo; **yolk sac**, present even though the human oocyte has no yolk granules; and **allantois**, whose most important function is probably the formation of the umbilical blood vessels. (Refer to Figures 35.20 and 35.21 for the 4 structures printed in boldface.)

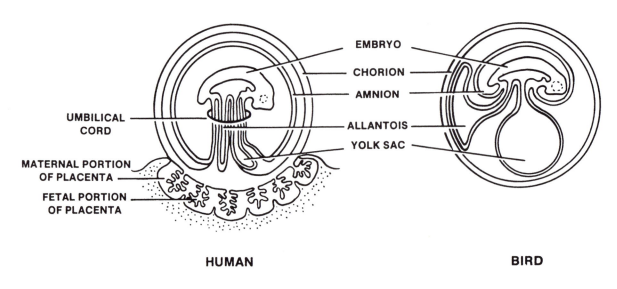

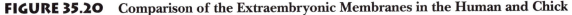

FIGURE 35.20 Comparison of the Extraembryonic Membranes in the Human and Chick

A few days after entering the uterus, the blastocyst undergoes the process of **nidation** (implantation), during which it buries itself in the endometrium from which it will acquire nutrients for the rest of its development.

Gastrulation occurs in somewhat the same manner as in the chick since a primitive streak is formed. The **embryo** with its 3 germ layers (ectoderm, endoderm, and mesoderm) is formed within 2 weeks after ovulation. By the end of 8 weeks of development, all of the body parts are present and it is now called a **fetus,** rather than an embryo. After this time, the organs develop further to eventually become functional and the fetus grows rapidly in size. To accommodate the increased nutritional demands for this growth, the **placenta** serves as the site of nutrient and waste exchange between fetus and mother.

To give you a better appreciation of the growth involved, the following chart indicates the "crown-rump" length (i.e., sitting height) of the human fetus at different stages of development:

2 weeks	— 0.23 mm	5 month	— 7 in.
1 month	— 3/8 in.	6 month	— 9 in.
2 month	— 1 in.	7 month	— 11 in.
3 month	— 3 in.	8 month	— 14 in.
4 month	— 5 in.		

Examine the materials on display in the lab. The models display the relationship of the fetus to the uterus, and the charts illustrate the process of birth.

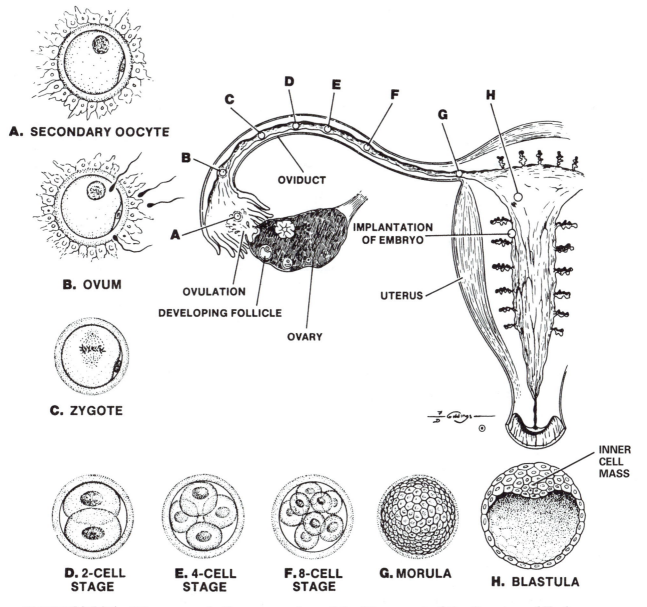

A. SECONDARY OOCYTE

B. OVUM

C. ZYGOTE

OVIDUCT

OVULATION

DEVELOPING FOLLICLE

OVARY

IMPLANTATION OF EMBRYO

UTERUS

INNER CELL MASS

D. 2-CELL STAGE

E. 4-CELL STAGE

F. 8-CELL STAGE

G. MORULA

H. BLASTULA

FIGURE 35.21 Diagrammatic Representation of the Movement of the Oocyte and Embryo

REVIEW QUESTIONS

1. Why is the maintenance of the corpus luteum so important during human pregnancy? _____
_____

2. What are the maternal and fetal portions of the placenta? _____
_____

3. How do substances actually get from the blood stream of the mother to the child? _____
_____

4. The pregnancy test is based on the presence of what hormone? _____

5. List the 3 germ layers and name two structures which develop from each.
_____
_____
_____

6. Define the following terms:

   Gastrulation _____

   Morphogenesis _____

   Fertilization _____

   Cleavage _____

   Blastopore _____

   Archenteron _____

   Invagination _____

   Blastocyst _____

   Nidation _____

   Neurulation _____

7. How does embryonic nutrition in the chick differ from that in a mammal? _____
_____

8. Examine the placenta. After the first eight weeks, do the chorionic villi entirely surround the embryo or are they confined to an area of attachment?_____

9. The chemical composition of the amniotic fluid is similar to that of blood. Is this of any significance?
_____

10. How does embryonic respiration differ in the chick and human embryos?_____
_____
_____

11. At what age do you begin to see the formation of eyes in the chick embryo? _____

12. At about what age do you see development of segmentation of the chick embryo body? _____

Excretory System

OBJECTIVES

After completion of this exercise, the student should be able to do each of the following:

- Identify the parts of the human urinary system.
- Identify the structural components of the nephron and describe where they are located.
- Describe where filtration and reabsorption occur in the nephron and explain what is involved in each process.
- Explain why increased salt intake lowers the rate of urine formation.
- Describe the relationship between the rate of urine formation and specific gravity (from lab discussion of results).
- Explain exactly how alcohol or caffeine increases urine output (applicable if lab instructor uses test solution E).
- Describe the normal range for the pH of the urine and list several factors that might alter the pH.
- Answer the review questions at the end of this exercise.

INTRODUCTION

The amount and chemical composition of the body fluids are controlled by **input** (food, fluid, etc.), **metabolism**, and **output** (urine).

The kidneys control output because they regulate the cellular environment by selective excretion of those substances in excess or toxic to the body and by a reabsorption of those substances beneficial to the body.

In this lab we will study the anatomy of the kidney and also experimentally test the effect of intake of different substances on kidney function.

KIDNEY ANATOMY

Each kidney is composed of approximately 1 million microscopic units known as **nephrons**. These tubular structures are the functional units of the kidney. Each one is associated with two capillary networks. Because of the pressure of blood in the **glomerulus** (first capillary network), dissolved substances and water are filtered out of the blood into the cavity of the **Bowman's capsule**. This solution in the capsule is basically blood plasma minus plasma proteins. When the solution reaches the tubular portion of the nephron, certain of the dissolved substances and most of the water are reabsorbed into a second capillary network and re-enter the bloodstream. Specifically, sodium is reabsorbed into the capillaries from the proximal convoluted tubule by active transport. As more sodium is reabsorbed, the water concentration in the capillaries drops, allowing water from the tubule to move into the capillaries by osmosis. Thus, if more sodium is available to be reabsorbed, then more water will enter the capillaries by osmosis. Thus, a combination of **filtration** and **reabsorption** maintains the proper balance of water and salts within the body and also rids the body of metabolic wastes (especially nitrogenous wastes). Fluid remaining in the tubules after reabsorption is **urine**, the color of which is dependent on the concentration of bile compounds present. The fluid from all of the nephrons drains into a common cavity.

The kidney is divided into three general regions: **cortex**, **medulla**, and **pelvis**. The cortex is the outer region of the organ, containing mainly the glomeruli and convoluted tubules of the nephrons. The medulla has pyramid-shaped groupings of collecting tubules and Henle's loops. The pelvis is the region into which the contents of the nephrons are emptied. From this point the urine is drained into the **urinary bladder** via a **ureter**. After temporary storage, urine is passed to the outside via the **urethra**.

Observe the charts, models, and preserved sheep's kidney on display, using Figure 36.1 as a guide.

URINARY PHYSIOLOGY — OPTIONAL

Analysis of urine can yield valuable information about the condition of the body. Certain diseases are characterized by the presence of substances in the urine which normally should not be present. Presence of glucose, for example, indicates that there is so much glucose in the bloodstream that the kidney tubule is incapable of completely reabsorbing it, and is indicative of diabetes mellitus. Presence of ketones in urine is also a symptom of diabetes. These chemicals result from excessive metabolism of lipids due to the cells' inability to obtain and use glucose. Presence of blood or protein may indicate damage to the urinary tract by a bacterial infection. Volume of urine produced indicates the level of fluid intake. A small amount of highly concentrated urine indicates a state of body dehydration. The kidney has the ability to adjust the pH of the urine over a wide range. It can excrete or reabsorb ions to maintain a constant blood pH of 7.4. A drastic change in pH may indicate malfunction in the nephron. Thus, analysis of

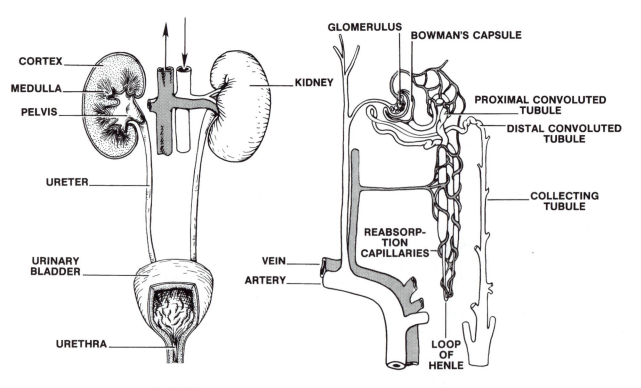

FIGURE 36.1 The Urinary System and the Nephron

urine composition as well as of volume is a valuable tool in determining the general state of health of the individual as well as the specific condition of the kidneys.

In the following experiment we will demonstrate the effect of intake of various substances on the volume and composition of urine produced.

EXPERIMENTAL PROCEDURE

At least 4 test groups will be set up (more, if the instructor so desires). Volunteers should be chosen in the previous lab period so that their food and water intake can be restricted for two hours before the experiment. Record the last time of urination prior to the experiment. At the beginning of the lab period, each subject should empty the bladder as completely as possible. This urine should be saved and labeled with the subject's name. This specimen will serve as the **control**.

Each subject will drink **one** of the following solutions. Drink only as much of the test solution as is comfortably possible.

Solution A: 700 ml distilled water
Solution B: 700 ml 0.9% NaCl
Solution C: 250 ml 1.5% NaCl
Solution D: 400 ml 1.0% $NaHCO_3$
Solution E: 500 ml coffee or cola beverage

Begin timing the subject as soon as the fluid is consumed. A urine specimen should be collected *every 20 minutes* for a total period of *two* hours. On the data sheet for the test solution to be analyzed by your group, indicate the times at which the specimens are to be collected. All specimens, including the control, will be analyzed separately.

NOTE

No other fluids should be consumed during the experimental period. IF A SUBJECT IS UNABLE TO PRODUCE ENOUGH URINE AT ANY OF THE COLLECTING PERIODS, THE URINE SHOULD BE RETAINED AND COMBINED WITH THE NEXT COLLECTION.

Record all the information for your test solution on the appropriate data sheet at the back of the lab write-up. At the end of the lab period, pool the necessary information from all of the groups.

ANALYSIS OF SPECIMENS

Description

Normal urine will vary in color from light straw to amber, due to a pigment resulting from hemoglobin breakdown. Certain colors may indicate pathological conditions — e.g., milky (due to pus, bacteria, fat), red or smoky brown (blood and blood pigments), green or brownish yellow (bile; also indicated by the presence of yellow foam). A fresh specimen should be clear, but may become cloudy after standing awhile. Cloudiness indicates the presence of pus, blood, bacteria, or salts.

Volume

Pour the specimen into a graduated cylinder and measure the volume in ml.

Specific Gravity

This measures the relative amounts of solutes present. In a 24-hour specimen, normal urine will have a value between 1.015–1.025. Individual specimens may range from 1.002–1.030. Values below or above this range may indicate kidney damage, diabetes insipidus, or diabetes mellitus.

Fill the urinometer cylinder 3/4 full of urine. Lower the float *gently* into the center portion of the cylinder, being sure not to touch the sides of the cylinder. Read the specific gravity value from the level indicated by the bottom of the urine meniscus. Be sure to wash, rinse, and dry the apparatus carefully after *each* use.

pH

Freshly voided urine has a normal **pH** range of 4.8–7.5. If the pH is very low, this may indicate acidosis, fever, or high protein diet; if the pH is very high, this may indicate urine stagnation in the bladder, anemia, or cystitis.

To use pH indicator paper, dip into urine 3 times. Tap off excess urine and, after the proper interval of time, compare color with the color chart. To use the Corning pH meter, follow these directions:

1. With the knob of the machine on "standby" (STDBY) lift the electrodes from the water and wipe them dry with Kim-wipes.

2. Lower the electrodes into a buffer of known pH.

3. Turn the top knob to the "pH" position and turn the calibration knob (lower knob) until the

needle on the scale is set for the pH of the buffer (usually 7, but ask your instructor).

4. Place the top switch back to "standby"; lift the electrodes and place them in water to wash off the buffer.

5. Dry with Kim-wipes and then place electrodes in the urine sample.

6. Turn the knob to "pH" and read sample pH from the scale.

7. Turn knob back to STDBY, then lift the electrodes and immerse again in water.

CAUTION

Always leave top knob in STDBY position unless you are actually testing the buffer or the urine. NEVER remove electrodes from buffer or urine when knob is in the "pH" position!

Do *not* adjust the middle knob (temperature control). Electrodes should be immersed in *water* when you are not using the machine.

Chloride Estimate

Sodium chloride (NaCl) is the main form of chloride present in the urine and accounts for about half of the inorganic substances excreted through the urine.

Measure 10 drops (1/2 ml) of urine into a test tube. Add 1 drop of 20% potassium chromate (K_2CrO_4). Using the labeled dropper, add 1 drop at a time of 2.9% silver nitrate ($AgNO_3$) solution, shaking the test tube constantly. Count the number of drops necessary to change the solution from yellow to brownish tan. Each drop of silver nitrate represents about 1 g of NaCl per liter of urine.

1. Record number of drops used under NaCl g/1.

2. Determine the total grams NaCl in the entire sample using the following equation:

$$\text{Total NaCl (grams)} = \frac{\text{ml of sample}}{1000 \text{ ml}} \times \text{Number of drops } AgNO_3 \text{ used}$$

The first figure for chloride estimate (g/1) is based on a one liter volume even though the sample voided is much less than this. The actual amount of salt excreted (total NaCl) depends on what fraction of one liter is voided at a particular time (ml of sample / 1000). Measurement in g/1 gives a relative figure that can be compared with other samples and among subjects, whereas "total NaCl" is a measurement of how much salt was physically present in a given sample.

Presence of Glucose
(control sample only)

To make a "pathological" urine sample, add 10 drops of a glucose solution to 5 ml of the urine sample in a test tube. Use 5 ml of the urine sample without the added glucose in another tube to represent the normal condition. Add 5 ml of Benedict's reagent to each tube and heat both of them in a water bath. The appearance of an orange to brick-red color is an indication that glucose is present. Was there any glucose in the normal control sample?

Heller Ring Test for the Presence of Albumin
(control sample only)

Make a "pathological" sample by adding 10 drops of albumin solution to 5 ml of the urine sample in a test tube. As before, use another tube with 5 ml of urine without albumin added to serve as a control. Place 2–3 ml of concentrated nitric acid into 2 empty test tubes. Slant the tube and carefully pour 2–3 ml of the normal and pathological urine down the sides of each tube. Two distinct layers should form and the contents of the tubes should not be mixed. The formation of a white ring at the junction of the two liquids indicates the presence of albumin.

REVIEW QUESTIONS

1. During the 2 hour period was more fluid excreted than was consumed by any of the subjects? Which ones? Why would this be so?_____

_____

_____

2. What were the differences in the results from the intake of the hypotonic, isotonic, and hypertonic solutions? (Consider only solutions A, B, and C.) _____

_____

_____

3. Did bicarbonate solution cause more, or less urine output than did water? What effect did this solution have on the pH of the urine? _____

_____

_____

4. When would the urine of a normal individual contain glucose? _____

_____

_____

5. Why is the presence of blood or protein in urine considered abnormal?_____

_____

_____

6. Compare the location of the proximal convoluted tubule to that of the distal convoluted tubule. _____

_____

_____

DATA SHEET A

SOLUTION: _____

Volunteer's Name _____

Time of last voiding before lab period _____

Volume of test substance drunk _____

Urine Specimen	Time	Volume Voided (ml)	Rate of Urine Formation (ml/min)	Specific Gravity	pH	NaCl g/l	Total g NaCl	Description of Sample
Control								
1 (20 min)								
2 (40 min)								
3 (1 hour)								
4 (1 hour, 20 min)								
5 (1 hour, 40 min)								
6 (2 hours)								

Conclusions:

DATA SHEET B

SOLUTION: _____

Volunteer's Name _____

Time of last voiding before lab period _____

Volume of test substance drunk _____

Urine Specimen	Time	Volume Voided (ml)	Rate of Urine Formation (ml/min)	Specific Gravity	pH	NaCl g/l	Total g NaCl	Description of Sample
Control								
1 (20 min)								
2 (40 min)								
3 (1 hour)								
4 (1 hour, 20 min)								
5 (1 hour, 40 min)								
6 (2 hours)								

Conclusions:

DATA SHEET C

SOLUTION: _____

Volunteer's Name _____

Time of last voiding before lab period _____

Volume of test substance drunk _____

Urine Specimen	Time	Volume Voided (ml)	Rate of Urine Formation (ml/min)	Specific Gravity	pH	NaCl g/l	Total g NaCl	Description of Sample
Control								
1 (20 min)								
2 (40 min)								
3 (1 hour)								
4 (1 hour, 20 min)								
5 (1 hour, 40 min)								
6 (2 hours)								

Conclusions:

DATA SHEET D

SOLUTION: _____

Volunteer's Name _____

Time of last voiding before lab period _____

Volume of test substance drunk _____

Urine Specimen	Time	Volume Voided (ml)	Rate of Urine Formation (ml/min)	Specific Gravity	pH	NaCl g/l	Total g NaCl	Description of Sample
Control								
1 (20 min)								
2 (40 min)								
3 (1 hour)								
4 (1 hour, 20 min)								
5 (1 hour, 40 min)								
6 (2 hours)								

Conclusions:

DATA SHEET E

SOLUTION: _____

Volunteer's Name_____

Time of last voiding before lab period _____

Volume of test substance drunk _____

Urine Specimen	Time	Volume Voided (ml)	Rate of Urine Formation (ml/min)	Specific Gravity	pH	NaCl g/l	Total g NaCl	Description of Sample
Control								
1 (20 min)								
2 (40 min)								
3 (1 hour)								
4 (1 hour, 20 min)								
5 (1 hour, 40 min)								
6 (2 hours)								

Conclusions:

Ecology

OBJECTIVES

After completion of this exercise, the student should be able to do each of the following:

- Define ecology, ecosystem, producer, primary consumer, secondary consumer, decomposer, biomass, and trophic level.
- Describe the flow of energy through an ecosystem.
- Describe the forest ecosystem in terms of: (1) species present in each trophic level; and (2) species present in the horizontal layers.
- List all examples of ecological interaction that were seen on the field trip.
- Describe the process of plant succession leading to a climax community. List all examples of succession that were seen on the field trip.
- Complete any additional objectives dealing with any of the three optional sections.
- Answer all questions throughout this exercise and the review questions at the end of this exercise.

INTRODUCTION

Ecology is the study of the interactions between organisms and their environment. Early studies in ecology involved purely descriptive studies of the species of organisms present and a description of their physical environment. Later on, ecology was studied as functional units which emphasized organisms interacting among themselves and with their physical environment.

An **ecosystem** is an ecological unit which is composed of a number of populations of organisms which interact with each other and with their physical environment. Since the primary interactions involve the flow of energy, an ecosystem can be viewed as interacting populations of organisms through which energy passes.

In most ecosystems, the primary source of energy is sunlight (there are a few exceptions to this statement), the energy of which is trapped in food molecules produced by photosynthesis by organisms having chlorophyll. Such organisms, mostly plants and some protists and monerans, are referred to as **producers.** All other organisms must get their energy by feeding on producers or their remains, or on other organisms which feed on producers. Organisms which feed directly on the producers are known as **primary consumers**. Organisms which feed

341

on primary consumers are known as **secondary consumers**, and so forth. In the end, the dead remains of all of these organisms are decomposed for energy by such **decomposer organisms** as bacteria or fungi. These four groups of organisms are referred to as the **trophic** (feeding) **levels** of the ecosystem.

The chemicals making up the bodies of organisms can then be recycled, but the energy is finally lost from the ecosystem in the form of unusable heat energy. Thus energy does not cycle in an ecosystem but makes a one-way trip through it.

Usually, the total amount of **biomass** (total mass of the living organisms in a trophic level) is reduced with each successive trophic level because of inefficiency of energy transfer and heat loss at each exchange. Therefore, one would expect the biomass of producers in an ecosystem to be greater than the biomass of primary consumers, and the biomass of primary consumers to the greater than the biomass of secondary consumers.

In this lab, your instructor has the option to study one of several kinds of ecosystems. You will also have the opportunity to study many different types of ecological interactions in nature.

THE FOREST ECOSYSTEM

Your instructor will take you on a field trip to a forest ecosystem, and will guide you through answering the following questions.

Name the most prominent producers in the forest ecosystem._____

_____

_____

The producers in the forest exist in several horizontal layers. The canopy of the forest is defined as consisting of those trees whose crowns are at the top forming a continuous layer of the forest. What were the major canopy trees in the ecosystem you studied? _____

_____

_____

_____

The understory is composed of those plants whose diameters measured 5 cm at the height of 4.5 feet above ground level and whose crowns are below the top layer of the forest. List the major understory trees that you saw in the ecosystem.

_____

_____

The shrub and ground layers are composed of those tree seedlings and saplings less than 5 cm in diameter. Shrubs and herbaceous plants would also be in this layer. List several plants that you saw in this layer.

_____

_____

Although they might not be seen on the field trip, name some of the primary consumers that you would expect to find in the forest ecosystem.

_____

_____

Name some of the secondary consumers.

_____

_____

Along the way, your instructor will point out a fallen tree, which is being decomposed by bacteria and fungi (the decomposer trophic level). What effect should this decay process have on:

1. The mineral content of the soil?

_____

_____

2. The amount of stored energy within the ecosystem? _____

_____

_____

3. Possible changes within the ecosystem?

_____

_____

ECOLOGICAL INTERACTIONS

On the field trip, your instructor will point out examples of different ecological interactions which may include the following among others:

1. lichens which are examples of mutualism between an alga and a fungus where both organisms benefit from the association;

2. leaf galls in which certain species of insects will lay their eggs in the soft leaf tissue of only certain tree species — the gall is the tree's growth response to the presence of the eggs;

3. ecological succession (described in the next section);

4. examples of parasitic plants

5. competition among plants for sunlight

PLANT SUCCESSION

Succession refers to the process of replacement of one group of species by another group of species. Usually the first group of plants will change the environment so that it is less suitable for themselves and more suitable for another group of species. This cycle is repeated many times in a given area over a long period of time. Succession, called **primary succession**, can start with areas of bare rock or open water. Over time the rock develops soil and the open water areas will fill in. Different plant groups will then occupy these areas in a succession until a final stable plant community called a **climax community** is developed which will maintain itself. More common examples of succession, called **secondary succession**, begin on areas that already have soil but vegetation has been removed by agricultural uses or by fire. In the first case (called "old field" succession) an agricultural field is abandoned and soon grows up in weeds and grasses. The grasses become invaded by trees such as cedar and pine and an evergreen forest develops. This forest eventually gives way to a hardwood forest in most areas.

Incidentally, succession also occurs among animals but replacement occurs because the animals are adapted to the particular form of vegetation in an area. As plant groups develop a succession sequence, the animal groups change along with them.

The instructor will point out early stages of old field succession if these areas are present on your field trip location.

If the evergreen forest stage of old field succession is available, answer the following questions:

1. Describe the physical environment in which these conifer trees are growing.

_____

_____

_____

2. Describe the appearance of the lower branches of the conifers in this forest. What would cause this effect? _____

_____

_____

_____

3. What species of plants are growing in the shade of these conifers? _____

_____

_____

4. Were the same canopy tree species represented as seedlings or small trees in the lower layer?

_____

_____

_____

5. Based on your answer to question 4, what do you think would be the future course of plant succession in this forest?

_____

_____

_____

OPTION 1 — ECOSYSTEMS AT A SANDY SEASHORE

Four ecosystem zones are usually found more or less parallel to the shoreline of a sandy shore. Certain parts of this area will not fit into one of the four zones because the zones plus the intermediate stages between zones demonstrate the concept of "ecological succession" in which there is no clearcut boundary between the zones, but rather a continuous transition from one type of community to another. The shore is dynamic, always changing in its plant and animal life. Use the diagram in Figure 37.1 as a guide to understand these zones better.

BEACH ZONE

The **intertidal zone** or **lower beach** area, is exposed only at low tide. Because of the tremendous variation in conditions — waves, scorching sunlight, and nighttime coolness — no vegetation can live here. The only animals which can survive here are those which burrow into the sand and are filter feeders or intermittent scavengers.

What is the primary source of food energy in this zone? _____

_____

_____

List the organisms found here and try to determine their trophic levels. _____

_____

_____

The **middle beach** is that area of the beach stretching between the high tide mark and the first sand dunes with vegetation. Here will be found the

debris left behind by the ocean tides — sea shells, egg cases, ocean plants, driftwood, and washed-up bodies of ocean animal forms. Some of the animals here live in homes tunnelled into the sand and scavenge dead animal material not already claimed by the shore birds. They are often protected in their environment by the color pattern simulating the sandy coloration of the shore.

Birds inhabit the beach area looking for food whenever it can be found. Some dive into the waves head-first to grab fish, while others fly along the surface of the water, skimming small fish into their lower bills. On the beach, some dodge incoming waves in their search for dead animal matter or insects. Others wander about on the middle beach looking for food.

What is the primary source of food energy in this zone? _____

_____

_____

List the organisms found here and try to determine their trophic levels. _____

_____

_____

The **upper beach** is the area where vegetation first really establishes itself. Exposure to salt spray and wind permits only grasses to exist here. Not only do these plants serve to anchor the sand dunes, but

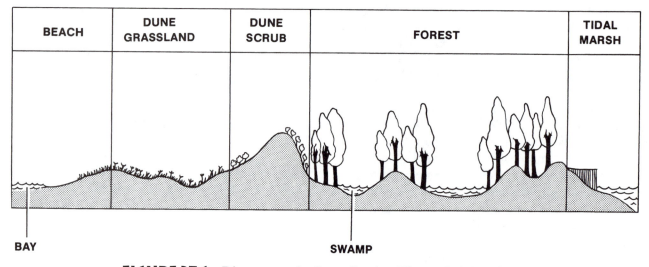

FIGURE 37.1 **Diagrammatic Cross-Section Through A Seashore**

they also protect areas behind the beach from wind erosion, wave action, and salt spray. Living among the grasses are various invertebrates.

Name the producers in this zone. _____

_____

_____

_____

List some primary consumers seen here.

_____

_____

_____

Are there any secondary consumers that you can see? _____

_____

_____

What would be the eventual expected result of building houses and hotels in this zone?

_____

_____

DUNE GRASSLAND ZONE

This zone is represented by the ridge of sand dunes. It is separated from the upper beach by a sandy bluff, marking the inner boundary of erosion caused by recent storms. Beyond this area, inland plants and animals are not subjected to regular invasion of sea water. However, the winds and salt spray still control the type of vegetation that is able to exist here, by bringing in new sand and handicapping the growth of woody plants.

The most common plants in this zone are still grasses, although they are now interspersed with some hardy herbaceous plants. Again, various types of animals are found in this region — mostly insects and animals which feed upon insects.

Name the producers in this zone.

_____

_____

_____

Primary consumers? _____

_____

Secondary consumers? _____

_____

_____

_____

DUNE SCRUB ZONE

This zone is located directly behind the dune grasslands and is lower in elevation than the ridge forming the dunes. It is characterized by the presence of woody plants which are able to grow in this area due to the decreased amount of salt spray reaching them and to the increased stability of the soil. However, periodic increases in the amount of salt spray during storms controls the height of the plants, creating a bushier type of growth. In this area, woody thickets are interspersed with more open spaces containing a variety of grasses, vines, and sometimes, cacti.

Between the sandy hills with their woody plants may be some low lying grassy meadows in which the soil is deep and relatively moist. It is in these areas that different species of rushes, grasses, weeds, and wildflowers may be found. Why would the soil be deeper here? _____

_____

_____

_____

Because of the abundance of plant life, a variety of animals will also be found in the dune scrub zone. Birds of various kinds inhabit the trees. In addition, many types of reptiles and amphibians (toads, lizards, skinks, and snakes) as well as a whole range of insects and arachnids inhabit this zone. Few mammals actually live in this region, though occasionally mice, squirrels, rabbits, etc., may be seen.

List several species of producers found in this zone. _____

_____

_____

_____

Which type of producers predominate (think in terms of mass)?

_____

_____

List several species of primary consumers found here.

_____

_____

_____

_____

Secondary consumers?

_____

_____

_____

_____

FOREST ZONE

The forest is the most complex zone found adjacent to a sandy beach. It is located behind the dune scrub zone. The forest zone encompasses a variety of areas, each of which differs and has its unique characteristics. Each subzone differs in the type of vegetation growing among the dominant trees, usually pines. There are a larger number of animal species here than in any other ecosystem associated with the sandy beach. More detail on the forest ecosystem is found in Part III of this exercise. Any unique features of this particular forest will be mentioned by your instructor.

OPTION 2 — STUDYING SMALLER ECOSYSTEMS

Your instructor may wish to have you study smaller, more available ecosystems.

Sampling can be done in a small lake or pond. Organisms can be identified (with a microscope when necessary) and placed in the appropriate trophic levels.

A small plot of ground can be sampled by removing the leaf litter and the upper few inches of soil. Organisms can be separated from the soil/litter and identified. Most of these organisms will be insects, arachnids, and other arthropods. After identification, these organisms can be listed by their appropriate trophic levels.

Your instructor (if this option is chosen) will give you specific instructions on sampling and on identification of the organisms.

List the organisms that you found by trophic levels below.

Producers

_____

_____

_____

Primary consumers

_____

_____

_____

Secondary consumers

_____

_____

_____

Decomposers

_____

_____

_____

OPTION 3 — COMPUTER MODEL OF TUNDRA SOIL-LITTER ECOSYSTEM

Ecological interactions in most ecosystems are so complex that it is difficult to determine what would happen if changes are made in the system. The science of computer modeling seeks to construct a computer program that takes all of these interactions into account and, when run, can simulate the behavior of an ecosystem over a period of years. Of course a model is only as good as the available information on the ecological interactions. This section utilizes a computer model for the tundra soil-litter ecosystem.

Soil-litter ecosystems (see Option 2) also occur in cold areas such as the tundra. The tundra is characterized by a climate so cold that the soil is frozen most of the year (permafrost) and only thaws a few inches for only 2 months of the year. Producers include lichens, mosses, grasses, sedges (similar to grasses) and a few tiny shrubs and stunted willows. Like most soil-litter ecosystems, the energy supply is in the dead organic matter that collects on the soil surface. This level is the "available soil carbon" (organic matter in the litter) or trophic level #1. Each year as the producers die, the organic material collects on the soil and some of it is broken down by bacteria and fungi (the microflora or trophic level #2), partially decomposed, and returned to the environment. Because of the short season of warm temperature, much of the organic matter is not used by the system since the microflora have such a short time to accomplish decomposition. Trophic level #3 is composed of the smaller animals (usually arthropods) that feed upon the microflora (level #2) and the organic matter in level #1. A group of larger animals (again, mostly arthropods) compose trophic level #4 and feed only on the organic matter in level #1. Trophic level #5 are the predators (carabid beetles, some mites, and some insect larvae) that feed on levels #3 and #4 as carnivores. Interestingly, level #1 is composed mostly of plant organic matter, but since it is "available soil carbon" it contains the dead remains of the other trophic levels. Another unusual feature of this system is that the animals in levels #3 and #4, as they feed on the organic matter of level #1, break the remaining material into smaller and smaller pieces and thus increase the

ability of the bacteria and fungi (microflora) to utilize it for energy.

For a small ecosystem, the tundra soil-litter system is very complex with many interactions among trophic levels. The system is diagrammed in Figure 37.2. The trophic levels are shown in boxes and the arrows represent the movement (flow) of mass and energy from one level to another.

The computer model of this ecosystem simulates the behavior of the system over a certain number of years. The model incorporates values for all of the energy flows shown on the diagram but for simplicity, the only flows that can be changed in the computer program are: (1) the flows from each trophic level to the others; and (2) the effect of levels #3 and #4 on increasing the amount of organic matter utilized by the bacteria and fungi. Students can manipulate these variables up or down and see how the changes affect the ecosystem. Research with this model has indicated that the system is more sensitive to INCREASES in the stated values than to decreases in them.

If your instructor chooses to use this option, additional computer instructions and data sheets will be provided.

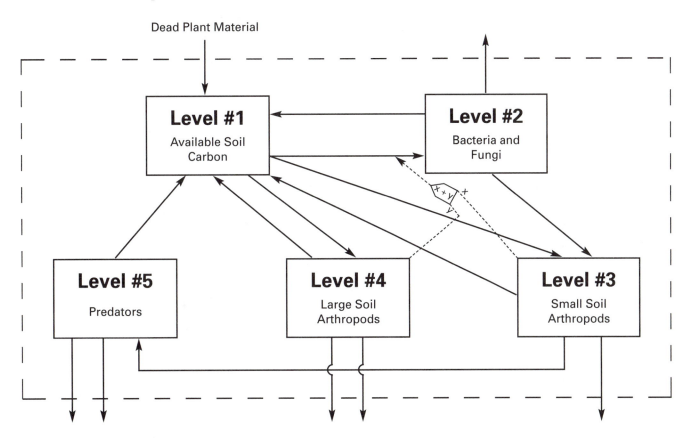

FIGURE 37.2 Diagram of tundra soil-litter ecosystem. Trophic levels are in boxes. Arrows between boxes represent the flow of energy and biomass between levels. Arrows to the outside represent energy and biomass losses in respiration.

REVIEW QUESTIONS

1. Name the trophic levels described below:

   _____ uses sunlight energy to make its own food molecules by photosynthesis.

   _____ breaks down dead organic material so that chemical elements can be recycled.

   _____ animals that feed directly on plants.

   _____ animals that feed on the level above.

2. _____ is never recycled and used again in the ecosystem.

3. The three layers in most forest ecosystems are the _____, _____, and _____.

4. The end of plant succession is a _____ community which maintains itself with no further changes.

5. List the different stages in "old field" plant succession. _____

   _____

   _____

6. Describe both primary and secondary succession. _____

   _____

   _____

Making Metric Conversions

Converting within the metric system involves multiplying by factors of 10 or 1/10. One way of looking at the conversion factors is to visualize a set of stair steps where each step is either 10 times or 1/10 as much as the next step (see Figure A-1). Since some metric measurements in biology require the kilo- unit (1000), the top step is 10^3 ($10 \times 10 \times 10$). The next steps are 10^2 ($10 \times 10 = 100$), 10^1 ($10 \times 1 = 10$), and 10^0 (1). Notice on these exponent numbers that 10 to the third power is a 1 followed by 3 zeroes, 10 to the second power is a 1 followed by 2 zeroes, and so on. By comparison (but not necessarily by logic) 10^0 is 1 with no zeroes after it. As the exponent number goes down, the decimal place is moved to the left. Following this progression, 10^{-1} is 0.1, (deci- or one tenth), 10^{-2} is 0.01 (centi- or one hundredth), and 10^{-3} is 0.001 (milli- or one thousandth). Micro- units are 10 with the exponent -6 (one millionth or 6 decimal places to the left of 1). Nano- units are 10 with the exponent -9 (one billionth or 9 decimal places to the left of 1). Each step is one exponent number from the next one. Thus one kilometer is at the top step, and one meter is 3 steps down at 10^0. One millimeter is 3 steps below the 1 meter step, 1 micrometer is 6 steps below, and 1 nanometer is 9 steps below the 1 meter step. Comparison can also be made between other units as well. A micrometer is 3 steps below the millimeter and 4 steps below the centimeter.

The first step in metric conversions is to determine the magnitude of the difference between the two units. The number of steps separating the two units will be the number of decimal places that must be moved to the right or left in making the conversion. For example, micrometers and nanometers are 3 steps apart, so in making conversions, the decimal point will be moved either 3 places to the right or 3 places to the left.

The second step is to determine which way the decimal point will be moved — right or left. This involves a little bit of logic and common sense. If you are converting micrometers to nanometers, you are converting larger units into smaller units (there will be a lot of small nanometers in a micrometer). Thus the decimal point should be moved to the right giving a larger whole number. If you are converting nanometers to micrometers, the opposite is true. Since a nanometer is smaller than a micrometer, one small nanometer will be some fraction of the larger micrometer unit. So logically, the decimal point will be moved to the left.

In summary: when converting from larger units to smaller — decimal point moves right; when converting from smaller units to larger — decimal point moves left. Therefore, since micrometers and nanometers are 3 steps apart, then: 1 micrometer = 1000 nanometers (decimal point 3 places to right of 1); 1 nanometer = 0.001 micrometers (decimal point 3 places to left of 1).

Thirdly, if the conversion requires multiples of the first unit, find the conversion factor and then multiply by the number in front of the first unit.

For example:

59 micrometers = _____ nanometers.

Since 1 micrometer equals 1000 nanometers, then 59 micrometers will be 59 $\times$ 1000 or 59,000 nanometers.

And finally, take one long look at your answer and see if it makes sense. For example, you know that a meter stick is a little longer than a yard and its smallest units are millimeters. If you complete a conversion and come up with "1 millimeter = 1000 meters", logic will tell you that you goofed and moved the decimal point the wrong way!

Using these 4 principles, fill in the blanks in the following conversions:

1. One centimeter   = _____ micrometers.

2. One kilogram   = _____ milligrams.

3. One milliliter   = _____ liter.

4. 55 milliliters   = _____ microliters.

5. 105 millimeters   = _____ meters.

6. 1500 nanometers   = _____ millimeters.

7. 0.2 millimeters   = _____ micrometers.

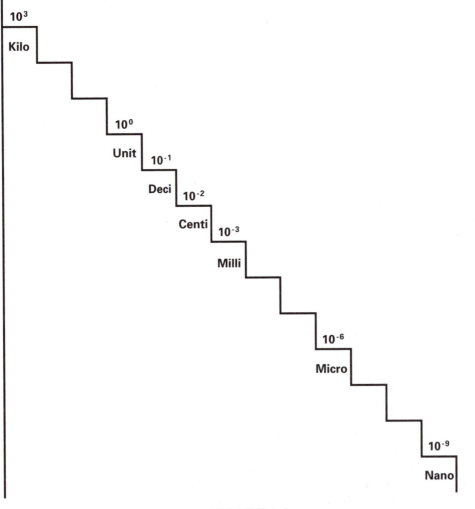

FIGURE A-1.

Presenting Data in Graph Form

B

Data collected in an experiment is usually presented in the form of a graph, a diagram showing the relationship between an independent variable and a dependent variable. The independent variable is what the student varies in the experiment, such as the temperature, time, or pH. The dependent variable is what the student measures, counts, or records and it is the thing that is being affected in the experiment. The following procedure should be followed in presenting data from an experiment:

- Use graph paper.

- Plot the independent variable on the x-axis (horizontal) and the dependent variable on the y-axis (vertical)

- The intervals labeled on each axis should be appropriate for the range of the data so that it will fit on the graph.

- The intervals labeled on the graph paper should be evenly spaced.

- Label each axis with the name of the variable and the appropriate with the correct units.

- Choose the type of graph (such as line or bar) which represents the data the best.

- For a line graph, plot the points from the data onto the graph. Mark with a dot where each value of the independent variable on the vertical line intersects with the value of its dependent variable on the horizontal line. Connect the points to create a line.

- For a bar graph, use the points in step #7 to create a vertical rectangular bar that spans from the points to the zero line at the bottom.

- Create a title for the graph that properly represents what it is showing.

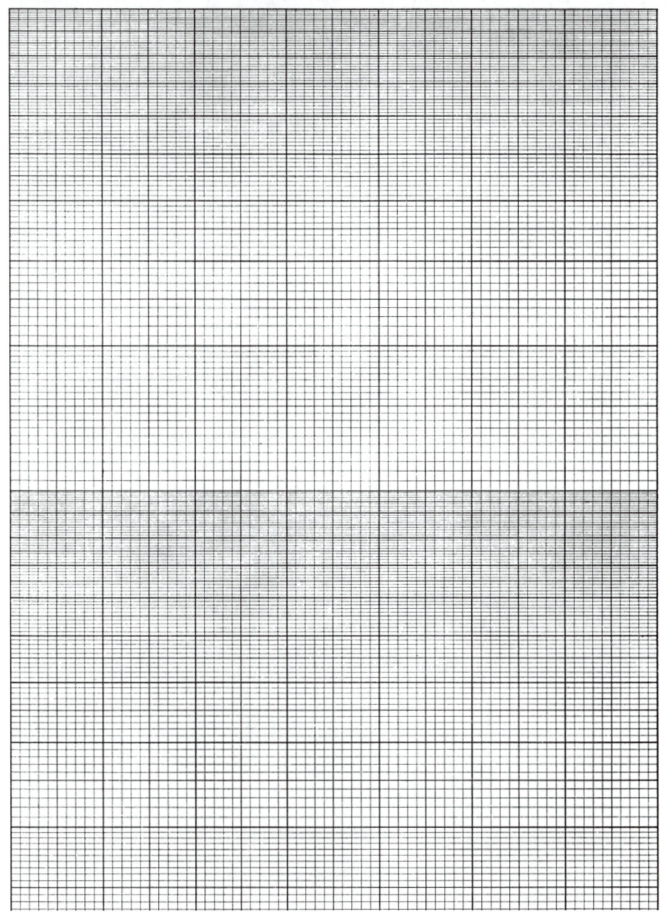

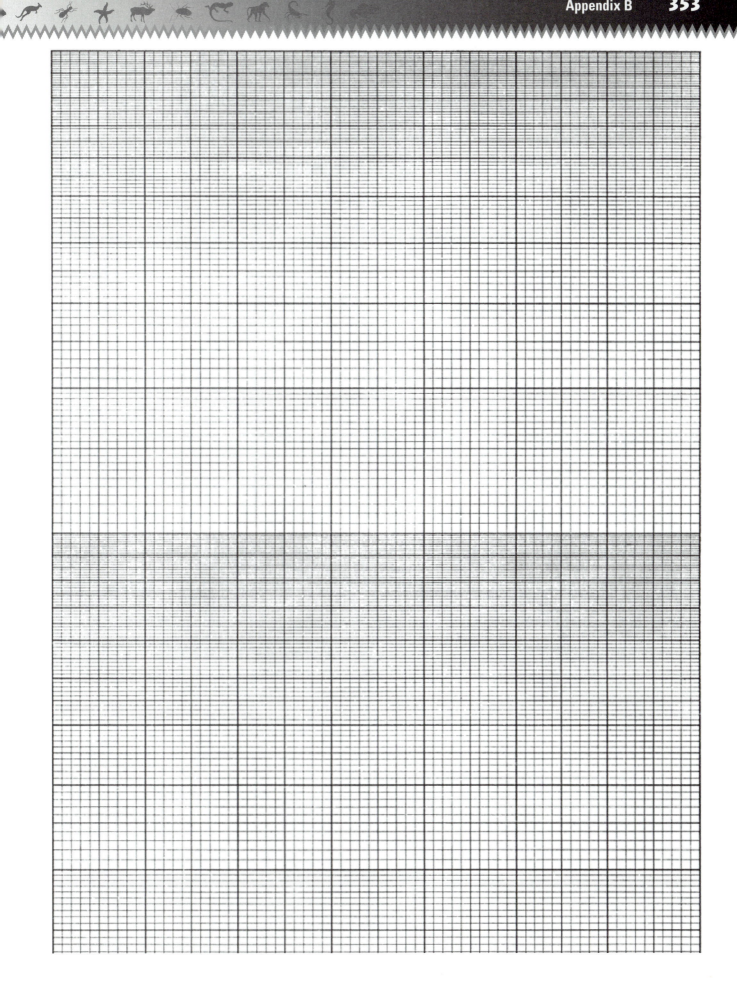

Glossary

A

Abdomen — the region of the mammalian body located between the diaphragm and the pelvis, that contains the abdominal cavity and its visceral organs; one of the three principal body regions (head, thorax, and abdomen) of many animals.

Abducens nerve — the cranial nerve that terminates in the lateral rectus muscle of the eye and consists of mostly motor fibers.

Abduction — a movement away from the axis or midline of the body; opposite of adduction.

Aboral — opposite to or away from the mouth.

Absorption spectrum — the range of a pigment's ability to absorb wavelengths of light.

Accessory pigment — an example would be carotenoids that can absorb wavelengths of light that chlorophyll cannot.

Acoelomate — an organism lacking a coelom.

Acorn worm — worm belonging to Enteropneusta.

Actinomycetes — bacteria that form colonies of branching chains that resemble fungi.

Adaptation — structural, physiological, or behavioral traits of an organism that promotes its survival and contributes to its ability to reproduce under environmental conditions.

Adduction — a movement toward the axis or midline of the body; opposite of abduction.

Adductor muscles — muscles that adduct or press the thighs together.

Adenine — one of the four nitrogenous bases found in DNA and RNA.

Adipose — fat, or fat-containing, such as adipose tissue.

Adrenal glands — endocrine glands; one superior to each kidney; also called *suprarenal glands.*

Adventitious root — supportive root developing from the stem of a plant.

African Sleeping Sickness — a human sleeping sickness caused by *Trypanosoma gambiense* which is carried by the tsetse fly.

Afterimage — the perceived persistence of light after cessation of the light stimulus, which may be positive (of the same color) or negative (of the complementary color).

Agglutination — clumping of cells; particular reference to red blood cells in an antigen-antibody reaction.

Agranulocytes — a nongranular lymphocyte or monocyte.

Air bladder — the swim bladder in fish or an air-filled chamber in some seaweeds that functions as a float.

Akinete — a stage in some green algae in which the newly formed cell fuses to the wall of the parent cell.

Albumen — the "white" of the egg that surround the ovum. It functions as a water reservoir for the embryo and a food source for later development.

Algae — simple plants of the Monera and Protista categories that possess chlorophyll and carry on photosynthesis.

Allantoic duct — a small vessel in the umbilical cord which serves to carry some of the small amount of urine formed by the kidneys away from the fetus.

Allantois — an extraembryonic membranous sac that forms blood cells and gives rise to the fetal umbilical arteries and vein. It also contributes to the formation of the urinary bladder.

Allele — an alternative form of a gene occurring at a given chromosome site, or locus.

Alternation of generations — two-phased life cycle characteristic of many plants in which there are sporophyte and gametophyte generations.

Alveolus — an individual air capsule within the lung. Alveoli are the basic functional units of respiration.

Ambulacral groove — the podia or tube feet of a sea star are contained on the ventral surface of each ray in this groove.

Ambulacral ridge — the skeleton running down the center of the ray of an echinoderm.

Amino acid — a unit of protein that contains an amino group (NH_2) and an acid group (COOH).

Ammocoetes larvae — sea lamprey larva. This lamprey developes in freshwater and then migrates to the sea or inland lakes.

Amnion — a membrane that surrounds the fetus to contain the amniotic fluid.

355

Amniotic egg — an egg of a reptile or bird. It contains four extraembryonic membranes: yolk sac, amnion, allantois, and chorion.

Amoebocyte — a primitive cell of connective tissue present in the tissues or body fluids of some invertebrates that is capable of independent movement.

Amphioxus — a marine chordate belonging to the Cephalochordata subphylum.

Amplexus — a sexual embrace without the occurrence of intercourse.

Ampulla — a sac or vesicle, such as that on the leaves or roots of some aquatic plants.

Amyloplast — a membrane-bound plastid containing starch grains.

Anaphase — a stage in mitosis or meiosis characterized by the movement of chromatids or halves of chromosome pairs to their respective poles from the equatorial planes.

Androgen — a male sex hormone, a masculinizing substance or one that promotes development and function of male genital organs and secondary sexual characteristics.

Animal pole — the darkly pigmented portion of the unfertilized egg of the frog.

Annulus — a row of cells down the back of the helmet-like sporangium of a fern.

Antenna — a sensory appendage on many species of invertebrate animals.

Antennule — the smaller of the two antennae in certain crustaceans such as the crayfish.

Anther — the portion of a plant stamen in which pollen is produced.

Antheridial receptacle — stalks which are topped with finger-like projections produced by female plants of liverworts.

Antheridium — male reproductive organ in certain nonseed plants where motile sperm are produced.

Anthocyanin — a blue or purple pigment, water-soluble, present in the sap of various plants. They appear red in acid solutions, and blue in alkaline solutions.

Antibody — an immune substance that can occur naturally or be formed as the result of the presence of an antigen.

Antigen — a foreign material, usually a protein, that triggers the immune system to produce antibodies.

Anus — the terminal end of the GI tract, opening of the anal canal.

Aortic arch — paired blood vessels encircling the pharynx connecting the ventral and dorsal aortae in vertebrate embryos or one of a series of paired afferent and efferent branchial arteries in aquatic vertebrates.

Apical — at or near an apex, summit, or tip.

Apopyle — opening of the radial canal into the spongocoel of sponges.

Appendicular portion — that part containing, consisting of, or pertaining to appendages.

Aqueous humor — the watery fluid that fills the anterior and posterior chambers of the eye.

Archegonial receptacle— stalks with discs on top produced by male plants of liverworts.

Archegonium — female reproductive organ in certain nonseed plants where eggs are produced.

Archenteron — the principle cavity of an embryo during the gastrula stage. Lined with endoderm, the archenteron develops into the digestive tract.

Archeocyte — an amoebocyte in the mesenchyme of the sponge.

Areolar connective tissue — loose connective tissue.

Articular cartilage — a hyaline cartilaginous covering over the articulating surface of bones of synovial joints.

Asci — little sacs of sexually-produced spores of ascomycetes.

Ascocarp — fruiting body of a spherical or cup-shape that contains asci found in Ascomycetes.

Ascospore — spore that an ascus produces.

Asexual reproduction — reproduction without the union of gametes or sexual union between two individuals, such as fission, fragmentation, the formation of spores, budding or gemmule formation.

Association neuron — a connector neuron; one transmitting an impulse between two neurons in a neural pathway.

Aster — minute rays of microtubules at the ends of the spindle apparatus in animal cells during cell division.

Astigmatism — this is due to unequal curvatures of the lens of the cornea which leads to a blurred vision problem.

Atlas — the first cervical vertebra that the skull articulates with.

Atriopore — where an atrial cavity opens to the outside.

Atrium — either of two superior chambers of the heart that receive venous blood.

Auditory nerve — the statoacoustic or vestibulocochlear nerve.

Auricle — an ear or a structure that is earlike; a receiving chamber of the heart; a ciliated projection in ctenophores and rotifers near the mouth.

Autotroph — an organism capable of synthesizing its own organic molecules (food) from inorganic molecules.

Axial skeleton — comprised of the skull, vertebrae, sternum, and ribs.

Axillary bud — a group of meristem cells at the junction of a leaf and stem which develops branches or flowers; also called *lateral bud*.

Axis — the second cervical vertebra; the stem of a plant.

Azygous vein — a single vein that drains the thorax and enters the superior vena cava just before it joins the heart.

B

Bacillus — a bacterium that is rod-shaped.

Bacteria — prokaryotes within the phylum Monera, lacking the organelles of eukaryotic cells.

Ball and socket joint — a freely movable joint such as the hip and shoulder.

Basal disc — the adhesive disc, flattened, of various coelenterate polyps or some sessile animals.

Base — a substance that contributes or liberates hydroxide ions in a solution; a solution in which the pH is greater than 7; alkaline.

Basement membrane — a thin sheet of extracellular substance to which the basal surfaces of epithelial cells are attached.

Basidia — club-shaped reproductive structures of club fungi that produce basidiospores during sexual reproduction.

Basidiocarp — fungi's (of the Basidiomycetes) fruiting body.

Basilar membrane — a membrane at the bottom of the cochlear duct. The organ of Corti rests there.

Basophil — a granular leukocyte that readily stains with basophilic dye.

Biceps brachii — the most familiar muscle of the forearm because it bulges when the elbow is flexed.

Biceps femoris — one of the hamstring muscles forming the muscle mass of the posterior thigh.

Bicuspid valve — a valve with two leaflets located between the left atrium and left ventricle.

Bilateral symmetry — the morphologic condition of having similar right and left halves.

Bile duct — formed by the union of the cystic duct from the gall bladder and the hepatic duct from the liver. It receives the pancreatic duct before entering the duodenum.

Binary fission — a process of reproduction that does not involve a mitotic spindle.

Biomass The total mass of the living organisms in a trophic level.

Bipolar cell — a cell (neuron) located in the retina of the eye.

Blade — the leaf of a plant; the expanded, thin part of a leaf, or the flat and leaflike thallus of certain algae.

Blastocoel — the enlarged central cavity in the blastula created by the arrangement of cells and their division.

Blastoderm — the protoplasm disc where cleavage takes place in eggs undergoing meroblastic cleavage. The surface layer of cells enclosing the yolk in centrolecithal ova.

Blastopore — the opening of the gastrocoel to the outside of the blastula, later to the become the anus.

Blastula — an early stage of prenatal development between the morula and embryonic stages.

Blind spot — the optic disk of the retina devoid of rods and cones, where optic nerve fibers make their exit from the eye. It is insensitive to light.

Blood — the fluid connective tissue that circulates through the cardiovascular system to transport substances throughout the body.

Bowman's capsule — the double-walled proximal portion of a renal tubule that encloses the glomerulus of a *nephron;* also called *glomerular capsule.*

Brachiocephalic artery — also the innominate artery which supplies blood to the right forelimb and head.

Brachial plexus — cervical and thoracic nerve network that ennervates the forelimb of vertebrates.

Bronchiole — a small division of a bronchus within the lung.

Bronchus — a branch of the trachea that leads to a lung.

Brownian movement — the vibration or jiggling of microscopic particles, especially those suspended in a fluid.

Buccal cavity — the mouth, or oral cavity.

Buccal cirri — in a lancelet, the oral tentacles that surround the vestibule.

Buccopharyngeal cavity — refers to the cheeks and pharynx.

Budding — asexual reproduction from an outgrowth or bud.

Bulbourethral gland — Cowper's gland, one of two, located near the bulb of the urethra, that secretes a viscid, alkalinizing fluid into the urethra.

Bursa — a saclike structure filled with synovial fluid, which occurs around joints.

C

Calcareous material — that which contains, consists of, or has the nature of calcium carbonate.

Calyptra — a hooklike covering enclosing a developing sporangium or a caplike structure that partially enclosed the developing capsule in mosses or the covering of certain fruits in angiosperms.

Calyx — a cup-shaped portion of the renal pelvis that encircles renal papillae.

Cambium — a layer of tissue in the roots and stems of dicotyledonous plants that later becomes secondary tissues, such as xylem, phoem, or parenchyma.

Canaliculi — small canal, such as those extending from lacunae into the matrix of bone.

Cancellous tissue — spongy bone; bone tissue with a latticelike structure.

Carbohydrate — a sugar or starch, comprised of carbon, hydrogen, and oxygen. The hydrogen and carbon occur in the ratio of 2:1.

Carbonyl — a functional group of aldehydes and ketones.

Cardiac muscle — a striated, involuntary muscle tissue in the myocardium of the heart.

Cardiac sphincter — a ring of muscle forming the boundary between the esophagus and the stomach.

Cardiac stomach — the gastric mill of a decapod crustacean; the folded portion of the stomach of a sea star that is everted when feeding.

Carotene — the precursor of vitamin A, a hydrocarbon synthesized by plants. It is a yellow pigment.

Carotid artery — one of the two large arteries which supply blood to the neck and head regions of vertebrates.

Carpal bone — one of the bones at the wrist or carpus.

Carpel — a pistil, a division of a compound pistil, a modified ovule- bearing megasporophyll, or a scale leaf of a female or carpellate cone.

Cartilage — a type of connective tissue with a solid elastic matrix.

Casein — a phosphoroprotein in milk which is soluble but is converted into insoluble casein through the action of enzymes. The basis of curd milk.

Catalyst — a substance or agent that affects a chemical reaction or alters its velocity without being changed itself.

Cecum — the pouchlike portion of the large intestine to which the ileum of the small intestine is attached.

Cell membrane — a cell's or protoplasmic mass's surface layer which acts as a selective barrier to the passage of ions and molecules.

Cell plate — the structure in the equatorial plane of a plant cell developed during telophase of mitosis. Cellulose cell walls form on each side of the plate.

Cellulose — the chief constituent of plant cell walls; a complex carbohydrate.

Cell wall — the wall secreted by the cytoplasm of a plant cell that lies outside of it. It is rigid, nonliving and composed mostly of cellulose. In animals, the cell wall is a thin sheath lying outside the plasma membrane.

Cementum — the bony connective tissue that covers the root of a tooth.

Central nervous system — the brain and the spinal cord.

Centriole — an organelle usually located in the centrosome, considered to be the active division center of the animal cell.

Centromere — a portion of the chromosome to which a spindle fiber attaches during mitosis or meiosis.

Cephalization — head or concentration of nervous tissue development and mouth and food-getting apparatus development anteriorly, in animal evolution.

Cephalothorax — fusion of the head and thoracic regions characteristics of certain arthropods.

Cerebellum — the portion of the brain concerned with the coordination of movements and equilibrium.

Cerebral aqueduct — a tiny canal that travels through the midbrain and connects the third ventricle of the diencephalon to the fourth ventricle below.

Cerebral fissures — the grooves that divide the back cerebral hemisphere into a number of lobes, named for the cranial bones that lie over them.

Cerebral peduncles — known as little feet of the cerebrum, composed primarily of two bulging fiber tracts in the anterior of the midbrain converging ascending and descending information.

Cerebrum — the largest portion of the brain, composed of the right and left hemispheres.

Cervical vertebra — a vertebra of the neck.

Chalazae — dense cordlike structures attached to the egg yolk that suspend it in the albumen.

Chalk — marine origin soft limestone formed from shell deposition of foraminiferans.

Cheliped — front pair of pincerlike legs in most decapod crustaceans, adapted for seizing and crushing.

Chemotaxis — the turning in response to a chemical stimulus of a motile organism or cell.

Chitin — strong, flexible polysaccharide forming the exoskeleton of arthropods.

Chlorophyll — green pigment in photosynthesizing organisms that absorbs energy from the sun's rays.

Chloroplast — a membrane enclosed organelle which contains chlorophyll and is the site of photosynthesis.

Choanocyte — a flagellated cell with a transparent collar at the base of the flagellum, in the gastral epithelium of sponges.

Chondrocyte — a cartilage cell.

Chordae tendinae — tiny white cords, "heartstrings," that anchor the cusps to the walls of the ventricles.

Chorion — the membrane in the human embryo that initially produces the hormone HCG that keep the corpus luteum functioning and absorbs nutrients from the endometrium and later becomes the placenta.

Choroid coat — the pigmented middle layer of the vertebrate eye, continuous with the ciliary body and iris anteriorly. All three comprise the vascular tunic or uvea.

Chromatid — one of the two spiral filaments if a chromosome during prophase and metaphase of mitosis. One of the four elements of a tetrad in meiosis.

Chromatin — threadlike network of DNA and proteins within the nucleus.

Chromoplast — a plastid containing red, orange, or yellow, responsible for the colors of various plant structures, such as leaves, flowers, or fruits.

Chromosome — structure in the nucleus that contains the genes for genetic expression.

Cilia — microscopic, hairlike processes that move in a wavelike manner on the exposed surfaces of certain epithelial cells.

Ciliary body — a portion of the choroid layer of the eye that secretes aqueous humor and contains the ciliary muscle.

Cleavage — a series of mitotic divisions that the zygote undergoes. The pattern of cleavage varies from one species of animal to another.

Cleavage furrow — the first sign of cleavage in an animal cell; a shallow groove in the cell surface near the old equatorial plate.

Climax community — the final stable community which, after a series of successions, can maintain itself.

Clitellum — the anterior, swollen glandular region in certain annelids.

Clitoris — a small, erectile structure in the vulva of the female.

Cloaca — terminal portion of the digestive tract of many animals, that also may serve the excretory, reproductive, and respiratory systems.

Cnidoblast — a cell producing a nematocyst in coelenterates.

Coccus — a bacterium that is spherically shaped.

Coccygeal vertebrae — vertebrae in the coccyx.

Coccyx — in man and apes, the small bone that terminates the vertebral column.

Cochlea — the spiral portion of the inner ear that contains the organ of Corti.

Cochlear duct — a small chamber filled with a liquid called endolymph. It is located in the cochlea of the inner ear.

Cocoon — protective, or resting, stage of development in certain invertebrate animals.

Coelenteron — in coelenterates, the gastrovascular cavity.

Coelom — body cavity of higher animals, containing visceral organs.

Coeliac artery — the first major branch of the dorsal aorta which supplies the stomach, spleen and liver.

Coenocytic mass — a multinucleated mass of cytoplasm undivided by membranes or walls.

Collagen — a protein which forms the main ingredient of white fibers of connective tissue. When boiled or upon hydrolysis, it is converted into gelatin.

Collagenous fiber — the nonelastic, white, and wavy fibers in connective tissue, especially tendons and ligaments.

Collar cells — flagella-supporting cells in the inner layer of the wall of sponges.

Colony — a group of organisms that live in close association with each other and are of the same kind.

Colorblindness — is due to lack of all or partial loss of three cone types in the retina. Most common is the lack of red or green receptors which leads to two varieties of red-green colorblindness. It is a gender-linked condition.

Columnar epithelium — (simple) made up of a single layer of tall cells that fit closely together. It lines the entire length of the digestive tract.

Commensalism — two species living in close association with the smaller species, the symbiont, benefiting but the host species is neither helped nor harmed.

Companion cell — a type of plant cell that is connected to a sieve-tube.

Complete digestive tract — cells associated with sieve tube cells in phloem.

Compound microscope — a microscope with two lenses.

Condom — a thin, natural barrier membrane or latex rubber sheath that fits over the penis to collect the semen.

Cone — a photoreceptor cell in the retina.

Conidia — spores produced by fungi during asexual reproduction.

Conjugation — sexual union in which the nuclear material of one cell enters another; the temporary uniting or joining, especially of individuals, where a transfer of nuclear material from one cell to another occurs; the union of isogametes in algae.

Conjugation tube — a protoplasmic bridge that forms between two organisms, such as *Paramecia*, while they are conjugating.

Connective tissue — one of the four basic tissue types within the body. It is a binding and supportive tissue with abundant matrix.

Contractile vacuole — in certain protozoans, an excretory vesicle which fills with fluid and then discharges its contents through a plasma membrane.

Control — a sample in an experiment that undergoes all the steps in the experiment except the one being investigated.

Conus arteriosus — also known as the trunus arteriosus which leads forward from the anterior base of the ventricle. It divides into the left and right aortic arch in the frog.

Convergence — the occurrence of structures of similar appearance in unrelated organisms; the synapse of several axons with one or a few motor neurons; cell movement from the lateral areas toward the midline and resulting in a primitive streak, as in the chick.

Convergent evolution — the acquisition of similar adaptive structures in unrelated animals of a particular habitat.

Cornea — the transparent convex, anterior portion of the outer layer of the eye.

Corolla — the petals of a flower.

Coronal suture — separation between the frontal and parietal bones.

Coronary artery — an artery which supplies blood to the heart.

Corpora quadragemina — the four bulging nuclei of the midbrain which are reflex centers involved with vision and hearing.

Corpus albicans — the small white scar tissue that develops after the corpus luteum degenerates.

Corpuscle — a cell floating in a fluid, such as blood, coelomic fluid, or saliva; an encapsulated sensory nerve ending.

Corpus callosum — the connecting band of commissural fibers between two cerebral hemispheres.

Corpus hemorrhagicum — a ruptured follicle together with its blood clot.

Corpus luteum — the small yellow body that develops in a ruptured follicle and is the principal source of progesterone.

Cortex — the outer layer of an organ such as the convoluted cerebrum, adrenal gland, or kidney.

Cotyledon The first leaf that develops in the embryo sporophyte of a seed plant; in certain ungulates, a mass of villi on the placenta.

Cowper's gland — *see* bulbourethral gland.

Cranium — in vertebrates, the portion of the skull that encloses the brain.

Crenation — when a red blood cell is placed in a hypertonic solution the cells lose water and shrink.

Crop — a portion of the bird esophagus that is dilated and used to store and moisten food; the dilated region of the foregut where food is stored in invertebrates.

Crossing over — the exchange of corresponding chromatid segments of genetic material of homologous chromosomes during synapsid of meiosis I.

Cross wall — a characteristic of some fungi when their hyphae are septate.

Crown — in a seed plant, the junction between the stem and the root; the large bud of a rhizome used for propagation; the circle of processes that are found on the inner surface of a corolla; the branches and leaves of a tree; a crinoid body with its arms; a bird's crest; that portion of a tooth that is exposed but covered by enamel; the top of the head.

Crustose lichen — those that lie flat and close to the substrate.

Ctenidium — a mollusk gill; a comblike structure of rows of stiff bristles, as on a flea.

Cuboidal epithelium — (simple) it forms one layer of cuboidal (cube) cells resting on a basement membrane. It is common in glands and forms the walls of the kidney tubules.

Cupule — structure that is shaped like a cup, such as the involucre of an acorn; a liverwort structure that produces gemmae; a cup fungus's apothecium.

Cuticle — waxlike covering on the epidermis of nonwoody plants to prevent water loss.

Cyanobacteria — photosynthetic prokaryotes that have chlorophyll and release oxygen.

Cyclosis — the streaming of protoplasm. This is seen in the cells of various protozoans.

Cytokinesis — division of the cellular cytoplasm.

Cytoplasm — the protoplasm of a cell located outside of the nucleus.

Cytosine — a pyrimidine base found in nucleotides and nucleic acids that is water-soluble.

D

Daughter cell — a first generation offspring with regard to gender.

Daughter colony — a characteristic of a green algae known as *Volvox*. They are formed during asexual reproduction from a single vegetative cell of the parent.

Deciduous teeth — the first set of teeth, "baby" teeth, that are shed at a later stage of maturity in mammals.

Decomposer organism — organism that feeds on the dead remains of other organisms, thereby decomposing them for energy.

Deltoideus or deltoids — Fleshy, triangle-shaped muscles that form the rounded shape of your shoulders.

Denaturation — a process in which protein unravels and loses its conformation.

Dendrite — a nerve cell process that transmits impulses toward a neuron cell body.

Dentin — the principal substance of a tooth, covered by enamel over the crown and by cementum on the root.

Deoxyribonucleic acid (DNA) — a substance made up of phosphoric acid, a sugar, and nitrogenous bases that is found in the nuclei of cells. DNA is the key molecule in genes for giving its cell its hereditary characteristics and information. It is localized principally in the chromosomes and regulates protein synthesis.

Deoxyribose — a pentose sugar found in DNA.

Depth of focus — the thickness of an object that is all in sharp focus at the same time under the microscope.

Dermal branchia — the minute respiratory structures or skin gills located on the surface of an echinoderm.

Dermal papillae — *see* dermal branchia.

Dermis — the second, or deep, layer of skin beneath the epidermis.

Diaphragm — a flat dome of muscle and connective tissue that separates the thoracic and abdominal cavities.

Diaphysis — the shaft of a long bone.

Diastole — the sequence of the cardiac cycle during which the ventricular heart chamber wall is relaxed.

Diatoms — aquatic unicellular algae characterized by a cell wall composed of two silica impregnated valves.

Diatomaceous earth — the accumulation of siliceous skeletons of diatoms over millions of years, used as an abrasive, for filtration, and for insulation.

Dichotomous or forked-line method — a method of branching or forking repeatedly in pairs.

Dicot — a kind of angiosperm characterized by the presence of two cotyledons in the seed; also called *dicotyledon.*

Dicotyledonae — a member of the class of flowering plants with two coyledons in the embryo, leaves with netted venation, cylindrical vascular tissue, flower parts in groups of four or five, and cambium present. This taxon includes the most advanced and highly developed plants.

Diencephalon — in vertebrates, the portion of the brain between the cerebral hemispheres and the midbrain, encompassing the epithalamus, thalamus, hypothalamus, and neural lobe of the hypophysis.

Diffusion — movement of molecules from an area of greater concentration to an area of lesser concentration.

Dikaryon — mycelium of certain septate fungi that possess two separate haploid nuclei per cell.

Dinoflagellate — a protozoan with two flagella, one transverse and one longitudinal.

Dioecious — having separate sexes, unisexual; staminate and pistillate flowers on separate plants; sex organs on separate gametophytes.

Diploid — twice (2N) the number of chromosomes found in gametes.

Disaccharide — a carbohydrate that yields two monosaccharide molecules upon hydrolysis.

Divergent evolution — the tendency of an organism to diversify when it spreads to various habitats.

DNA Polymerase — an enzyme that catalyzes the synthesis of a new DNA strand.

Dominant gene — a hereditary characteristic that expresses itself even when the genotype is heterozygous.

Dorsal — pertaining to the back or posterior portion of a body part; the opposite of *ventral.*

Ductus arteriosus — a short vessel that connects the aorta and the pulmonary trunk during fetal circulation.

Ductus venosus — a vessel that carries the blood toward the heart of the fetus by-passing the immature liver.

Duodenum — the first portion of the small intestine.

Dura mater — the outermost meninx covering the central nervous system.

E

Ecology — the study of the relationship of organisms between themselves and the physical environment.

Ecosystem — a biological community and its associated abiotic environment.

Ectoderm — the outermost of the three primary embryonic germ layers.

Ectoplasm — the outermost layer of a cell's protoplasm, usually more rigid and lacking granules.

Elastic cartilage — cartilage found where a structure with elasticity is desired, such as the external ear.

Elastic fiber — fiber which is capable of resuming its original size, shape, or length after being deformed or stretched.

Elater — a structure in sporangia of liverworts or attached to the spores of horsetails that is sensitive to and changes shape in response to changes in humidity.

Embryo — a young organism in early stages of development or before hatching from an egg; the mammal between the stages of blastocyst and fetus, when organs are forming; the sporophyte of a vascular plant before its period of rapid growth; the rudimentary plant within the seed of a seed plant, consisting of plumule, cotyledons, hypocotyl and radicle.

Embryo sac — the female gametophyte of an angiosperm, which develops within the ovule of a flower.

Emulsion — a colloid consisting of droplets of a liquid suspended in another liquid.

Enamel — a hard, calcareous substance that makes up the outer layer of the crown of a tooth.

Endoderm — the innermost of the three primary germ layers of an embryo.

Endodermis — a plant tissue composed of a single layer of cells that surrounds and regulates the passage of materials into the vascular cylinder of roots.

Endoplasm — the inner protoplasm of a cell, more granular than the rest.

Endoskeleton — hardened, supportive internal tissue of echinoderms and vertebrates.

Endosperm — reserve nutrients formed in the ovule or embryo sac of seed plants, formed when a sperm unites with two endosperm (polar) nuclei.

Enteropneust — a member of the class Enteropneusta, phylum Hemichordata, an invertebrate, including the acorn worm.

Enzyme — a protein catalyst that activates a specific reaction.

Eosinophil — a type of white blood cell that becomes stained by acidic eosin dye; constitutes about 2%–4% of the white blood cells.

Epidermis — the outermost layer of the skin, composed of stratified squamous epithelium.

Epididymis — a coiled tube located along the posterior border of the testis; stores spermatozoa and discharges them during ejaculation.

Epiglottis — a cartilaginous leaflike structure positioned on top of the larynx that covers the glottis during swallowing.

Epiphyseal plate — a cartilaginous layer located between the epiphysis and diaphysis of a long bone and functions in longitudinal bone growth.

Epiphysis — the end segment of a long bone, distinct in early life but later becoming part of the larger bone.

Epithelial tissue — one of the four basic tissue types; the type of tissue that covers or lines all exposed body surfaces.

Equator — also known as the metaphase plate on which the double-stranded chromosomes line up during metaphase of mitosis or meiosis.

Erythrocyte — a red blood cell.

Esophagus — a tubular organ of the GI tract that leads from the pharynx to the stomach.

Estrogen — female sex hormone secreted from the ovarian (Graafian) follicle.

Ethmoid — a bone at the base of the cranium forming the supporting structure of the walls and roof of the nasal cavity.

Eukaryotic — possessing the membranous organelles characteristic of complex cells.

Eustachian aperture — an opening into the eustachian or auditory tube.

Excurrent siphon — an opening through which water and debris is carried from an animal, such as a clam.

Exoskeleton — an outer, hardened supporting structure secreted by ectoderm or epidermis.

Extension — straightening of a limb; the backward movement of the head, trunk, or limbs in man.

External ear — the outer portion of the ear, consisting of the auricle (pinna), external auditory canal, and tympanum.

External oblique — paired superficial muscles that make up the lateral walls of the abdomen.

Exteroceptor — a receptor that responds to outside stimulus, such as the eye, ear, or olfactory cells.

Extraembryonic membrane — members that are not a part of the embryo but are essential for the health and development of the organism.

Eyespot — a light-sensitive pigmented structure that various invertebrates have; a stigma; a fungus disease of plants.

F

Facial nerve — activates the muscles of facial expression and the lacrimal and salivary glands. It carries sensory impulses from the taste buds of the anterior tongue.

Fallopian tube — the tube through which the ovum is transported to the uterus and where fertilization takes place: also called the *oviduct*.

Farsightedness — a condition in which light rays focus behind the retina.

Fat cell — a cell found in adipose tissue composed of a greasy substance; a lipid soluble in alcohol or ether but not in water.

Fatty acid — an organic acid that is a component of neutral fat.

Feather — a specialized epidermal outgrowth forming the surface covering or plumage of birds. It consists of a proximal hollow portion (quill) and a distal rachis.

Femur — the thigh bone of vertebrates; the third segment of the leg distal to the trochanger in arthropods.

Fermentation — the decomposition of complex organic substances through enzyme action produced by microorganisms under anaerobic conditions.

Fertilization — the fusion of nuclei and other events associated with the union of sperm and egg, resulting in a zygote.

Fetus — the unborn offspring during the last stage of prenatal development.

Fibrin — a fibrous protein that develops in the formation of a blood clot from fibrinogen.

Fibrinogen — a protein in blood plasma that converts into fibrin in a coagulation.

Fibroblast — a connective tissue cell that functions in the formation of fibers and the amorphous ground substance.

Fibrocartilage — a type of cartilage that forms the cushion-like disks between the vertebrae of the spinal column.

Fibrous root — an intertwining mass of many roots of about equal lengths.

Fibula — the smaller of the two bones in the lower limb extending from ankle to knee, lying lateral to the tibia.

Field of view — the circular field you see when you look through the ocular lens.

Filament — a threadlike growth form, process, or structure; an anther stalk.

Filtration — the passage of a liquid through a filter or a membrane.

Fission — division, cleavage or splitting; an asexual method of reproduction.

Flagella — long slender locomotor processes characteristic of flagellate protozoans, certain bacteria, and sperm.

Flexion — a movement that decreases the angle between two bones of a joint; opposite of extension.

Fluoresced light — light absorbed at one wavelength and then emitted as a different color and wavelength.

Foliose lichen — a lichen having many leaves.

Follicle stimulating hormone (FSH) — a gonadotrophic hormone secreted by the pituitary that promotes maturation of ovarian follicles. It stimulates the seminiferous tubules in males.

Food vacuole — formed by many freshwater protists during the process of phagocytosis.

Foramen ovale — the opening through the interatrial septum of the fetal heart.

Forked foot — a charateristic of a rotifer. It is a narrow tail-like posterior foot usually movable and often ending in two slender toes which enable the animal to attach temporarily to some object.

Fragmentation — an asexual reproductive method in which new organisms are formed by the breakup into several parts of the original organism.

Frond — the leaf of a fern containing many leaflets.

Frontal — pertaining to the anterior portion of the body; a bone of the cranium.

Fruit — a mature ovary enclosing a seed or seeds.

Fruticose lichen — a lichen that is erect and branched.

Fungus — a simple plant that lacks chlorophyll, with a filamentous body, subsisting on organic matter.

G

Gallbladder — a pouchlike organ, attached to the inferior side of the liver, which stores and concentrates bile.

Gametangia — a structure in plants that produce gametes, unicellular in algae and fungi, multicellular in higher plants.

Gamete — a haploid sex cell, sperm or egg.

Gametogenesis — the origin of gametes and their development and maturation.

Gametophyte — a plant that produces gametes; the generation of a plant which produces gametes or sex cells; the sexual generation.

Ganglion — an aggregation of nerve cell bodies outside the central nervous system.

Ganglion cell — nervous tissue mass, principally nerve cell bodies; the brain or nervous tissue mass of the ventral nerve cord in invertebrates; nervous tissue mass lying outside the brain and spinal cord of invertebrates.

Gastric mill — a grinding structure that various crustaceans have in the stomach; an insect's gizzard.

Gastric vein — a vein that drains the right side of the stomach directly into the hepatic portal vein.

Gastrocnemius — a two-bellied muscle that forms the curved calf of the posterior leg.

Gastrocoel — the gastrula cavity.

Gastrodermis — the epithelium layer lining the gastrovascular cavity of a coelenterate.

Gastrovascular cavity — also known as the coelenteron which is the internal body cavity of a coelenterate.

Gastrulation — the migration inward of cells of the blastula resulting in a new cavity known as the gastrocoel.

Gemmae — buds or budlike bodies; an asexual reproductive body developing in gemmae cups on the thalli of liverworts and some mosses.

Gemmule — an internal asexual reproductive body that certain sponges produce; a spinelike, numerous process on the dendrite of a neuron.

Gene — one of the biologic units of heredity; parts of the DNA molecule located in a definite position on a certain chromosome.

Generative epithelium — an epithelium giving rise to sex cells; the cover of the surface of an indifferent gonad in the embryo; the lining of a seminiferous tubule or covering an ovary in mammals.

Generative nucleus — the structure produced when pollen undergo mitosis and it will form two sperm.

Genital artery — the artery supplying blood to the sex organs.

Genotype — the genetic makeup of an organism.

Germ hill — a region of the follicular wall that supports and surrounds the oocyte.

Germ layer — also called embryonic cell layers known as ectoderm, mesoderm, and endoderm from which all the organs of the new organism develop.

Gill — a gas exchange organ characteristic of fishes and other aquatic or semiaquatic animals.

Gill bar — in lancelets, the supporting structure separating gill clefts; a vertical supporting structure in a bivalve's gill filament.

Gill slit — a branchial cleft.

Gizzard — a thick-walled grinding structure between the crop and intestine in various annelids and arthropods; in birds, the second portion of the stomach.

Gladiolus — the sternum's middle portion; a plant of the Iris family.

Glans clitoris — also known as the clitoris. A small, erectile structure in the female, homologous to the penis in the male.

Glial cell — a conducting cell of the nervous sysem that provides support, insulation, and protection for the neruons.

Gliding joint — a diarthroses or freely movable joint also known as a plane joint such as the joints of the wrist.

Glomerulus — a coiled tuft of capillaries that is surrounded by the glomerular capsule and filters urine from the blood.

Glossopharyngeal nerve — supplies motor fibers to the pharynx that promote swallowing and saliva production.

Glottis — a slitlike opening into the larynx, positioned between the true vocal cords.

Gluteus maximus — a superficial muscle of the hip that forms most of the flesh of the buttock.

Glycerol — a hygroscopic and viscous liquid that is a by-product in the manufacture of soap as a result of the hydrolysis of fats.

Goblet cell — a unicellular gland within columnar epithelia that secretes mucus.

Gonad — a reproductive organ, testis or ovary, that produces gametes and sex hormones.

Gonangium — a reproductive polyp that consists of a blastostyle bearing gonophores enclosed with a gonotheca, in colonial coelenterates.

Gram negative bacteria — a type of bacteria whose walls have less peptidoglycan and are more complex in structure than gram positive bacteria.

Gram positive bacteria — bacteria that have simpler walls with a relatively large amount of peptidoglycan.

Gram staining technique — a valuable tool for identifying bacteria into two groups based on a difference in their cell walls.

Gravid — Bearing young or eggs; pregnant.

Gray crescent — the pigmented area of a fertilized frog egg between the yolk-filled and black portions.

Gray matter — the portion of the central nervous system that is composed of nonmyelinated nervous tissue.

Green gland — an excretory organ of the crayfish and other crustaceans located in the base of the antenna.

Ground parenchyma — also called ground tissue which makes up the bulk of a young plant. Functions include photosynthesis, storage and support.

Growth ring — one of the concentric rings apparent in a cross section of a stem of a dicot or gymnosperm.

Guanine — a crystalline substance present in DNA and RNA, fish scales, various animal and plant tissues, and in excreta. A purine, nitrogenous base.

Guard cell — an epidermal cell to the side of a leaf stoma that helps to control the stoma size.

Gullet — in vertebrates, the esophagus; in protozoans, the cytopharynx; in anthozoans, the tube from the mouth to the gastrovascular cavity; a food passageway.

Gymnosperm — a member of the plant class Gymnospermae, vascular plants producing naked seeds.

Gyrus — a convoluted elevation or ridge.

H

Haploid — the number (N) of unpaired chromosomes.

Hard palate — the bony partition between the oral and nasal cavities, formed by the maxillae and palatine bones.

Haversian canal — canals that run lengthwise through the bony matrix, carrying blood vessels and nerves to all areas of the bone.

Haversian system — see *osteon*.

Heartwood — the central portion of the stem that is darker, denser wood, consisting of dead elements that have lost their conducting ability.

Hematocrit — the volume percentage of red blood cells in whole blood.

Hemoglobin — the pigment of red blood cells that transports O_2 and CO_2.

Hemolysis — the act of liberating hemoglobin from red blood cells by their destruction or by the hemoglobin's diffusion through an altered cell membrane.

Hermaphrodism — the condition of possessing both male and female sex organs or a combined ovotestis; in flowers, possessing both stamens and pistil.

Heterocyst — a large empty cell found in the filaments in certain algae.

Heterogamy — producing unlike flowers or gametes.

Heterotroph — an organism that utilizes preformed food.

Heterozygous — having two different alleles (i.e., *Bb*) for a given trait.

Hilum — a concave or depressed area where vessels or nerves enter or exit an organ.

Hinge joint — a joint that works like a hinge, such as the elbow joint, allowing movement in one plane only.

Holdfast — an attachment organ or an anchoring structure.

Holoblastic cleavage — a type of total cleavage resulting in blastomeres or equal size.

Hollow nerve tube (dorsal) — one of four anatomical structures in the embryo stage of a chordate. It forms the central nervous system, the brain and spinal cord.

Homeothermic — warm-blooded; having a constant body temperature.

Homologous — similar in developmental origin and sharing a common ancestry.

Homozygous — a condition in which all sperm or eggs produced by an organism are genetically alike due to two members of a pair or a series of pairs of genes are alike.

Human chorionic gonadotropin (HCG) — the hormone produced by the chorion in the first 3 months of human development responsible for keeping the corpus luteum functioning.

Humerus — the upper arm bone between shoulder and elbow in man; the corresponding bone of the forelimb in other vertebrates.

Hyaline cartilage — the most common kind of cartilage in the body, occurring at the articular ends of bones, in the trachea, and within the nose, and forms the precursor to most of the bones of the skeleton.

Hydranth — a feeding polyp of a colonial coelenterate that is hydralike; the oral end of a hydrozoan polyp bearing the terminal mouth and surrounded by tentacles.

Hydrogen bond — a weak bond in which a hydrogen atom forms a bridge between two electronegative atoms such as nitrogen or oxygen. An important intra-molecular bond.

Hydrolysis — is the process in which water is used to split a substance into smaller particles.

Hydroxyl — a functional group in which a hydrogen atom is bonded to an oxygen atom which in turn is bonded to the carbon skeleton of the organic molecule.

Hyoid bone — a "u" shaped bone suspended by ligaments from the pointed styloid processes on the lower part of the temporal bones.

Hyperopia — *see* Farsightedness.

Hypertonic — a solution with greater osmotic pressure than an isotonic solution. Plasmolysis occurs when cells are placed in such a solution.

Hyphae — filaments of the mycelium or vegetative body of a fungus.

Hypocotyl — portion of plant embryo that contributes to stem development.

Hypodermis — in arthropods and invertebrates, the outermost layer of cells of the body wall; a supportive or protective cell layer immediately under the epidermis in plants.

Hypoglossal nerve — the nerve beneath or at the base of the tongue.

Hypostome — a conical elevated portion of the hydrozoan, bearing the mouth.

Hypothalamus — a structure within the brain below the thalamus, which functions as an autonomic center and regulates the pituitary gland.

Hypotonic — a solution with less solute concentration outside the cell than inside the cell. Hemolysis occurs when red blood cells are placed in such a solution.

I

Iliac artery — the artery located near the ilium.

Ilium — of the three bones comprising the os coxae, the more dorsal one.

Incomplete dominance — a type of inheritance in which F_1 hybrids have an appearance that is intermediate between the parents.

Incurrent canal — the canal permitting inflowing.

Incurrent siphon — an opening into which water and food enters an animal such as a clam.

Incus — the middle of the three ear bones, the anvil.

Inguinal canal — the circular passage through which a testis descends into the scrotum.

Initiation codon — the codon, AUG, of m-RNA which initiates translation.

Inner ear — the innermost portion or chamber of the ear, containing the cochlea and the vestibular organs.

Innominate artery — that artery proximal to the hipbone or os coxae.

Innominate bone — the os coxae or hipbone.

Inorganic compound — not containing carbon; not of animal or plant origin.

Intercalated discs — they are the gap junctions between cardiac muscle cells.

Interkinesis — the interphase between meiosis I and meiosis II.

Intermediate inheritance — *see* incomplete dominance.

Internode — a region between stem nodes.

Interoceptor — a sensory receptor in a visceral organ.

Interphase — the time between two successive mitotic divisions.

Intertidal zone — the part of the ecosystem of the beach that is exposed only at low tide. It is characterized by an extreme variation in conditions and no vegetation can live here.

Intestine — part of the alimentary canal extending from the stomach or crop the cloaca or anus. It consists of the duodenum, jejunum, ileum cecum, colon, rectum, and anal canal in man. In certain invertebrates, the digestive cavity.

Interstitial cell — epidermis cells of coelenterates that become sex cells; the cells of Leydig in vertebrates that secrete the male hormone testosterone, located between the seminiferous tubules of the testis.

Invagination — the unfolding process forming a pocket-like cavity.

Involucres — a structure for protecting or investing; a group of bracts surrounding and enclosing the base of a flower or fruit.

Iris — the pigmented muscular portion of the eye that surounds the pupil and regulates its diameter.

Iris diaphragm — in the eye of cephalopods and vertebrates, a pigmented opening containing the pupil, where light enters.

Ischium — of the three bones of the hipbone, the most posterior.

Isogamy — conjugation of isogametes method of reproduction, as with certain protozoans and algae.

Isotonic — Equal osmotic pressure.

J

Jellyfish — a free-swimming saucer-shaped coelenterate having a jellylike consistency; a medusa.

Jointed appendages — a characteristic of the arthropods.

Jugular — pertaining to the veins of the neck which drain the areas supplied by the carotid arteries.

K

Key — a samara or key fruit.

Kidney — one of the paired organs of the urinary system that contains nephrons and filters urine from the blood.

Kinetochore — a specialized region on the centromere that links each sister chromatid to the spindle.

Kingdom — one of the three major division of natural objects, the animal, plant, or mineral kingdom.

L

Lacrimal gland — a tear-secreting gland, located on the superior lateral portion of the eyeball underneath the upper eyelid.

Lacuna — a hollow chamber that houses an osteocyte in mature bone tissue or a chondrocyte in cartilage tissue.

Lamella — a concentric ring of matrix surrounding the central canal in an osteon of mature bone tissue.

Latent period — the time period between the application of a stimulus and a response; the time period between exposure to radiation and the appearance of effects; the time period between exposure to infection and the appearance of symptoms.

Lateral bud — a bud in the upper angle where a leaf joins a stem.

Latissimus dorsi — a large flat muscle pair that covers the lower back. It extends and adducts the humerus.

Leaf — green outgrowth of a plant stem where photosynthesis, transpiration, and respiration are concentrated and consisting of a blade, petiole, and stipules.

Leaf scar — a scar on a stem where a leaf was previously attached.

Leaf stalk — a petiole.

Lens — a transparent refractive structure of the eye, derived from ectoderm and positioned posterior to the pupil and iris.

Lenticel — an opening on a woody plant's root or stem that admits air to the underlying tissue.

Leucoplast — a colorless plastid functioning in starch storage and synthesis.

Leukocyte — a white blood cell; also spelled *leucocyte*.

Lichen — algae and fungi coexisting in a mutualistic relationship.

Ligament — a fibrous band of connective tissue that binds bone to bone to strengthen and provide support to the joint; also may support viscera.

Limb bud — that part of the chick embryo that will later become the wing.

Lip cell — two thin-walled cells that mark the point of rupture in the wall of a sporangium.

Lipid — an organic compound that is insoluble in water but soluble in organic solvents. Most yield fatty acids upon hydrolysis.

Longitudinal groove — a groove extending lengthwise along the long axis.

Lumbosacral plexus — a network of interlacing nerves which extend into the legs.

Lumen — the space within a tubular structure through which a substance passes.

Lung — one of the two major organs of respiration within the thoracic cavity.

Lymph — a clear fluid that flows through lymphatic vessels.

Lymphatic network — the vessels and associated organs which function in the return of tissue fluid to the bloodstream, filtration of lymph, manufacture of leukocytes, and formation of antibodies.

Lymphocyte — a type of white blood cell characterized by a granular cytoplasm.

Lysis — the destruction of cells by lysins or other lytic agents.

M

Macronucleus — the larger of the two nuclei, the vegetative nucleus, in protozoans.

Madreporite — Water enters the water-vascular system of various echinoderms through this porous, calcareous plate on the aboral surface.

Magnification — the appearance of enlargement of an object to the eye when seen through a series of lenses.

Malleus — the first of the three ear ossicles, the hammer.

Maltose — a disaccharide formed when hydrolysis of starch takes place.

Mammary gland — the gland of the female breast responsible for lactation and nourishment of the young.

Mandible — Either the upper or lower jaw; a mouth part in arthropods used for cutting, crushing, or grinding.

Mantle — fleshy fold that envelops the viscera of a mollusk.

Manubrium — a handlelike structure, such as the malleus of the ear, the mastax of a rotifer, or the furcula of a springtail; in the antheridium of certain algae, an inward-projecting cell; an elongated cylindrical base of some cymbas and spathes.

Mastax — in rotifers, the enlarged portion of the pharynx which contains jaws used for grinding food.

Matrix — the intercellular substance of a tissue.

Maxilla — the upper jaw bone; in arthropods, an appendage posterior to the mandible.

Maxillary teeth — in the frog, located along the margin of the upper jaw.

Maxilliped — an appendage located posterior to the maxillae in crustaceans; a centipede's poison claw.

Medulla oblongata — a portion of the brain stem between the pons and the spinal cord.

Medusa — a jellyfish; in coelenterates, a free-swimming sexual form.

Megasporangium — *see* Ovule.

Megaspore — the larger of two kinds of spores in heterosporous plants; a spore that later gives rise to a female gametophyte; a macrospore.

Megasporophyll — a leaf that bears sporangia; a carpel in higher plants.

Meiosis — cell division by which gametes, or haploid sex cells, are formed. There is a reduction in the number of chromosomes by one-half.

Meninges — a group of three fibrous membranes that cover the central nervous system.

Meristem tissue — Undifferentiated plant tissue that is capable of dividing and producing new cells.

Meroblastic cleavage — partial cleavage where cell division is limited to the region of the animal pole.

Mesencephalon — the midbrain located between the forebrain and hindbrain, one of the three primary vesicles of the vertebrate brain.

Mesentery — a fold of peritoneal membrane that attaches an abdominal organ to the abdominal wall.

Mesoderm — the middle germ layer that gives rise to connective tissue, muscles, blood, the circulatory system, urinogenital organs, serous cavities and their linings.

Mesoglea — the substance in coelenterates between the epidermis and gastrodermis.

Mesophyll — the centrally located tissue of a leaf composed of parenchyma cells.

Messenger RNA — one of the three types of nucleic acids which yield ribose upon hydrolysis. It is of nuclear origin.

Metabolism — the chemical changes that occur within a cell.

Metacarpal — pertaining to the region of the metacarpus in the hand and forefoot between the carpus and phalanges.

Metaphase — the stage of mitosis when the chromosomes arrange themselves with their centromeres in an equatorial plane.

Metatarsal — a bone of the metatarsus, between the tarsus and phalanges.

Micronucleus — the smaller of the two nuclei in some protozoans that functions in reproduction and rejuvenation.

Micropyle — a small opening of certain eggs where the sperm enters; an opening the shell of a sponge's gemmule through which cells exit; an opening in an ovule through which a pollen tube gains access to the embryo sac in certain plants.

Microsporangium — a sporangium which produces micro- or meiospores.

Microspore — a spore in seed plants that develops into a pollen grain.

Microsporophyll — a leaf bearing microsporangia such as the carpellate cone of a gymnosperm; a flower stamen.

Middle beach — the area of the beach between the high tide mark and the first sand dunes with vegetation.

Middle ear — the middle of the three ear chambers, containing the three ear ossicles.

Middle lamella — a thin layer of adhesive material, primarily pectins, found between primary walls of adjacent young plant cells.

Milk teeth — deciduous (temporary) teeth of a mammal, "baby" teeth.

Mimicry — a protective resemblance of an organism to another.

Mitochondria — a small organelle, spherical, rod-shaped or filamentous, present in all cells, containing enzymes of Kreb's citric acid cycle and the electron transport systems. They are of primary importance in the cell's metabolic activities.

Mitosis — the process of cell division, in which the two daughter cells are identical and contain the same number of chromosomes as the original cell.

Mitral valve — the left atrioventricular heart valve; also called the bicuspid valve.

Molting — periodic shedding of an epidermal derived structure.

Monocot — a type of angiosperm in which the seed has only a single cotyledon; also called *monocotyledon.*

Monocotyledonae — a class of the Angiosperms including vascular plants with embryos with a single cotyledon, flower parts in threes or sixes, leaves with parallel veins, and a stem with scattered vascular bundles.

Monocyte — a large leukocyte that is granular and has a large, oval or slightly indented nucleus.

Monoecious — hermaphroditic; both testes and ovaries in the same individual; in plants, both antheridia and archegonia are on the same plant, or both staminate and carpellate cones or staminate and pistillate flowers are on the same plant.

Monomer — a flower with only one member in each whorl.

Monosaccharide — a simple sugar or monose, that cannot be decomposed by hydrolysis.

Mons pubis — the elevated area over the pubic symphysis that is covered with pubic hair.

Morphogenesis — specific organ and structure development, resulting in an organism's characteristic size, form, and structure.

Motor neuron — a nerve cell that conducts action potential away from the central nervous system and innervates effector organs (muscles and glands); also called *efferent neuron.*

Mucosa — a mucous membrane that lines cavities and tracts opening to the exterior.

Muscle — an organ adapted to contract; three types of muscle tissue are cardiac, smooth, and skeletal.

Muscle fatigue — resulting from oxygen debt that occurs during prolonged muscle activity. When muscles lack oxygen, lactic acid begins to accumulate in the muscles; the muscle cannot contract.

Muscularis externa — one of the four layers of the intestine. It has two distinct layers of smooth muscle.

Mutualism — a beneficial relationship between two organisms of different species.

Mycelium — a mass of hyphal filaments making up the vegetative body or thallus of a fungus.

Mycoplasms — Bacteria lacking cell walls, the smallest known free-living organisms.

Myelencephalon — the part of the hindbrain from which the medulla oblongata develops.

Myelin sheath — a lipoprotein material that forms a sheath-like covering around nerve fibers.

Myxobacteria — gliding bacteria forming the most elaborate colonies of all prokaryotes.

Myotomes — a myomere; the part of a somite that becomes striated or skeletal muscle.

N

Nares — the nasal cavity openings, nostrils.

Nasal bones — the small rectangular bones forming the bridge of the nose.

Nasal conchae — the curved shell-like bones sloping inwardly, like shelves, forming the lateral walls of the nasal chamber.

Natural selection — the evolutionary mechanism by which better adapted organisms are favored to reproduce and pass on their genes to the next generation.

Nearsightedness — or myopia. When the image normally focuses in front of the retina.

Neck — the cervical region which connects the head with the main portion of the body; a constricted region resembling a neck; the portion of an archegonium which is elongated and through which sperm gain access to the venter; the point of juncture between the stem and root of a plant.

Neck (of the tooth) — the narrowed portion of the tooth.

Negative feedback mechanism — a chemical response to chemical regulation of glandular secretions that opposes the original change.

Nematocyst — a coelenterate's or ctenophore's stinging cell. It consists of a spherical or oval capsule with a coiled tube or thread capable of being everted or discharged, and functions as a protective, food-getting, and adhesive structure.

Nephridia — a tubular structure, simple or branched, that functions as an exretory organ in invertebrates. It opens to the outside through a nephridiopore.

Nephron — the functional unit of the kidney, consisting of a glomerulus, glomerular capsule, convoluted tubules, and the loop of the nephron.

Neural fold — a fold developing from the neural plate and forms the neural tube upon fusion.

Neural groove — a longitudinal median groove lying between the neural folds prior to their closing.

Neural plate — medullary plate; median sheet of ectoderm that develops into the neural tube and neural crest cells; one of the plates overlying or fused with the vertebrae in turtles.

Neural tube — the ectodermal longitudinal tube formed in the middorsal region of a vertebrate embryo that develops into the brain and spinal cord.

Neuron — the structural and functional unit of the nervous system, composed of a cell body, dendrites, and an axon; also called a *nerve cell.*

Neurulation — the development of the notochord, neural tube, and coelom. This only occurs in chordates.

Neutrophil — a type of phagocytic white blood cell.

Nictitating membrane — a transparent membrane functioning as a third eyelid in some vertebrates but is vestigial in man.

Nidation — to place an implant with an organ or tissue; the attachment of the blastocyst to the endometrium of the uterus in mammals.

Nipple — a dark pigmented, rounded projection at the tip of the breast.

Nitrogen fixation — a process carried out by certain organisms, such as by soil bacteria, whereby free atmospheric nitrogen is converted into ammonia or nitrate compounds.

Node — location on a stem where a leaf is attached.

Notochord — a flexible rod of tissue that extends the length of the back of an embryo.

Nuclear membrane — the double-layered surface of the nucleus of a cell, contining pores on the outside which communicate with the endoplasmic reticulum.

Nucleic acid — an organic molecule composed of joined nucleotides, such as RNA and DNA.

Nucleolus — a basophilic, densely staining body in the nucleus of a cell in the interphase, composed principally of RNA.

Nucleoplasm — a nucleus's protoplasm.

Nucleotide — a nucleoside whose phosphoric acid group is attached to a sugar; a nucleoside phosphate.

Nucleus — a spheroid body within a cell that contains the genetic factors of the cell.

O

Occipital — pertaining to the occiput, the back of the head.

Ocular — the eyepiece of a microscope or other optical instrument; pertaining to the eye.

Oculomotor Nerve — the third cranial nerve whose fibers innervate all eye muscles except the superior oblique and lateral recturs.

Odontoid process — the toothlike process on the axis. The atlas rotates around it.

Oil — a viscous liquid soluble in ether or other organic solvent but insoluble in water; a fluid lipid.

Olfactory lobe — Rhinencephalon; the anterior portion of the telencephalon.

Oocyte — a developing egg cell.

Oogamy — method of sexual reproduction involving gametes of unequal size.

Oogenesis — the process of female gamete formation.

Oogonium — it is the female stem cell resulting from a primordial germ cell undergoing mitosis. It develops into a primary oocyte.

Oosphere — a large female nonmotile gamete formed in an oogonium; egg cell.

Ootid — the mature ovum after the second maturation division.

Operculum — a covering, a lid; the lid over the discharge pore in the sporangium of certain fungi; a pysix lid; the cerebral fold concealing the insula; in coelenterates, the flap closing the hydrotheca opening; in gastropods, the covering of the shell aperture; the close of the egg opening through which an embryo or larve exits; a burrow closing after the occupant is retracted; in cicadas, the cover of the sound-producing structures; in the king crab, the cover of the book gills and genital openings; a respiratory structure in terrestrial isopods; the gill chamber cover in foshes; in amphibian larvae, the cover of the gills and developing legs; the second ear ossicle in urodeles and some anurans.

Optic chiasma — an X-shaped structure on the inferior aspect of the brain where there is a partial crossing over of fibers in the optic nerves.

Optic lobe — one of two rounded bodies, the corpora bigemina, that form the roof of the midbrain, concerned with vision. In amphibians, reptiles, and birds, it is well developed but is absent in mammals.

Optic nerve — a sensory nerve, the second cranial nerve, formed of fibers arising from ganglion cells in the retina.

Optic tract — a fiber band extending from the optic chiasma to the lateral geniculate body of the thalamus.

Oral groove — the surface groove leading to the cytostome in ciliates.

Oral sucker — the sucker surrounding the mouth in flukes.

Organ — a structure consisting of two or more tissues, which performs a specific function.

Organelle — a minute structure of a cell with a specific function.

Organic compound — a compound that contains carbon in the form of chains or rings.

Organism — an individual living creature.

Organ of Corti — hearing organ, a spiral elongated structure that rests on the basilar membrane of the cochlea, containing hair cell receptors for auditory stimuli.

Organogenesis — the differentiation and association of cells to form organ systems.

Organ system — group of structures functioning together as a unit.

Osculum — a sponge's excurrent opening.

Osmosis — the passage of a solvent, such as water, from a solution of lesser concentration to one of greater concentration through a semipermeable membrane.

Ossicle — one of the three bones of the middle ear.

Osteocyte — a mature bone cell.

Ostia — the opening to a cavity or tube; where water enters a radial canal or spongocoel in sponges; a septum opening in anthozoans; a heart opening where blood enters in arthropods; where water enters a water tube of a gill in bivalves; an oviduct or uterine tube opening in vertebrates.

Otolith — a body present in the statocyst of invertebrates or in vertebrates' saccule.

Oval window — a membrane-covered opening in the bony wall between the middle and inner ear, into which the footplate of the stapes fits.

Ovary — the female gonad in which ova and certain sexual hormones are produced.

Oviparous — the method of reproduction as in birds and insects, in which the female lays eggs which hatch outside the body.

Ovisac — the structure for storing eggs; in elasmobranchs and amphibians, the uterus; the fairy shrimp's egg sac.

Ovulate cone — is the female cone consisiting of many scales, each with two ovules.

Ovulation — the rupture of an ovarian follicle with the release of an ovum.

Ovule — the female reproductive organ in a seed plant that contains megasporangium where meiosis occurs and the female gametophyte is produced.

Ovum — a secondary oocyte after ovulation but before fertilization.

Oxygen debt — s*ee* Muscle fatigue.

Oxyhemoglobin — hemoglobin that has been oxygenated, formed in red blood cells on their passage through the lungs or gills.

P

Palatine — pertaining to or of the palate.

Palisade parenchyma — a leaf's photosynthetic tissue forming layers of chlorophyll-containing cells (palisade cells), which comprise the mesophyll of a leaf along with the spongy parenchyma.

Pancreas — organ in the abdominal cavity that secretes gastric juices into the GI tract and insulin and glucagon into the blood.

Pancreatic duct — a tube carrying fluids from the pancreas to the small intestine.

Paper chromatography — chemical analysis based on the selective absorption of parts of a mixture on a strip of paper.

Papillae — small nipplelike projections.

Parasite — an organism that resides in or on another from which it derives sustenance.

Paraphyses — a slender sterile filamentous structure in spore-producing or gamete-producing structures in bryophytes and fungi; a non-nervous portion of the roof of the telencephalon in lower vertebrates.

Parapodia — Fleshy appendages that extend out from the body wall of a polychaete, functioning as a locomotor and respiratory organ; in certain mollusks, a finlike projection from the foot.

Parietal — pertaining to a wall of an organ or cavity.

Parthenogenesis — ovum development without fertilization. In rotifers or aphids, it may occur naturally or be induced by mechanical or chemical stimus to eggs.

Passive transport — substance movement across a plasma membrane not involving energy expenditure by the cell.

Patella — a sesamoid bone in the quadriceps tendon at the knee, the kneecap.

Pectoral fin — an anterior lateral fin of a fish, one of a pair.

Pectoral girdle — the portion of the skeleton that supports the upper extremities.

Pectoralis major — is a large fan-shaped muscle covering the upper part of the chest.

Pedicel — a peduncle, a stalk or stem.

Pedicellaria — the small pincherlike organ of various echinoderms.

Pellicle — thin covering or membrane; a periplast.

Pelvic girdle — the part of the skeleton supporting the pelvic fins of fishes or the hind limbs of a tetrapod; three paired bones, the ilium, ischium, and pubis.

Pelvis — the cavity or basinlike structure; the ring of bones in man consisting of the two innominate bones, sacrum, and coccyx; the kidney cavity consisting of the expanded end of the ureter.

Penis — the external male genital organ, through which urine passes during urination and which transports semen to the female during coitus.

Pericardial cavity — the cavity in the pericardium surrounding the heart.

Pericardial sac — *see* Pericardium

Pericardium — a protective serous membrane that surrounds the heart.

Pericycle — parenchymous cells in the roots and stems of vascular plants located in a layer between the endodermis and phloem, or the xylem and phloem in some cases.

Peridontal membrane — a fibrous membrane (ligament) that holds the tooth in place in the bony jaw.

Periosteum — a fibrous connective tissue covering the surface of bone.

Peripheral nervous system — the nerves and ganglia of the nervous system that lie outside of the brain and spinal cord.

Peristalsis — rhythmic contractions of smooth muscle in the walls of various tubular organs, which move the contents along.

Peristome teeth — hygroscopic teeth surrounding a spore capsule opening in mosses.

Peritoneum — the serous membrane that lines the abdominal cavity and covers the abdominal viscera.

Petal — the leaf of a flower, which is generally colored.

pH — a symbol indicating the concentration of hydrogen ions or hydroxyl ions.

Phagocytosis — the surrounding and engulfing of substances by phagocytes, usually foreign bodies such as bacteria, dust particles, or colloidal dyes.

Phalanx — pl. *phalanges*: a bone of the finger or toe.

Pharynx — the region of the GI tract and respiratory system located at the back of the oral and nasal cavities and extending to the larynx anteriorly and the esophagus posteriorly; also called the *throat.*

Phenotype — the appearance of an organism caused by the genotype and environmental influences.

Phloem — vascular tissue in plants that transports nutrients.

Phosphate — a salt of phosphoric acid.

Photosynthesis — the process of using the energy of the sun to make carbohydrate from carbon dioxide and water.

Phycocyanin — a blue pigment that is light-absorbing, found in blue-green and other algae.

Phycoerythrin — a red pigment that is light-absorbing, in red and other algae.

Pinacocyte — a contractile cell, flattened in a sponge's epidermis.

Pinna — a protruding structure; an external ear auricle; a pinnate leaf's primary division.

Pistil — the reproductive structure of a flower that contains the stigma, style, and ovary.

Pith — a centrally located tissue within a dicot stem.

Pith ray — are located between the vascular bundles in a dicot stem.

Pituitary gland — a small, pea-shaped endocrine gland situated on the inferior surface of the brain that secretes a number of hormones; also called the *hypophysis* and commonly called the *master gland.*

Pivot joint — a freely movable joint (diarthroses) such as between the axis and atlas.

Placenta — the organ of metabolic exchange between the mother and the fetus.

Placoid scale — Found in elasmobranchs and consisting of a bony plate embedded in the dermis. A dentine-covered spine projects through the epidermis from this plate.

Plankton — aquatic, free-floating microscopic organisms.

Planula — Free-swimming larva, ciliated, of some coelenterates and a genus of ctenophorans; a parasite.

Plasma — the fluid, extracellular portion of circulating blood.

Plasmodesmata — tiny protoplasmic threads connecting adjoining cells through cellulose cell walls of plants.

Plasmodium — an ameboid mass of naked protoplasm with more than one nucleus, forming the vegetative body of a slime mold; a member of the *Plasmodium* genus.

Plasmolysis — the shrinking from the cell wall of the protoplasm, resulting in turgidity loss from exosmosis.

Plasmolyze — the process of plasmolysis.

Platelets — fragments of specific bone marrow cells that function in blood coagulation: also called *thrombocytes.*

Pleural membranes — serous membranes that surround the lungs and line the thoracic cavity.

Pleuron — the lateral body wall of an arthropod segment.

Pleuroperitoneal cavity — located in the frog's coelom in which respiratory, digestive, reproductive and excretory organs are located.

Plumule — the primary bud of a plant embryo; the epicotyl; a down feather.

Pneumatic — of air or gas; operated by air pressure; containing air or conducting it.

Poikilothermy — having a varying body temperature, dependent upon the environment; cold-blooded.

Polar body — a minute cell produced at oogenesis, a polocyte.

Pollen — male gametophyte generation of seed plants.

Pollen tube — the slender tube developing at germination of a pollen grain. Two sperm nuclei gain access to the female gametophyte and its ovum through it.

Pollinate cone — the male pine cone.

Polymer — the resulting substance of the combination of two or more molecules of the same type.

Polymorphonucleocyte — cell whose nucleus consists of two or more lobes joined by slender strands.

Polyp — a sessile coelenterate whose cylindrical body is attached at its basal end, the free end bearing a mouth surrounded by tentacles; a neoplastic, pedunculated structure that develops on a mucous membrane.

Polysaccharide — a complex carbohydrate that yields more than two molecules of a monosaccharide upon hydrolysis.

Pons — the portion of the brain stem just above the medulla oblongata and anterior to the cerebellum.

Primary consumer — organisms which feed directly on the producers of the ecosystem.

Primary root — the first structure to emerge from a sprouting seed.

Primary succession — the process of replacement of one group of species by another group of species, starting with bare rock or open water and developing soil and filling in open water areas.

Primitive streak — the enlongated blastopore in the chick embryo.

Primordia — the first indication of organ or structure development.

Proboscis — Elephant's trunk; an elongated snout of certain animals; an elongated and usually tubular structure extending from the body or head of some invertebrates.

Producer — an organism that synthesizes organic compounds from inorganic constituents within an ecosystem.

Proglottid — a segment of the body of a tapeworm (strobila).

Prokaryotic cell — cell lacking distinct membrane-bound structures organelles; monerans.

Prophase — the initial phase of mitosis when chromatin organizes into chromosomes, centrosomes move to opposing poles, spindle fibers develop and attach to the chromosomes, and the nuclear membrane disappears.

Prospopyle — the opening between the incurrent and radial canal in sponges.

Prostate — a walnut-shaped gland surrounding the male urethra just below the urinary bladder that secretes an additive to seminal fluid during ejaculation.

Prostomium — that part of the the head projecting forward and overhanging the mouth, as in various annelids.

Protein — a macromolecule composed of one or several polypeptides.

Prothallus — a heart-shaped structure that is the gametophyte generation of a fern.

Prothrombin — a substance in blood plasma essential for clotting.

Protonemata — a branched structure in a moss developed from a spore which gives rise to leafy, gametophyte moss plants by budding; a similar algal structure developing from a zygote.

Protoplasm — the viscid substance comprising a living cell, with which living processes are associated, such as metabolism, growth, irritability, and reproduction.

Protoplast — the living, organized protoplasm of a cell.

Protozoan — Unicellular or acellular animals of the phylum Protozoa. Some form colonies but lack differentiation of tissue, including rhizopods, flagellates, sporozoans, and ciliates.

Pseudocoelom — invertebrates of the Pseudocoelomata group, of the phyla Entoprocta, Aschelminthes, or Acanthocephala. They are characterized by the possession of a pseudocoel.

Pseudopodia — a usually temporary, blunt extension of a cell's cytoplasm; a slender stalk forming the axis of the gametophyte in some bryophytes and mosses.

Pubic symphysis — the ventral union of two pubic bones.

Pubis — the pubic bone or os pubis, the most ventral or anterior of the three bones forming the os coxae (hipbone).

Pulmocutaneous arch — found in the frog carrying blood to the lungs and skin for gas exchange.

Pulmonary artery — the artery supplying blood to the lungs.

Pulmonary trunk — a large vessel carrying blood from the right ventril of the heart.

Pulp — soft, spongy tissue filling the cavity of a tooth; the central substance of an invertebral disk; the substance of the spleen; the succulent part of a fruit; the pith in certain stems; plant fiber mixture from which paper is made.

Pupillary opening — or pupil. The opening in the center of the iris of the eye which allows light to enter the eye.

Purine — a substance parent to many biologically significant substances such as uric acid, caffeine, and adenine, but not occuring naturally, such as uric acid and adenine.

Pus — a yellowish semifluid consisting of serum, leukocytes, bacteria, and tissue debris that is usually a result of inflammatory processes.

Pyloric sphincter — pertaining to the pylorus.

Pyloric stomach — a circular ring of muscles around the alimentary canal at the pylorus regulating the passage of food from the stomach to the intestine, also known as the pyloric valve.

Pyrenoid — a highly refractive protein body in the chloroplast of lower organisms, serving as a center for the deposition of starch.

Pyrimidine — a nitrogenous base in nucleotides.

Q

Quadriceps femoris — the large muscle that comprises the rectur femoris, vastus lateralis, vastus medialis, and vastus intermedius in man. It forms the anterior portion of the thigh, and extends the leg and flexes the femur.

R

Radial canal — a canal that extends outward from a central cavity in sponges or medusae; a canal of the water-vascular system in sea stars which extends peripherally in each ray.

Radial symmetry — a condition in which a number of similar parts radiate outward from or are uniformly distributed around a central axis.

Radicle — a small root as a nerve root; the part of a plant embryo developing from the primary root.

Radius — one of the bones in the forelimb of a tetrapod; the bone that lies laterally to the ulna in man; one of the parts in a radially symmetrical structure or organism; the third longitudinal vein in an insect's wing.

Radula — a structure bearing rows of chitinous teeth in the floor of the pharynx of most mollusks but not including bivalves. It is a chewing, rasping organ.

Raphe — a fusion or union of a structure's two halves like a seam, externally marked by a ridge or groove; a groove or cleft lying longitudinally in the valve of a diatom; a ridge longitudinally located on a seed that develops in an anatropous ovule, marking fusion of the funiculus and integument.

Receptacle — a structure for receiving and retaining something; the swollen tips of branches bearing sex organs in rockweeds; a disc-shaped structure bearing the archegonium or antheridium in liverworts; the tip of a peduncle or pedicel on or around which the floral parts of seed plants develop.

Recessive gene — one producing its effects only in a homozygous individual or in the absence of a dominant gene.

Rectal caeca — *see* Cecum.

Rectum — the terminal portion of the GI tract, from the sigmoid colon to the anus.

Rectus abdomines — the paired strap-like muscles of the abdomen. Their main function is to flex the vertebral column.

Rectus femoris — one of the four muscles that makes up the quadriceps. It helps to extend the knee and flex the hip.

Red blood cells (erythrocytes) — they carry oxygen and carbon dioxide.

Red tide — Enormous numbers of dinoflagellates (*Gonyaulax* and *Gymnodinium*) in waters off the Florida and California coasts. This occurrence imparts a reddish hue to the waters by day and luminescence by night.

Reducing sugar — such as a monosaccharide which have a free carbonyl group in close proximity to a hydrogen group.

Reflex — a reflex action.

Refractory period — the time following stimulation when a muscle or nerve fiber is incapable of responding.

Regeneration — regrowth of tissue or the formation of a complete organism from a portion.

Renal arteries — they branch off the abdominal aorta (dorsal aorta) and serve the kidneys.

Respiratory tree — a treelike structure in holothuroids attached to the cloaca into which water can be pumped; the trachea, bronchi, and bronchioles as a unit in man.

Retina — the inner layer of the eye that contains the rods and cones.

Rhizoid — Rootlike; a slender filament attaching the mycelium of some fungi or the gametophyte of mosses, liverworts, and ferns to the substrate.

Rhizome — a rootstock or stem growing on or under the ground and producing stems and roots; a stolon.

Ribonucleic acid (RNA) — one of the nucleic acids which yield ribose upon hydrolysis, important in the synthesis of proteins. The three types are messenger RNA, transfer RNA, and ribosomal RNA.

Ribose — a pentose sugar found in riboflavin, ribonucleic acid, and various nucleotides.

Ribosomal RNA — a type of RNA transcribed from DNA in the nucleolus and found in the ribosomes.

Ribosomal subunit — one of the ultramicroscopic bodies in the cytoplasm of cells, usually associated with the endoplasmic reticulum, that function in protein synthesis.

Rickettsia — a microorganism intermediate between viruses and bacteria, of the group Rickettsiales. They are obligate, intracellular parasites common in various arthropods. Some are pathogenic and cause various diseases.

Ring canal — a circular canal in an echinoderm's water-vascular system or the tentacle of an ectoproct.

RNA polymerase — an enzyme that links together the growing chain of ribonucleotides during transcription.

Rod — a photoreceptor in the retina of the eye that is specialized for colorless, dim light vision.

Root cap — end mass of parenchyma cells which protects the apical meristem of a root.

Root hair — minute epidermal projection from the root of a plant, which functions in absorption.

Root meristem — plant tissue that remains embryonic as long as the plant lives.

Root (tooth) — one of two regions of a tooth. It is that portion embedded in the jaw bone.

Rostellum — a small projection at the anterior end of certain flagellates; the beak of certain sucking insects; a round process on the scolex of certain tapeworms, possibly retractile and surrounded by recurved hooks; a sterile, beaklike lobe on the stigma of an orchid.

Rotation — is the movement of a bone around its longitudinal axis. A common movement of the ball-and-socket joints.

Round window — one of two membrane-covered holes in the middle ear. It is the more inferior membrane.

Rugae — elevations or ridges as in the mucosa of the stomach.

S

Sacral vertebrae — those vertebrae in the region of the sacrum.

Saprophyte — a plant whose nourishment is obtained from dead organic matter.

Sapwood — the softer outer wood (xylem) providing mechanical support and functioning in conduction and food storage.

Sartorius — the long slender muscle of the inner thigh, a flexor of the leg; an external rotator and flexor of the thigh.

Scanning electron microscope — an electron microscope that uses an electron beam focused to a fine point and passed back and forth over the surface of the specimen.

Scapula — the pectoral girdle bone that the forelimb articulates with; the shoulder blade in man.

Sciatic nerve — the nerve near the ischium.

Sclera — the outer white layer of connective tissue that forms the protective covering of the eye.

Sclerenchyma — supporting tissue in plants composed of hollow cells with thickened walls.

Scolex — head region of a tapeworm.

Scrotal sac — see Scrotum.

Scrotum — a pouch of skin that contains the testes and their accessory organs.

Sea walnut — a ctenophore.

Secondary consumer — organisms which feed on the primary consumers of an ecosystem.

Secondary succession — the process of replacement of species on areas where vegetation has been removed by agricultural uses or fire.

Seed coat — the outer seed covering, consisting of an outer tegmen and an inner testa, which develops from the integuments.

Segmentation — Being divided into segments; metamerism; cleavage.

Selectively permeable — a property of membranes that allows some substances to cross more easily than others.

Semen — the secretion of the reproductive organs of the male, consisting of spermatozoa and additives.

Semicircular canals — tubular channels within the inner ear that contain the receptors for equilibrium.

Seminal fluid — the fluid found in semen containing nutrients and mucus.

Seminal receptacle — in various invertebrates, a saclike structure storing sperm received from the male.

Seminal vesicles — a pair of accessory male reproductive organs lying posterior and inferior to the urinary bladder, which secrete additives to spermatozoa into the ejaculatory ducts.

Seminiferous tubule — the vertebrate testis's sperm-producing tubule.

Semipermeable membrane — a cell or plasma membrane partially permeable allowing certain molecules or ions to readily pass but not other molecules.

Semitendinosus — one of the muscles that makes up the hamstring group which form the muscle mass of the posterior thigh.

Sensory neuron — a nerve cell that conducts an impulse from a receptor organ to the central nervous system; also called afferent neuron.

Sensory receptor — a structure responding to stimulus, such as a sensory end organ or peripheral nerve ending; a neuroepithelial cell.

Sepal — leafy division of the calyx at the base of a petal of a flowering plant.

Septa — membranes or walls dividing a cavity or structure into two or more parts.

Septate hyphae — hyphae which are cross-walled such as in fungi.

Serosa — a serous membrane; a membrane enveloping an insect embryo; reptile and bird chorion.

Sertoli cell — in a mammalian testis, a sustenacular cell in the epithelial lining of a seminiferous tubule.

Sessile — organisms that lack locomotion and remain stationary, such as sponges and plants.

Setae — stiff bristles or bristlelike structures; the support stalk of the capsule in a bryophyte.

Sexual reproduction — the method of reproduction involving the union of sex cells or gametes.

Sheath — a covering or protective envelope; a fold or sac into which an organ that is protrusible is retracted; the covering of a blood vessel, nerve, muscle, or tendon, for protection; nerve fiber covering; the gelatinous covering of certain filaments or algal cells; a leaf base enveloping a stem.

Sieve tube cell — a cell of the conducting structure in phloem, joined together end to end and separated by sieve plates.

Simple squamous epithelium — a single layer of flattened cells resting on a basement membrane. It is in the air sacs of lungs and it forms the walls of capillaries.

Sinus venosus — a heart chamber of lower vertebrates that receives venous blood and conducts it to the atrium; a temporary structure in the embryos of higher vertebrates.

Skeletal muscle — a type of muscle tissue that is multinucleated, occurs in bundles, has crossbands of proteins, and contracts either in a voluntary or involuntary fashion.

Smooth muscle — a type of muscle tissue that is nonstriated, composed of fusiform, single-nucleated fibers, and contracts in an involuntary, rhythmic fashion within the walls of visceral organs.

Soft palate — the velum palati, a musculomembranous partition forming the posterior of the roof of the mouth, closing the posterior openings to the nasal cavity during swallowing.

Solute — a substance dissolved in a solvent to form a solution.

Solvent — a fluid such as water that dissolves solutes.

Somite — Bump-like structures on either side of the spinal cord that will later give rise to muscles and vertebrae.

Sperm — a spermatozoan; the mature male gamete; seminal fluid or semen.

Spermaries — organs producing male gametes; testes or antheridia.

Spermatid — a haploid cell from the division of a secondary spermatocyte, transforming into a functional spermatozoan.

Spermatocyte — one of two cell generations occuring in the development of spermatozoa.

Spermatogenesis — the production of male sex gametes, or spermatozoa.

Spermatogonia — primordial germ cells giving rise to a primary spermatocyte.

Spermatozoon — a sperm cell, or gamete.

Sperm duct — see Vas deferens.

Sphenoid — a butterfly-shaped bone which spans the width of the skull and forms part of the floor of the cranial cavity.

Sphygmomanometer — an instrument used to measure the blood pressure in the artery of the upper arm.

Spicule — a pointed, sharp silicious or calcareous body, as the exoskeleton of sponges, corals, certain protozoans, or formed in bone development; a curved copulatory structure in a male nematode.

Spinal nerve — one of the thirty-one pairs of nerves that arise from the spinal cord.

Spinal accessory nerve — the eleventh cranial nerve which travels to the muscles of the neck and back.

Spindle fiber — a structure of a spindle shape; a mass of delicate, curved fibrils in mitosis extending between the two poles or asters.

Spirillum — a bacterium that is spiral-shaped, such as *Vibrio* and *Spirillum.*

Spirochete — a bacterium that is spiral-shaped and flexible, of the order *Spirochaetales.*

Spleen — a large, blood-filled organ located in the upper left of the abdomen and attached by the mesenteries to the stomach.

Spongocoel — a sponge's central cavity, opening to the outside through the osculum.

Spongy parenchyma — a layer of leaf cells lying below the palisade layer. It is loose-textured, containing air spaces, and part of the mesophyll.

Sporangium — an organ within which spores are produced.

Sporangiophore — a structure in fungi that is stalklike and bearing spores or sporangia.

Spore — a structure for asexual reproduction consisting of one or a few cells that can give rise to a new organism with gametic union; a cell that results from multiple fission of a sporont; a dormant body into which a bacterium may be transformed that is resistant to drying, sunlight, and heat.

Sporophyll — a leaf or structure like a leaf bearing sporangia.

Sporophyte — the diploid plant that produces haploid meiospores in those plants that exhibit alternation of generations.

Springwood — the inner layer of a dicot's annual ring.

Stalk — the principal supporting plant stem or axis; a supporting structure in an animal; a structure that connects.

Stamen — the portion of a flower which is composed of a filament and an anther, where pollen grains are produced.

Stapes — an ear ossicle in the middle ear, also known as the stirrup.

Starch — a polysaccharide, present in leucoplasts and chloroplasts of plants and the cotyledons and endosperm of seeds, composed of many glucose units.

Statoacoustic nerve — the eighth cranial nerve, from the internal ear, composed of two branches. Also called the auditory nerve.

Stem — a vascular plant's main axis that bears buds; a part that supports a structure; the main stalk.

Stereomicroscope — a microscope with two lenses.

Sterigmata — one of the slender stalks at the tip of a basidium, bearing a basidiospore.

Sternum — a tetrapod breastbone; a structure of bone and cartilage with the pectoral girdle or ribs articulate with.

Stigma — the upper portion of the pistil of a flower.

Stipe — a supporting structure or short stalk; the stalk supporting the pistil in seed plants.

Stolon — a horizontal branch at a plant's base which gives rise to new shoots; in some fungi, a horizontal hypha which gives rise to new plants at the place where it comes into contact with the substratum.

Stomate — an opening in a leaf or stem that gases pass through; the pharynx or buccal capsule of a nematode.

Stomochord — a cartilaginous rod providing support in the hemichordates.

Stone canal — the S-shaped canal in various echinoderms leading from the madreporite to the ring canal.

Stratified squamous epithelium — is the only common stratified epithelium in the body. It usually consists of several layers of cells and is found in the esophagus, mouth, and the outer portion of the skin.

Striation — marked by stripes, cross bands, or fine parallel lines, grooves, or ridges.

Strobila — a chain of similar structures, such as the body of a tapeworm with its chain of proglottids.

Style — the long slender portion of the pistil of a flower.

Subclavian artery (left) — the third branch of the aortic arch. It gives rise to the vertebral artery, axillary artery and brachial artery.

Subcutaneous layer (tissue) — it is located under the dermis and contains many fat cells. It anchors the skin to the underlying organs.

Submucosa — a connective tissue layer lying beneath a mucous membrane.

Substrate — the substance that an enzyme acts upon; a solid material to which an organism is attached or upon which it lives.

Sucker — a cup-shaped structure for getting food or for attachment; a shoot developing from the roots or lower part of the stem.

Sulcus — a furrow or groove on the brain. Less deep than a fissure.

Summerwood — the annual ring's outer layer consisting mostly of small, thin-walled elements.

Suspensory ligament — delicate fibers encircling the eye lens that attach it to the processes of the ciliary body.

Swimmeret — an unspecialized, biramous appendage of various crustaceans, a pleopod.

Symbiosis — the relationship between two organisms of different species in which they live in intimate association with each other.

Synapse — a narrow gap between a synaptic knob of an axon and the dendrite of another neuron or effector cell.

Synapsis — homologous chromosomes coming together in pairs during prophase I of meiosis.

Synovial membrane — a membrane that forms the inner lining of the capsule of a freely moveable joint.

Systemic arch — it is found in the frog. One of the three branches of the aortic arch, carrying blood to most of the body.

Systole — the muscular contraction of the ventricles of the heart during the cardiac cycle.

T

Tactile receptor — a cell sensitive to stimulus by touch.

Tapetum lucidum — a light-reflecting layer in the choroid of the eye, such as in elasmobranchs. Found in cats and sheep.

Taproot — a plant root system in which a single root is thick and straight.

Tarsal — pertaining to the tarsus.

Taste buds — a sensory receptor containing gustatory cells and located principally on the tongue.

Taxon — *pl.* taxa. The named taxonomic unit at any given level.

Teat — a protuberance that is elongated by which the young of mammals draw milk from the udder or mammary gland.

Telencephalon — the portion of the embryonic forebrain that gives rise to the rhinencephalon, corpora striata, and cerebral cortex develop.

Telophase — the last stage of mitosis or meiosis when two nuclei are formed and cytokinesis occurs.

Telson — a portion of the abdomen in crustaceans; the terminal spine of the king crab; a scorpion's sting or curved, dorsal spine; a primitive insect's terminal abdominal segment.

Temporal — pertaining to the temporal bone or temple.

Tendon — a band of dense regular connective tissue that attaches muscle to bone.

Tentacle — a slender, flexible, elongated, unsegmented process located at the oral or anterior end of an animal, that serves as a sensory, food-getting, defensive, or attachment structure.

Tergum — as in arthropods, the dorsal surface or a body segment or somite; a plate that covers the anterior lateral surface of a barnacle.

Terminal bud — the bud at the tip of a plant stem.

Terminal bud scar — an area on a twig formed each spring when the scales covering the terminal bud fall off.

Termination codon — one of three codons of m-RNA, UAA, UAG, and UGA which do not code for amino acids but instead act as signals to stop translation.

Testis — the primary reproductive organ of a male, which produces spermatozoa and male sex hormones.

Testosterone — a male hormone produced by the mammalian testis.

Tetanus — the contracted state of a muscle; a disease caused by *Clostridium tetani* that is infectious, also known as lockjaw.

Tetrad — a group of four; two pairs if bivalent chromosomes formed during meiosis; a group of four cells.

Thalamus — a number of nuclei in the diencephalon functioning as relay centers for sensory impulses passing to the cerebral cortex and for crude, uncritical sensations; the receptacle of a flower.

Thallus — an undifferentiated plant body, such as algae and fungi.

Thoracic vertebrae — those in the region of the thorax.

Threshold — the lowest limit that a certain phenomenon will occur at.

Thrombin — in shed blood, a substance formed from prothrombin to induce clotting by converting fibrinogen to fibrin.

Thrombocyte — a blood platelet in mammals; a spindle cell of the blood in vertebrates below mammals.

Thromboplastin — a substance released from injured tissue and disintegrating blood platelets which converts prothrombin to thrombin in the presence of calcium ions.

Thymine — a pyrimidine base in DNA.

Thymus gland — a bi-lobed lymphoid organ positioned in the upper mediastinum, posterior to the sternum and between the lungs.

Thyroid gland — a large endocrine gland in the neck alongside the larynx and trachea producing the hormone thyroxine and other iodine-containing compounds that regulate metabolic rate, electrolyte balance, and other functions. It controls metamorphosis in amphibians.

Tibia — the larger and innermost bone of the two of the leg, lying medial to the fibula; the shinbone.

Tissue — an aggregation of similar cells and their binding intercellular substance, joined to perform a specific function.

Tongue — a protrusible muscular organ on the floor of the oral cavity.

Tonoplast — a layer of protoplasm around a water vacuole of a cell; a vacuolar membrane.

Totipotency — referring to embryonic cells that retain the potential to form all parts of animals.

Trachea — the airway leading from the larynx to the bronchi; also called the *windpipe.*

Tracheid — an elongated dead cell in vascular plants with a pronounced cavity or lumen forming an element of xylem. Its walls are usually thick and pitted and it functions in support and water conduction.

Trait — a distinguishing feature studied in heredity.

Transfer RNA — a type of RNA that carries a particular amino acid to m-RNA at the ribosome in protein synthesis.

Translation — the process combining animo acids into polypeptide chains as directed by messenger RNA.

Translocation — substance movement from one place to another; ion, atom, or molecule movement through cell membranes; a chromosome aberration in which a portion is transferred to a nonhomologous chromosome.

Transmission electron microscope — a microscope that uses electron beams rather than light and electromagnetic lenses rather than glass lenses.

Transpiration — the evaporation of water from a leaf, which pulls water from the roots through the stem to the leaf.

Transverse fission — *see* Binary fission.

Trapezius — muscles that form a diamond or kite-shaped muscle mass. The most superficial muscles of the posterior neck and upper trunk.

Triceps brachii — a muscle with three heads. Only muscle fleshing out the posterior humerus.

Trichocyst — an oval or rod-shaped structure in the ecto-plasm of ciliates functioning as a defensive or anchoring structure; in certain cryptomonads and chloromonads, a hairlike structure capable of being discharged.

Tricuspid valve — the heart valve between the right atrium and the right ventricle.

Trigeminal nerve — the fifth cranial nerve having three main division: the ophthalmic, maxillary, and mandibular.

Triglyceride — a fat, composed of a glycerol molecule bonded to three fatty acids by ester linkages.

Triploblastic — having three primary germ layers: ectoderm, mesoderm, and endoderm.

Trochlear nerve — the fourth cranial nerve which supplies the superior oblique muscle of the eye.

Trophic level — one of the four divisions of the ecosystem based upon how an organism feeds.

True ribs — the first seven pairs which attach directly to the sternum by costal cartilages.

Trunk — a tree's main stem; the body minus head or limbs; the torso; an arthropod's thorax and abdomen; the main portion of a structure; a proboscis, as on an elephant.

TseTse fly — a bloodsucking African fly of the genus *Glossina.*

Tubal ligation — occurs in women as a means of sterilization which involves cutting a short section out of the oviduct to prevent eggs from traveling into the uterus.

Tube feet — tubular extensions of the water-vascular system functioning in locomotion and grasping in echinoderms.

Tube nucleus — a nucleus at the tip of a pollen tube. It func-tion in the development and growth of the tube.

Turgid — abnormally inflated, distended, or swollen; tightly stretched by pressures from within.

Turgor pressure — osmotic pressure that provides rigidity to a cell.

Tympanic membrane — the membranous eardrum posi-tioned between the outer and middle ear; also called the *tym-panum,* or the *ear drum.*

Typhlosole — the infolding longitudinally of the intestinal wall in annelids or in isopods; a cyclostome or elasmobranch spiral valve.

U

Ulna — in man, the medial forearm bone and the corre-sponding bone in the forelimb of other tetrapods.

Umbilical arteries — two of the three vessels in the umbilical cord. They carry carbon dioxide and debris-laden blood from the fetus to the placenta.

Umbilical cord — a cordlike structure containing the umbili-cal arteries and vein, which connects the fetus with the placenta.

Umbilical vein — one of the three blood vessels in the umbil-ical cord. It is largest and carries blood rich in nutrients and oxygen to the fetus.

Uncinate process — a thin, flat process extending backward and upward from a rib overlying the next succeeding rib, strengthening the rib cage.

Undulating membrane — a thin, vibrating organelle in cili-ates of fused cilia located in the region of teh cytopharynx; a thin finlike, cytoplasmic process in trypanosomes through which flagellum runs.

Upper beach — that part of the beach ecosystem where it be-comes possible for vegetation to establish itself.

Uracil — a pyramidine base in nucleic acids (RNA).

Urea — a crystalline white substance derived from proteins. It is the principal solid excreted in the urine of most mam-mals and is formed in the liver from ammonia following the deamination of amino acids.

Ureter — a tube that transports urine from the kidney to the urinary bladder.

Urethra — a tube that transports urine from the urinary bladder to the outside of the body.

Uric acid — a crystalline white substance in the urine of car-nivorous animals and the end product of nitrogen metabolism in uricotelic animals.

Urinary bladder — a distensible sac in the pelvic cavity which stores urine.

Urine — the fluid or semifluid product of the urinary or ex-cretory organs; in man it contains water, salts, pigments, hor-mones, and waste products of metabolism.

Urogenital opening — one that functions as both an excre-tory and reproductive structure.

Urogenital sinus — or vestibule. It is the area in a female where the urethra from the urinary bladder enters.

Uropod — an appendage located on either side of the telson in various crustaceans.

Uterus — a hollow, muscular organ in which a fetus develops. It is located within the female pelvis between the urinary bladder and the rectum.

V

Vacuole — a membrane-bound sac in the cell. Examples would be food and contractile vacuoles in protists and the central vacuole in plants.

Vagina — a tubular organ that leads from the uterus to the vestibule of the female reproductive tract and receives the male penis during coitus.

Vagus nerve — the tenth cranial nerve whose fibers are dis-tributed to the head, neck, thorax, and abdomen.

Valve — a structure permitting flow in only one direction; a movable pedicellaria jaw; a part of an ovipositor of an insect; one of the halves of the cell wall of a diatom; a partially de-tached anther lid.

Vascular bundle — connecting tissue consisting of primary xylem and phloem and perhaps procambium.

Vascular cambium — the cambium of a vascular cylinder giv-ing rise to secondary xylem and phloem.

Vascular tissue — the xylem, phloem, and vascular cambium in tracheophytes.

Vas deferens — the principal duct conveying sperm from the testis to the outside.

Vasectomy — the removal of part of the vas deferens.

Vastus lateralis — one of the four muscles of the quadriceps group which flesh out the anterior thigh. The group as a whole acts to extend the knee.

Vastus intermedius — see Vastus lateralis.

Vastus medialis — see Vastus lateralis.

Vegetal pole — the lightly colored, yolk-filled portion of the unfertilized frog egg.

Vegetative reproduction — nonsexual as opposed to sexual reproduction.

Vein — a blood vessel that conveys blood toward the heart.

Velum — a thin membranous structure resembling a curtain or veil; the soft palate; a protective membrane covering the sporangium in quillworts.

Vena cava(s) — large vein(s) that return blood to the sinus venosus or right atrium of the heart.

Ventricle — the cavity of an organ; the cavity of the vertebrate brain; a recess between the true vocal folds and false vocal folds of the larynx.

Vessel — a canal or tube that fluid passes through; a conducting tube or duct in the xylem of vascular tissue of cylndrical dead cells joined end to end with perforated intervening walls.

Vestibular membrane — a membrane which lines the vestibular cord in the cochlea of the inner ear.

Villi — fingerlike, minute processes or papillae; a straight, soft hair, such as those found on variosu plant structures.

Visceral body mass — it consists of the organs in a clam.

Vitreous humor — the transparent gell that occupies the space between the lens and retina of the eye.

Viviparous — giving birth to living young that have developed within, nourished through the uterus; seed germination while still attached to the parent plant.

Vocal fold — true vocal folds, on the inner surface of the larynx extending from the thyroid cartilage to the arytenoid cartilages. Their vibration results in sound.

Vocal sac apertures — openings into two vocal sacs in the throat of many species of male frogs.

Vomer — a membrane bone in mammals in the nasal region forming part of the nasal septum.

Vomerine teeth — the teeth borne on the vomerine bone, which is located in the roof of the mouth. They are used to hold food. They are found in frogs.

W

Walking leg — legs found in arthropods such as crayfish, used for locomotion, handling food, and cleaning the body.

Water-vascular system — a system of canals and structures functioning in locomotion and food-getting, in echinoderms.

White blood cell — a leukocyte, which is a nucleated blood cell that protects the body from disease-causing organisms such as bacteria and viruses.

White matter — in the brain, it is located in the inner region where there are pathways going to the cell bodies in the outer gray matter. It is reversed in the spinal cord.

Wing — a flight organ; a movable appendage possessed by flying animals.

Working distance — it is the space between the slide and the objective lens.

X

Xanthophyll — a yellow-orange accessory pigment used in photosynthesis.

Xiphoid process — the sword-shaped tip of the sternum or breastbone.

Xylem — vascular tissue in plants that transports water and minerals.

Y

Yolk plug — small circular area of cells within the blastopore opening.

Yolk sac — one of four extra-embryonic membranes that supports embryonic development.

Z

Zoospores — a swimming or motile spore; a spore with flagella or cilia.

Zygosporangium — a thick-walled structure resulting from the fusion of gametangia.

Zygospore — the zygote that results from the union of isogametes.

Zygote — a fertilized egg cell formed by the union of a sperm and an ovum.

Index